Stochasticity and Intramolecular Redistribution of Energy

NATO ASI Series

Advanced Science Institutes Series

A series presenting the results of activities sponsored by the NATO Science Committee, which aims at the dissemination of advanced scientific and technological knowledge, with a view to strengthening links between scientific communities.

The series is published by an international board of publishers in conjunction with the NATO Scientific Affairs Division

A Life Sciences	Plenum Publishing Corporation
B Physics	London and New York
C Mathematical and Physical Sciences	D. Reidel Publishing Company Dordrecht, Boston, Lancaster and Tokyo
D Behavioural and Social Sciences	Martinus Nijhoff Publishers
E Engineering and Materials Sciences	Dordrecht, Boston and Lancaster
F Computer and Systems Sciences	Springer-Verlag
G Ecological Sciences	Berlin, Heidelberg, New York, London, Paris, and Tokyo

Series C: Mathematical and Physical Sciences Vol. 200

Stochasticity and Intramolecular Redistribution of Energy

edited by

R. Lefebvre

Laboratoire de Photophysique Moléculaire du CNRS,
Orsay, France

and

S. Mukamel

Department of Chemistry, University of Rochester,
New York, U.S.A.

D. Reidel Publishing Company

Dordrecht / Boston / Lancaster / Tokyo

Published in cooperation with NATO Scientific Affairs Division

Proceedings of the NATO Advanced Research Workshop on
Stochasticity and Intramolecular Redistribution of Energy
Orsay, France
June 23 - July 4, 1986

Library of Congress Cataloging in Publication Data

NATO Advanced Research Workshop on Stochasticity and Intramolecular Redistribution
    of Energy (1986: Orsay, France)
    Stochasticity and intramolecular redistribution of energy.

    (NATO ASI series. Series C, Mathematical and physical sciences; vol. 200)
    Includes index.
    1. Excited state chemistry—Congresses. 2. Stochastic processes—Congresses.
I. Lefebvre, R. (Roland), 1928–   . II. Mukamel, S. (Shaul), 1948–   . III. Title.
IV. Series.
QD461.5.N357   1986   541.2'24   87–4598
ISBN-13: 978-94-010-8208-2   e-ISBN-13: 978-94-009-3837-3
DOI: 10.107/978-94-009-3837-3

Published by D. Reidel Publishing Company
P.O. Box 17, 3300 AA Dordrecht, Holland

Sold and distributed in the U.S.A. and Canada
by Kluwer Academic Publishers,
101 Philip Drive, Assinippi Park, Norwell, MA 02061, U.S.A.

In all other countries, sold and distributed
by Kluwer Academic Publishers Group,
P.O. Box 322, 3300 AH Dordrecht, Holland

D. Reidel Publishing Company is a member of the Kluwer Academic Publishers Group

CONTENTS

PREFACE

This volume contains the invited papers presented at the NATO Advanced Research Workshop on "Stochasticity and Intramolecular Redistribution of Energy" held in Orsay (France) from June 23 to July 3, 1986. The Workshop brought together leading researchers involved in the experimental and the theoretical studies of vibrational energy flow and relaxation in activated polyatomic molecules. The recent experimental developments in this area include the study of ultracold molecules in supersonic beams and the development of high resolution (frequency domain) and ultrafast (time domain) spectroscopic techniques. On the theoretical side the introduction of statistical methods (random matrix theory, reduced equations of motion) and efficient numerical algorithms provide an adequate framework for the interpretation of vibrational dynamics in large polyatomic molecules. Classical, semiclassical and quantum calculations on simple model systems show the existence of regular and chaotic regions in the phase space. The articles in this volume provide an updated review of the current status of experimental studies and the relevance of the recent theoretical developments to their interpretation.

We wish to thank the organizations which made this workshop possible. NATO provided the basic grant. We acknowledge the essential contribution of the late Dr Mario di Lullo in providing pertinent advice. The generous support of the U.S. Air Force Office of Scientific Research, the U.S. Navy Research Office and the Exxon Corporation made it possible to organize the planning meeting for this workshop which was held in Rochester, N.Y., October 3-5, 1985. Financial support was awarded by CECAM to facilitate the participation of European Colleagues belonging to the CECAM member countries. Dr Carl Moser, Director of CECAM helped us in many ways, in particular in providing the site for the meeting. Madame M. A. Duchassaing has been very efficient in managing the practical aspects of the meeting and of the preparation of the proceedings.

Roland Lefebvre
Laboratoire de Photophysique
Moléculaire du CNRS
Orsay, France

Shaul Mukamel
Department of Chemistry
University of Rochester
New York, U.S.A.

RELEVANCE OF CHAOS TO QUANTUM MECHANICS

Giulio Casati, Giorgio Mantica
Dipartimento di Fisica dell'Università di Milano
Via Celoria 16, 20133 Milano, Italy

Italo Guarneri
Dipartimento di Fisica Teorica e Nucleare
Università di Pavia, Italy

ABSTRACT

The relevance of the concept of chaos to Quantum Mechanics is discussed in this paper. We consider the problem of subthreshold ionization of Rydberg Hydrogen atoms, showing what relations are to be expected between classical chaotic dynamics and quantum motion. The phenomenon of Quantum Localization plays a central role in our description, which is completed by numerical computations confirming the theoretical results.

INTRODUCTION

One of the leading theoretical problems of today physics is commonly addressed as "Quantum Chaos", and is related to the manifestation in Quantum Mechanics of any erratic, unpredictable kind of motion [1]. The theory of Chaos was originally developed in Classical Mechanics, and it led to the recognition (among other things) of a large category of systems characterized by truly chaotic dynamics, the so called K-systems. The importance of these systems stands far beyond that of exotic dynamical models: their existence shows indeed the basic instability of classical dynamics. The theory of classical chaos can correctly address and eventually solve many long-standing theoretical problems [2], like the dynamical justification of the methods of Statistical Mechanics, the Fourier law of heat transport, and so on.

Quantum Mechanics has an inherent statistical nature related to the measurement process and to the probabilistic interpretation, in the line of the school of Copenhagen. Nevertheless, the Schroedinger equation provides a dynamical evolution for the wave function that is deter-

1

R. Lefebvre and S. Mukamel (eds.), Stochasticity and Intramolecular Redistribution of Energy, 1–13.
© *1987 by D. Reidel Publishing Company.*

ministic, and can be studied with the techniques of modern dynamics.

A basic question arises if we consider the implications of Classical Chaos to the "correspondence principle": precisely, let's focus our attention on the quantized version of a classical K-system, a billiard, for example. Loosely speaking, the correspondence principle states that as the Planck constant goes to zero, the predictions obtained by quantum mechanics should converge to the classical analogues. Now, these latter are determined by the chaotic trajectories of the system. We are so left with the expectation that also in Quantum Mechanics some "chaotic" process could possibly exist.

The above considerations generated a new field of research, concerned with the relation between Classical Chaos and Quantum Mechanics. The first systems to be analyzed were the quantized chaotic billiards. Being these systems bounded and autonomous, their energy spectrum is forced to be pure point. Since chaotic motion is necessarily related to a continuous spectrum of the evolution operator, <u>stricto sensu</u> the evolution of these quantum systems cannot be chaotic. Some interesting spectral properties have however been detected for these systems, and related to the theory of Random Matrices Ensembles[3]. We shall not discuss this aspect here, however.

Much richer spectral properties are possible in a different context: the class of externally perturbed quantum systems, were dynamical characteristics resembling chaos might possibly manifest. We shall discuss in this paper this latter kind of systems, and particularly we shall expose our results concerning a single interesting example: a Rydberg Hydrogen atom in the electric field of a microwave cavity. Since the early work of Bayfield and Koch[4], much attention has been devoted to the study of this problem, both theoretically and experimentally[5-15], as a consequence of its interesting features. In fact, a high underthreshold ionization is detected, which cannot be explained by the standard techniques of perturbation theory. On the contrary, a classical picture based on chaotic diffusion in phase space was shown to provide estimates in agreement with experimental findings[10-12].

We shall review in this paper a key feature of the ionization process: the phenomenon of "Quantum Localization". On such theoretical basis one can predict that the agreement between classical and quantum mechanics is to be expected only in a particular range of the external parameters, whose boundaries can be determined theoretically[7,9]. Away from this region, quantum motion is localized over a finite number of quasi-energy eigenfunctions for a long time, and ionization can occur only through a direct multiphoton ionization of low probability. Finally, we address the problem of the chaoticity of the quantum evolution by describing a time-reversal numerical experiment[16]. It indicates a dynamical stability of the motion even in the range of quantum delocalization.

We are convinced that the relevance of these results goes far beyond the present application. Detecting regions of localized quantum evolution may be a key feature in many atomic and molecular problems. To mention only one, the possibility of selective excitation in molecules might as well be related to localized characteristics of the motion, where good control can be gained dealing with a finite number of quasi-energy eigenfunctions.

THE HYDROGEN ATOM IN A MICROWAVE FIELD

The hydrogen atom in a microwave cavity can be described by the following one-dimensional Hamiltonian:

$$H = P^2/2 - 1/x + \mathcal{E} x \cos(\omega t) , \qquad x > 0 \qquad (1)$$

provided that the atom is initially in a very elongated quantum state – cigar state – having parabolic quantum numbers $n_1 = n - 1 \gg 1$, $n_2 = 0$, and magnetic quantum number $m = 0$ [8]. The dipole approximation for the linearly polarized monochromatic field has also been employed. $\mathcal{E}$ and ω are the microwave field strength and frequency respectively, in atomic units. Since the matrix elements for transitions with $\Delta n_2 \neq 0$ are small, the above Hamiltonian is significative of the real experimental situation. Moreover, it provides a unique opportunity to compare the classical and the quantum description in the semiclassical region.

CLASSICAL ANALYSIS

We recall first the results of the classical analysis [5,11]. The introduction of the external perturbation in the motion of the free atom renders the system not integrable. However, for small values of the parameter $\mathcal{E}$, K.A.M. theory grants us that at least some of the tori – regular orbits – of the Kepler motion are still present, although if in a somewhat deformed fashion. Such tori divide the phase space – that is they completely encircle closed regions – and in this way they prevent any enclosed orbit to reach a region of positive potential energy and ionize. We can summarize this by saying that there exist regions of stable motion.

As the perturbation gets stronger and stronger, most tori are destroyed, and adjacent chaotic regions merge together. By applying the

"Chirikov Resonance Overlapping Criterion", one can predict that when

$$\varepsilon n_c^4 \equiv \varepsilon_o > \varepsilon_{cz} \simeq 1 \big/ \big(50 \, \omega_o^{1/3} \big),$$
$$\omega n_o^3 \equiv \omega_o > \omega_{cz} \sim 1 \tag{2}$$

holds, all the tori above the value $n=n_o$ of the unperturbed action are destroyed. In the above relation ω_{cz} is a threshold frequency, and rescaled values for ε and ω have been introduced. Their meaning is the following: $\omega_o = 1$ is the frequency of the Kepler rotation of the electron, and $\varepsilon_o = 1$ is the intensity of the local Coulomb field, at $n=n_o$. When Eq. (2) is satisfied, the trajectories starting at $n=n_o$ wander randomly in phase space till they eventually ionize. It is important to stress that the presence of chaos renders the concept of trajectory meaningless: close trajectories part abruptly in the course of the motion, causing a high sensitivity to initial conditions. This prevents the calculation of the actual trajectories, both practically - on a computer[2] - and theoretically, according to algorithmical complexity theory[2]. In these circumstances a statistical description is most appropriate, and may be obtained considering the distribution function $f(n,\tau)$, of the statistical population at the value n of the action, after τ periods of the external wave. The evolution of f is determined by a Fokker-Planck equation[11,16]:

$$\partial f / \partial \tau = \frac{1}{\tau} \, \partial / \partial n \big(D(n) \, \partial f / \partial n \big), \tag{3}$$

where D(n) is the appropriate diffusion coefficient. We now agree to express the action in atomic units, so that integer n will represent the action of the nth quantum state. This is to be done here to render the correspondence with the quantum case more transparent. The diffusion coefficient can now be written as:

$$D(n) = 2 \varepsilon_o^2 \, n^3 \big/ \big(\omega_o^{7/3} \, n_o \big). \tag{4}$$

An exact solution for the diffusion equation (3) can be obtained[16]; however, we shall keep all our discussion at an informal level. A distribution peaked originally at n_o will diffuse in action space over a width Δn that can be estimated from (3) and (4).

Because D(n) increases with n, the diffusive process can complete to ionization ($n=\infty$) in a finite time τ_I. A rough estimation for τ_I can be obtained assuming it to be comparable to the time required to have $\Delta n \sim n_0$:

$$\tau_I \sim n_o^2 / D(n_o) \simeq \omega_o^{7/3} / (2\varepsilon_o^2).$$ (5)

Synthetically, the classical ionization mechanism is a chaotic diffusive process determined by eqs. (3)-(6). This opens the question whether the correspondent quantum motion is also chaotic. Before moving to this latter point, we must notice that richer details are offered by the classical description [11,12], which we do not report here because they are not of primary use for our next development.

THE QUANTUM LOCALIZATION PHENOMENON

We now consider the quantum motion given by Hamiltonian (1).Because the external field depends on time in a periodic way, we can follow the convenient choice of considering the period evolution - Floquet - operator U defined by [17]:

$$(U\psi)(x,t) = \psi(x, t + 2\pi/\omega).$$ (6)

U is the period evolution operator generated by the time-dependent Hamiltonian (1).

The methods of functional analysis [18] applied to (1), (6) prove that the spectrum of U is continuous, which is a necessary condition for chaos [19] - no matter how small ω and ε. It is easy to interpret physically this otherwise puzzling mathematical result: what renders the spectrum continuous is the fact that, due to multiphoton transitions to positive energy states, the stationary states of the unperturbed atom become quasi-stationary, with a finite lifetime depending on ε. This obviously quantum phenomenon represents a first difference with the classical case, where regions of stability are possible for small values of ε. It must be noticed, however, that the probability of direct multiphoton transitions is surely negligible for small ε.

Perturbation theory can be employed to resolve the dynamics of the system as far as the intensity of the external field is sufficiently small. Much less is known in general about the behaviour of strongly perturbed systems. As discussed in the introduction, the correspondence principle should suggest that, if Eq. (2) is verified, ionization should also occur in the semiclassical region n $\gg$ 1. However, this prediction does not take into account the phenomenon of quantum localization, which we are going to discuss now.

Quantum localization was hypothesized as a phenomenon related to the Anderson localization in disordered materials [20]. Since the basic assumption of randomness on which such description rests turned out to be too strong and inadequate to describe properly a variety of situations [21,22], we describe here a different picture of the localization mechanism [23].

Let's suppose that the quasi-stationary states of a time-dependent Hamiltonian are characterized by sufficiently long lifetimes as compared to the relevant experimental time-scale. Therefore, we can conveniently speak, even if not rigorously, of quasi-energy eigenfunctions of the operator U. Such 'eigenfunctions' actually exist, in the precise mathematical sense of 'resonances' [17]. Let's also suppose that the conditions for chaos in the related classical system are verified. In this regime, the evolution of a quantum state initially peaked on the unperturbed eigenstate n_o, will follow the classical evolution of $f(n, \tau)$, at least as long as the quantum nature of the spectrum becomes evident, say up to a time τ_D. Up to that time, the spreading on the unperturbed states will be approximately given by the classical spread in action $\Delta n(\tau_D)$.

The occurrence of the 'break-time' τ_D is described by the theory of 'Transient Stochasticity' [23], which relates it to the number N of quasi-energy states significantly involved in the evolution:

$$\tau_D \sim N \qquad (7)$$

It is easy to give an intuitive picture of the estimation (7): $2\pi/N$ is the mean quasi-energy spacing, which can be related to the time necessary to the system to 'realize' the discreteness of the spectrum via the uncertainty relation between energy and time. τ_D is the time required to quantum interference effects to set up and stop the 'classical' diffusion. In our case, the number N occurring in (7) is of the same order as $\Delta n(\tau_D)$, that is the spreading from the initial delta distribution. Should the diffusion coefficient be a constant, the estimate $\Delta n(\tau_D) \simeq (D \tau_D)^{1/2}$ would hold. Then, using (7), one would find:

$$\ell \equiv \Delta n(\tau_D) \simeq \tau_D \simeq D \qquad (8)$$

The above relation shows the essence of the localization phenomenon: Quantum Mechanics limits the classical diffusion over a finite number of states. The quantity ℓ is then called 'localization length', and is directly related to the classical diffusion coefficient. This picture has been verified for the case of the quantum kicked rotator in Ref. (24).

THE QUANTUM DELOCALIZATION BORDER

A different situation is met if the diffusion coefficient is a function of n, like in our case. What happens then, is a competition between two different effects: the inhibitory action of quantum interference, against the enhanced diffusion induced by the growing D(n). Depending on which process is dominant, we can have localization,

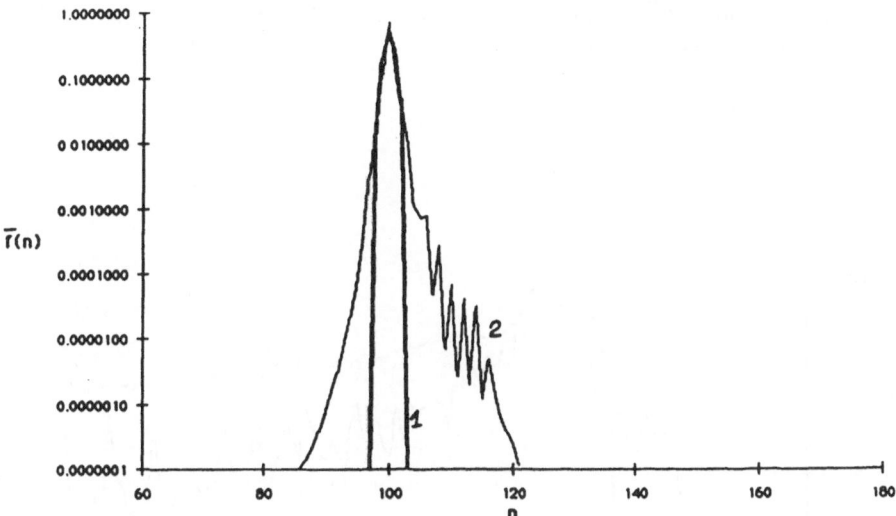

Fig. 1 Classical (1) and quantum (2) distribution functions $\bar{f}$ (n), averaged over 40 values of τ within the interval $80 < \tau < 120$. Here ω_o =1.5, $\mathcal{E}_c$=.01, n_o =100. Notice the small quantum tunneling through the classical Kolmogorov invariant curves.

or complete diffusion till ionization. It is possible to roughly decide such alternative, if we recall the classical ionization time τ_I. If this latter is less than the break-time τ_D at n_o, or D(n_o), according to Eq. (8), then ionization will surely occur before the occurrence of interference effects. In this way, using Eqs. (4), (5) and (8) it was found that localization is prevented and complete diffusion takes place [7] if:

$$\mathcal{E}_o > \mathcal{E}_q = \omega_o^{7/6} / (6 n_o)^{1/2} \qquad (9)$$

The above equation expresses the 'Quantum Delocalization Border'. It is of paramount importance in determining the behaviour of the excited atom, that turns out to be more detailed than naively predicted by the correspondence principle. As a matter of fact, in the semiclassical region $n \gg 1$, the relation $\mathcal{E}_\alpha < \mathcal{E}_q$ holds, such that the relative amplitude of $\mathcal{E}_0$ determines three possible dynamical situations, depicted in Figures 1-3. There, we compare the classical distribution function $f(n, \tau)$ with the quantum analogue, given by the square coefficients $|c_n(\tau)|^2$ of the projection of the wave function on the Hydrogen bound state eigenfunctions. The details of these numerical calculations are given in Ref. 9.

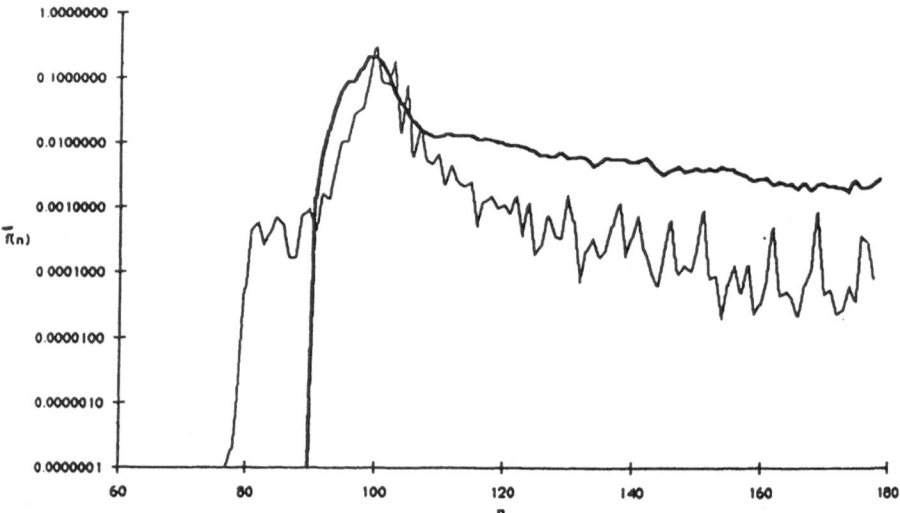

Fig. 2 Same as Fig. 1 but with $\mathcal{E}_0 = .03$. The distribution functions are averaged over the interval $440 < \tau < 480$. In this case $\mathcal{E}_\alpha < \mathcal{E}_0 < \mathcal{E}_q$, and the quantum packet is localized -except for a small resonant plateau. On the contrary, the classical packet is strongly diffuse.

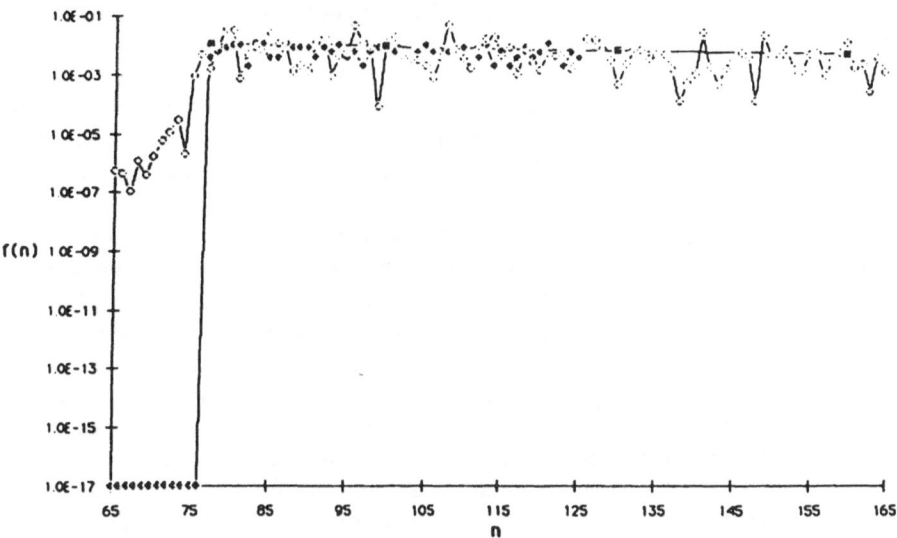

Fig. 3 Same as Fig. 1, but with $\mathcal{E}_o$ =.08. Classical (solid lozenges) and quantum (open lozenges) distributions after $\mathcal{T}$=60 cycles of the microwave, compared to the solution of Eq. (3)(■).

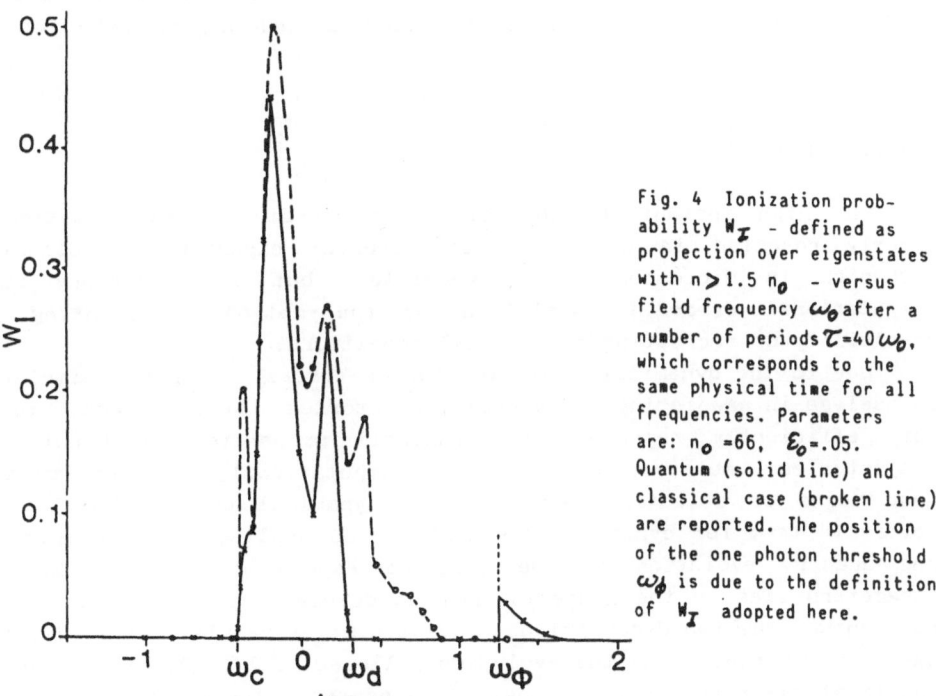

Fig. 4 Ionization probability W_I - defined as projection over eigenstates with $n > 1.5\, n_o$ - versus field frequency ω_o after a number of periods $\mathcal{T}$=40 ω_o, which corresponds to the same physical time for all frequencies. Parameters are: n_o =66, $\mathcal{E}_o$=.05. Quantum (solid line) and classical case (broken line) are reported. The position of the one photon threshold ω_ϕ is due to the definition of W_I adopted here.

Fig. 2 clearly illustrates the effect of the localization phenomenon in quantum mechanics: diffusion is inhibited and any 'correspondence' with the classical is lost. In Fig. 3, the two distribution functions agree between themselves and with the solution of the Fokker-Planck diffusion equation. However, this does not necessarily imply chaos in the quantum evolution, as we shall see in the next paragraph.

The above theory of quantum localization not only offers an important theoretical picture, but also provides a useful method for predicting laboratory results. Let's come back to the classical threshold given in Eq. (2). Since classical chaos is due to the interaction between the external field and the harmonics of the unperturbed motion, its onset requires a minimum allowed value ω_{cr} of the frequency of the microwave. The quantum delocalization border - Eq. (9), can be equivalently interpreted as a threshold in frequency, if we keep the external field intensity as well as the other variables fixed. Precisely, quantum localization is effective above a critical fre-quency ω_q. When this latter is bigger than ω_{cr} a window of underthreshold ionization is open in frequency. This condition is met in the case of Fig. 4, where the numerically detected ionization rates are reported versus the frequency of the external field. A direct comparison between diffusive and one-photon ionization shows a much higher rate for the former.

QUANTUM STABILITY

The high instability characterizing classical chaotic motion is clearly revealed in numerical time-reversal experiments. Classical mechanics is microscopically reversible, but any arbitrary small computational error is amplified by the motion, preventing the possibility of recreating the initial conditions.

Due to the dynamical amplification of errors, computer simulation is useless in predicting individual trajectories. Nevertheless, it can very efficiently reproduce the ensemble properties of bundles of trajectories; moreover, the lack of computational reversibility is a good means of detecting the onset of dynamical chaos. This "time-reversal" test for dynamical instability was applied to both classical and quantum evolution in the case corresponding to Fig. 3. The parameters lies in the range of quantum delocalization, where, as we have shown, the two descriptions provide similar results. At $\tau=60$, time has been reversed, and the evolution followed till $\tau=0$. Fig. 5 reports the final distribution functions. As expected, the classical does not recover the initial condition, but proceeds according to the diffusion equation. On the contrary, the quantum initial distribution is reproduced with high accuracy. We interpret this fact in the sense of a strong stability of the quantum motion. Its full relevance is to be left, however, to a more complete investigation [16].

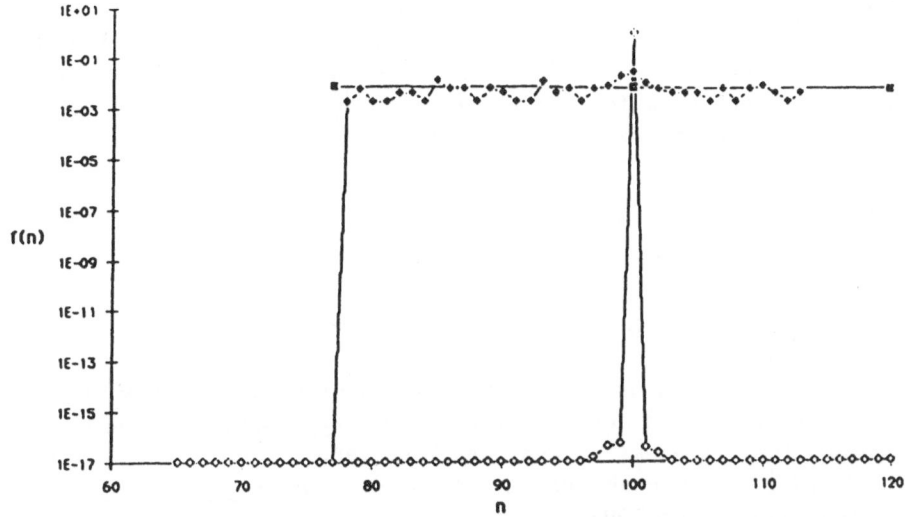

Fig. 5 Same as Fig. 3. At time τ =60 velocities have been reversed, and the evolution followed for other 60 periods. Distribution functions at this latter time are plotted, and compared with the solution of the Fokker-Planck equation at τ =120.

CONCLUSIONS

This paper has been concerned with the relevance of the concept of chaos to Quantum Mechanics. We have focused our attention on the class of the externally perturbed systems, which includes the interesting example of a Rydberg Hydrogen atom in a microwave cavity, where new theoretical predictions have been made possible by a semiclassical analysis accounting for the presence of chaos in the classical system.

Starting from the possible implications of the correspondence principle towards a 'chaotic' quantum motion, we have seen that the phenomenon of quantum localization may jeopardize any correspondence between classical and quantum evolution, no matter how far in the semiclassical region. On the other hand, we have also seen that Quantum Mechanics may indeed reproduce classically chaotic results - like the subthreshold ionization of Rydberg atoms - provided that a condition for quantum delocalization is met. The relevance of these results to the case of atomic ionization and more generally of molecular physics could be of primary importance.

The existence of Chaos in Quantum Mechanics, as well as its very definition, are still matter of debate among the physics community. Indeed a number of "pseudo-chaotic" quantum processes, like the ionization process we discussed in this paper, can now be predicted, but - we believe - not be defined as chaos. Indeed, there seems to be no room in Quantum Mechanics for the extreme instability properties related to the classical characterization of this word $^{(2)}$.

REFERENCES

(1) Proc. Int'l Conf.on Quantum Chaos, Como 1983, G. Casati Ed. (Plenum, 1985).

(2) Joseph Ford, 'Chaos, Solving The Unsolvable, Predicting The Unpredictable!', in Chaotic Dynamics and Fractals, M. Barnsley and S. Demko Eds. (Academic Press, New York 1986).

(3) G. Casati I. Guarneri and G. Mantica, in Proc. Int'l Conf.on Quantum Chaos, Como 1983; 'Random Matrices as Models for the Statistics of Quantum Mechanics', Physica D, to appear.

(4) J.E. Bayfield, P.M. Koch, Phys. Rev. Lett. 33 (1974) 258.

(5) B. Delone, V.P. Krainov, D.L. Shepelyansky, Usp. Fiz. Nauk 140 (1983); Sov. Phys. Uspeky 26 (1983) 551.

(6) D.L. Shepelyansky, in Proc. Int'l Conf.on Quantum Chaos, Como 1983 G. Casati Ed. (Plenum, 1985).

(7) G. Casati, B.V. Chirikov and D.L. Shepelyansky, Phys. Rev. Lett. 53 (1984) 2525.

(8) J.E. Bayfield, L. Pinnaduwage, Phys. Rev. Lett. 54 (1985) 313.

(9) G. Casati, B.V. Chirikov, I. Guarneri, D.L. Shepelyansky, Phys. Rev. Lett. 57 (1986) 823.

(10) J.G. Leopold, I.C. Percival, Phys. Rev. Lett. 41 (1978) 944. J. Phys. B 12 (1979) 709.

(11) R.V. Jensen, Phys. Rev. A 30 (1984) 386.

(12) K.A.H. Van Leeuven et al. Phys. Rev. Lett. 55 (1985) 2231.

(13) R. Blumel and U. Smilansky, submitted to Phys. Rev. Lett.

(14) J.N. Bardsley, B. Sundaram, Phys. Rev. A 32 (1985), 689.

(15) P.M. Koch, in Fundamental Aspects of Quantum Theory, V. Gorini and A. Frigerio eds. Plenum Press (1987).

(16) G. Casati, B.V. Chirikov, I. Guarneri and D.L. Shepelyansky, Phys.

Rev. Lett. 56 (1986) 2437.

(17) K.J. Yajima, Comm. Math. Phys. 87 (1982) 331.

(18) S. Graffi, V. Grecchi and H.J. Silverstone, Ann. Inst. Poincare 42 (1985) 215.

(19) J. Bellissard, 'Stability and Instability in Quantum Mechanics', in 'Trends in The Eighties', Lect. Notes in Math. P. Blanchard ed. 1985.

(20) S. Fishman, D.R. Grempel and R.E. Prange, Phys. Rev. Lett. 49 (1982) 509. Phys. Rev. A 29 (1984) 1639.

(21) G. Casati and I. Guarneri, Comm. Math. Phys. 95 (1984), 121.

(22) G. Casati, J. Ford, F. Vivaldi and I. Guarneri, 'On the Search for Randomness in The Kicked Quantum Rotator', Phys. Rev. A, to appear.

(23) B.V. Chirikov, D.L. Shepelyansky, F.M. Izrailev, Soviet Scient. Reviews 2C (1981) 209.

(24) B.V. Chirikov, D.L. Shepelyansky, Preprint 85-29, INP Novosibirsk 1985.

CHAOS IN MOLECULAR SYSTEMS?

Stavros C. Farantos
Department of Chemistry,
University of Crete, and
Institute of Electronic Structure and Laser,
Research Center of Crete,
711 10 Iraklion, Crete, Greece

and

Jonathan Tennyson
Department of Physics and Astronomy,
University College London,
Gower Street, London WC1E 6BT, England.

ABSTRACT. Theoretical studies on the floppy molecules KCN, LiCN and ArHCl, and on the more conventional molecules HCN and HCCH suggest that regular/irregular states can be identified in vibrationally excited molecules. Excitation of the low frequency mode (bending for triatomics) is the favoured route to chaotic behaviour. In accord with Percival's conjecture, spectra of LiCN associated with regular states are characterized by sharp, high intensity lines whereas chaotic states result in broad, low intensity spectra. Stable periodic trajectories are found in the phase space of HCN and HCCH at energies above the barrier to isomerization which provide bottlenecks to the reaction.

1. INTRODUCTION

In classical mechanics the generic behaviour of a multidimensional system with N degrees of freedom (N⩾2) and a non-separable potential is to have quasiperiodic trajectories at low energies where the anharmonicities and the coupling among the degrees of freedom is small whereas at very high energies the phase space is mostly occupied by chaotic trajectories[1]. Quasiperiodicity means that there are N constants of motion in contrast to chaotic trajectories where the only constants of motion are the total energy and angular momenta. At intermediate

R. Lefebvre and S. Mukamel (eds.), Stochasticity and Intramolecular Redistribution of Energy, 15–30.

energies, quasiperiodic and chaotic trajectories coexist.
For dynamical systems with $N \geqslant 3$ degrees of freedom, it is
quite possible to have "restricted" chaos which means that
there are M constants or approximate constants of motion
with $1 < M < N$ [2,3]. The behavior of classical systems described
above is the subject of one of the most celebrated theorems
in classical mechanics, the Kolmogorov-Arnold-Moser (KAM)
theorem [4].
       Is there and what is the "quantum analogue" of
classical chaos? These questions have been the subject of
a plethora of numerical investigations on model 2-D
potentials such as the stadium [5] and the Henon-Heiles
potentials [6,7]. The latter represents two harmonic
oscillators coupled through an anharmonic cubic term [8]. By
calculating the eigenfunctions and eigenvalues of the
hamiltonian operator with an expansion of the wavefunctions
in a suitable basis set and diagonalizing the resulting
hamiltonian matrix, it is possible to investigate the
behaviour of the eigenfunctions and eigenvalues as the
energy increases and classical mechanics shows a transition
from quasiperiodic to chaotic dynamics. Although by
defining a few indicators, one can see a transition from
regular (mode-localized) to irregular (mode-delocalized)
states as the energy increases [9], in a way similar to
classical mechanics, a precise definition of what is meant
by chaos in a quantum mechanical system is still awaited.
       In ro-vibrational spectra regular/irregular quantum
states should have their fingerprints in observable
quantities such as energy distributions, lifetimes and
intensities. Percival [10] extrapolating from the classical
results, was the first to conjecture that molecules at low
excitations have regular spectra (assignable transitions)
and at high energies, irregular spectra, where the use of
good quantum numbers to characterize the states involved is
not possible. Indeed molecules, being quantum objects, are
the best candidates for an experimental manifestation of
"quantum chaos". At the moment there are a limited number
of such studies. Acetylene [11] and hydrogen cyanide [12] are
the only molecules whose vibrational spectroscopy has
directly been associated to the present theories of chaos.
Attempts to understand the complexity of NO_2 [13] and H_2CO [14]
spectra in terms of chaos have also been made.
       In order to stimulate further experimental work on the
investigation of "quantum chaos", calculations on realistic
potential energy surfaces are needed. In recent years we
have undertaken such a study using classical dynamics and
the time-independent Schrodinger equation for a few
triatomic systems. The results have immediate implications
for spectroscopy and intramolecular dynamics. In the
following sections of this article we survey the results of
this work (15-19,46).

Full ro-vibrational quantum mechanical calculations are restricted to triatomic and a few tetraatomic molecules and to low and intermediate energies of excitation[20]. On the other hand classical mechanics remains a powerful tool for investigating the non-linear properties of molecular systems. In this article we discuss the significance of periodic orbits, which are found embedded in the chaotic regions of phase space of HCN and HCCH at energies above the barrier to isomerization, to spectroscopy. The importance of periodic orbits has been emphasized by Gutzwiller[21] who proposed a semiclassical theory for the construction of eigenvalues in the chaotic regions of phase space. Berry and Tabor[22] used periodic orbits to estimate the spectral density of states. More recently Heller[23] has shown that unstable periodic orbits in the stadium problem* act as attractors of the wavefunction. In the author's vocabulary periodic orbits "scar" the eigenfunctions.

To find possible experimental manifestation of "quantum chaos" in molecules is one of the goals of our work. As a second goal we try to associate the non-linear mechanical properties with the potential energy topography.

2. POTENTIAL ENERGY SURFACES AND COMPUTATIONAL METHODS

2.1 Potential Functions

Triatomics are the simplest molecules which can display chaos. Nonetheless we are still dealing with a system of three degrees of freedom in contrast to most theoretical studies which use 2-D model potentials. However it is possible on physical grounds to separate one degree of freedom in some classes of triatomic molecules. Examples of this are the floppy molecules KCN and LiCN, and van der Waals systems such as ArHCl. It can be shown that an adiabatic separation of the high frequency CN or HCl mode is possible[18]. Furthermore ab initio calculations on the ground electronic state of KCN [24] and LiCN [25] have demonstrated a weak coupling of the CN bond length, r, and the distance of alkalide atom from the center of mass of CN, R, and the angle θ between $\underline{r}$ and $\underline{R}$. Thus we first present the results of 2-D calculations in order to compare classical and quantum mechanics[16].

Another difficulty encountered in calculations with real potential energy surfaces is the representation of the potential function at high energies where experimental and theoretical data are scarce. It appears that the method

*It has been rigorously proved that the stadium model is ergodic and therefore only unstable periodic orbits can exist.

of Murrell and coworkers[26] is able to produce molecular
potential functions which reasonably interpolate the
regions between minima and dissociations channels. We use
this type of functions for HCN[27] and the 6-D surface of
acetylene[28], constructed from a mixture of experimental and
ab initio data.

In table I the main topographical characteristics of
the potential energy surfaces of cyanide compounds are
shown. LiCN and HCN each have two stable isomers which
are both linear but have different barriers to
isomerization, X-CN->X-NC. The potential function of KCN
supports one minimum with a triangular geometry but with a
small barrier to linearization of the molecule.

Table I

X	ref.	Absolute minimum E/cm^{-1}	θ	Metastable minimum E/cm^{-1}	θ	Barrier E/cm^{-1}	θ
K	24	-39064	105	-	-	503	180
Li	25	-53681	180	2280	0	3370	55
H	27	-45670	0	3920	180	12140	67

Table I. Energies of the minima and barriers to
isomerization for X-CN compounds. The metastable minima
and barrier heights are given relative to the energies of
the absolute minima. θ gives the angle of each feature,
$\theta=0$ corresponding to the linear XCN cyanide structure.

2.2 Clasical mechanical calculations

The classical hamiltonian for a rotationless triatomic
molecule in scattering coordinates is:

$$H = \frac{P_R^2}{2\mu_1} + \frac{P_r^2}{2\mu_2} + \left(\frac{1}{2\mu_1 R^2} + \frac{1}{2\mu_2 r^2}\right)P_\theta^2 + V(R,r,\theta) \tag{1}$$

where

$$\mu_1^{-1} = m_X^{-1} + (m_C + m_N)^{-1}, \quad \mu_2^{-1} = m_C^{-1} + m_N^{-1}$$

m_X, m_C and m_N denote the atomic masses. The six
first-order Hamilton's equations are integrated by using
several algorithms with constant or variable integration

step. We have found Shampine and Gordon's algorithm[29] which is a variable step and order procedure to be the most efficient for the type of potential functions used here. The energy is conserved to seven significant figures when integrating the trajectories for 16 ps. In order to characterize the trajectories as quasiperiodic (regular) or chaotic (irregular) several diagnostics are used. These are:

a) Inspection of the projection of the trajectory onto a coordinate plane; regular and irregular patterns can be assigned by appearance.

b) Poincare surfaces of section. For systems with two degrees of freedom the distinction between quasiperiodic and chaotic trajectories is given by closed curves as opposed to shotgun type surfaces of section. However for systems with more than two degrees of freedom, the computation of the Poincare surfaces of section is not easy. Recording passes of a trajectory through a plane in a certain direction gives an annular pattern instead of a curve. If a proper coordinate system has been chosen or in the case of resonant trajectories these generalized Poincare surfaces of sections are useful for 3-D systems.

c) The power spectrum of a dynamical variable[1]. The Fourier transform of the time series of a dynamical variable gives a finite number of discrete lines for quasiperiodic trajectories. These correspond to the fundamental vibrational frequencies and their combinations and overtones. Chaotic trajectories are characterized by a broad spectrum in which the number of lines increases with increasing spectral resolution.

d) The rate of exponential divergence of two initially neighboring trajectories[30]. For quasiperiodic trajectories this quantity tends to zero with time but for chaotic trajectories it converges to a positive number (the maximum Lyapunov exponent). This stochastic parameter can be evaluated for any multidimensional system. In fact for systems with $N > 2$ degrees of freedom $2N$ Lyapunov functions can be obtained[31]. The number of non-vanishing Lyapunov exponents give the dimensionality of the chaotic subspace or equivalently the number of zero Lyapunov exponents provides the number of constants of motion. According to Pensin's theorem[32] the average over the energy hypersurface of the sum of positive Lyapunov exponents is equal to the dynamical Kolmogorov entropy[33]. The Kolmogorov entropy provides the time scale for the randomization of the energy in the system. Although the computation of Lyapunov functions and exponents is very expensive for multidimensional systems, these quantities reveal the chaotic structure of phase space and give a local measure of its irregularity. Algorithms for the computation of the maximum or all Lyapunov functions have been published[34].

2.3 Quantum mechanical calculations

The quantum analogue of hamiltonian (1) is

$$\hat{H} = -\frac{\hbar^2}{2\mu_1 R^2} \frac{\partial}{\partial R}\left(R^2 \frac{\partial}{\partial R}\right) - \frac{\hbar^2}{2\mu_2 r^2} \frac{\partial}{\partial r}\left(r^2 \frac{\partial}{\partial r}\right) - \frac{\hbar^2}{2}\left(\frac{1}{\mu_1 R^2} + \frac{1}{\mu_2 r^2}\right)$$

$$\times \frac{1}{\sin\theta} \frac{\partial}{\partial\theta}\left(\sin\theta \frac{\partial}{\partial\theta}\right) + V(R,r,\theta) \tag{2}$$

A convenient method for finding the eigenvalues and eigenfunctions of $\hat{H}$ is via a basis set expansion. Use of Legendre functions for the angular coordinate allows kinetic energy integrals in this coordinate to be evaluated analytically. We have solved for the radial coordinates either by expanding in Morse oscillator-like functions[35] or, in the case of LiCN, by obtaining numerical functions which involves the solution of a model problem for some fixed $\theta = \theta_f$[36]. The former technique is available in program ATOMDIAT[37]. To gauge the size of these calculations, we mention that in the 2-D LiCN computations[16] convergence of the first 80 states to within 1 cm^{-1} was achieved by diagonalizing a 855 x 855 matrix.

Several diagnostics have been proposed to characterize the time-dependent or independent wavefunctions as regular or irregular. For time-independent solutions of the Schrodinger equation the most pertinent criteria are:
a) The nodal structure of the wave functions[38]. For regular states and a suitable choice of coordinates, the nodes of the wavefunction form a regular grid which allows quantum numbers for each coordinate to be assigned by inspection. This regularity can be associated with an approximate separability of coordinates. High energy states lose this structure so that quantization along qualitatively separable coordinates can no longer be achieved. It has been suggested that these irregular states are quantum chaotic. However this criterion must be used with caution since isolated resonances can make nodal patterns appear irregular[39].
b) Sensitivity of the eigenvalues to small perturbations in the hamiltonian[40]. The appearence of a dense web of overlapping avoided crossings has been considered as an indication of quantum chaos by Marcus and coworkers. A drawback with this criterion is the difficulty of finding the proper perturbative term (see ref.46).
c) Distribution of the nearest neighbor spacing of the energy levels[41]. Berry and Tabor [42] have shown from semiclassical arguments that regularity should be associated with a Poisson distribution. Numerical experiments with random matrices (Gaussian Orthogonal Ensembles) have led to the suggestion that chaotic spectra should obey a Dyson-Wigner distribution. Although a large

number of levels are required for reliable statistics, this criterion has special appeal as it can be applied to experimental data[11]. Recently Berry and Robnik[43] have shown that for systems without time reversibility (such as particles in a magnetic field), the local statistics of the energy levels change from the Gaussian Orthogonal Ensemble (GOE) of random-matrix theory to the Gaussian Unitary ensemble (GUE).

3. VIBRATIONAL CHAOS IN FLOPPY MOLECULES

Calculations on the regular/irregular behaviour on the 2-D KCN[15] and LiCN[16] show that the two metal cyanides display qualitative different behaviour . KCN, which has a small barrier to linearization, shows classical chaos even below the quantum ground state. Analysis of the nodal pattern of wavefunctions, avoided crossing diagrams, and level spacing distribution all point to an early onset of quantum chaos. In fact only the ground and the first two excited states can be assigned good quantum numbers by inspecting the nodal patterns. On the other hand LiCN, despite its large amplitude bending mode, is a more typical molecule. At low energies the phase space is mainly occupied by quasiperiodic trajectories, at intermediate energies quasiperiodic and chaotic trajectories coexist and at very high energies the phase space is occupied by irregular trajectories. This is a typical behaviour predicted by KAM theorem.

Analysis of the quantum diagnostics also shows a transition from regular to irregular motion. Fig. 1 shows a typical quasiperiodic and chaotic trajectory for LiCN. Fig. 2 displays typical regular and irregular quantum states.

For LiCN we were able to find for each type of classical trajectory an analogous quantum state[16]. Thus quasiperiodic trajectories localized below and above the minima of LiNC and LiCN as well as chaotic localized and delocalized trajectories have their quantum counterparts. It is anticipated that even the two free rotor or "polytopic" states [16] found in the quantum calculations should have their classical analogue, periodic orbits.

By computing the rate of exponential divergence of two neighboring trajectories for LiCN, we have located the onset of stochasticity at 1600 cm^{-1} above the absolute minimum. However irregularity in the quantum results is observed above about 2000 cm^{-1}. This "quantum sluggishness" in the appearance of chaos, which is also observed in KCN, is typical of quantum mechanics. It is due to the large spacing of the low-lying levels and hence the finite value of $\hbar$.

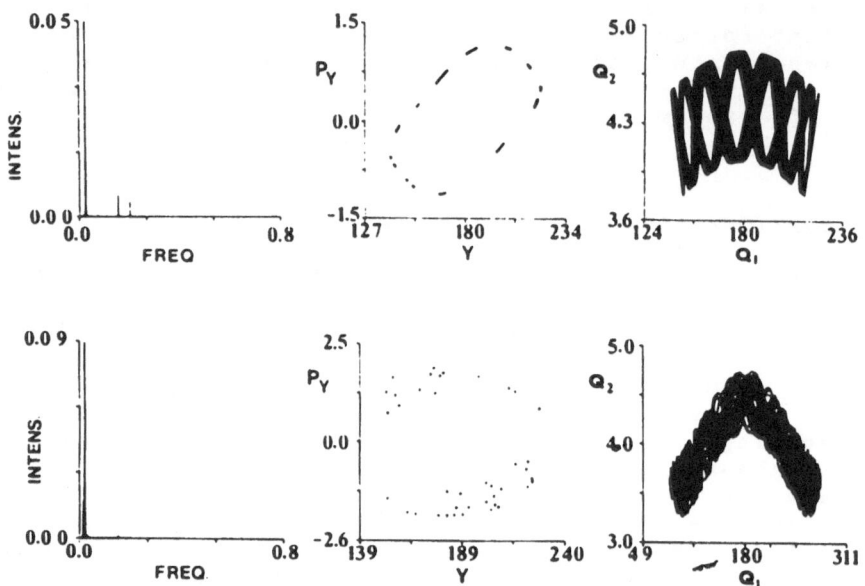

Figure 1.   Typical   quasiperiodic   and   chaotic   LiNC
trajectories.

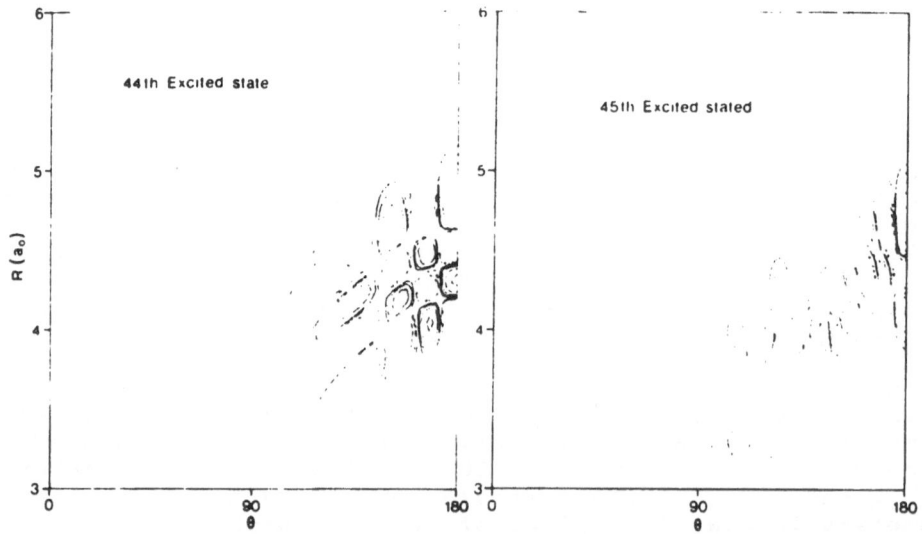

Figure 2.   Typical regular and irregular LiNC states.

Above the onset of chaos in LiCN, regular and irregular trajectories (states) coexist. An interesting problem related to chemical dynamics is the identification of regions of phase space which support either quasiperiodic or chaotic motion. Analysis of both classical and quantum results on LiNC and LiCN shows that the onset of vibrational chaos is strongly associated with the excitation of the bending mode[17]. On the other hand quasiperiodic stretching states persist above the barrier to isomerization. Hose and Taylor[7] were the first to analyze the phenomenon of the mode localization in highly excited vibrational states. They argued that overtone states have a stability not displaced by combination states and thus can exist as mode localized states embedded in the mode mixed quasicontinuum. In fact our results on LiCN and O_3[17] suggest that the excitation of the low frequency bending mode leads to an early onset of chaos.

Calculated fluorescence spectra of LiCN show that transitions from regular states are characterized by a few high intensity lines in contrast to the irregular states which display a broad spectrum with low intensity lines[19]. A profound distinction between regular and irregular states is found by plotting the fluorescence lifetimes of LiCN states as a function of energy (Fig.3). The fluorescence lifetimes of regular states lie in progressions with increasing bending quantum number; the progressions are labeled by the stretching quantum number. These types of graphs demonstrate that fluorescence properties provide a useful experimental probe for indentifying chaotic states.

The localization of the regular states of LiCN found above the barrier to isomerization is striking. This behaviour cannot be anticipated from the rather complicated potential. Similar states were found in the Henon-Heiles and stadium models by Aquilanti et al[44] and Taylor and coworkers[45]. These authors showed that the regularity can be understood by introducing adiabatic potentials. These are produced by a Born-Oppenheimer (BO) type separation of fast and slow vibrational modes. Taylor et al[45] have also computed the adiabatic action variables for the stadium problem showing that localization has the same origin in classical and quantum mechanics.

We have applied the Born-Oppenheimer approximation to 2-D potentials of KCN, LiCN and ArHCL with the slow coordinate (bending) as the adiabatic variable[18]. Comparison with fully coupled quantum mechanical calculations is good for those states which were assigned regular by inspection of their nodal patterns. The BO approximation breaks down at energies where the states become irregular and delocalized. One anticipates that this approximation will be useful in calculating highly excited regular states as well as in reducing the

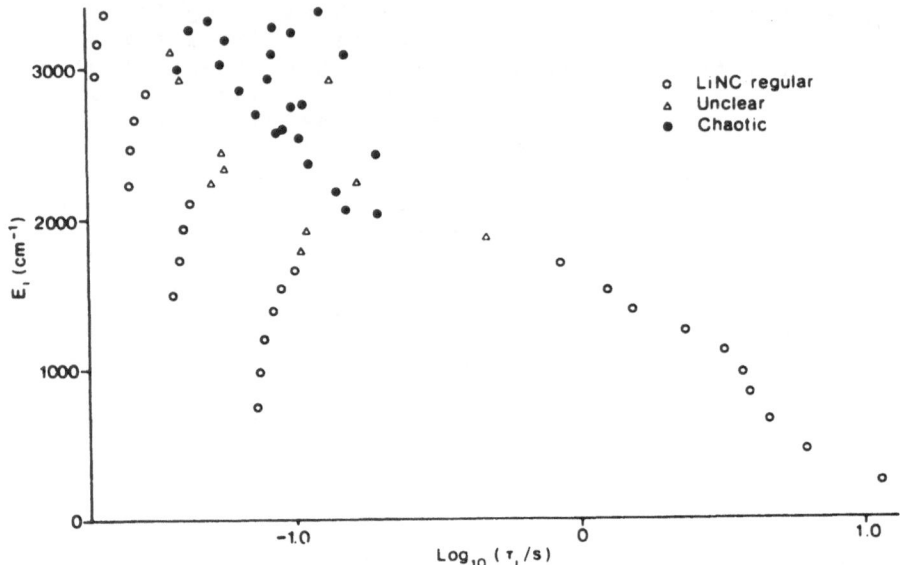

Figure 3. Fluorescence lifetime versus excitation energy for low-lying vibrational states of LiNC with J=0.

dimensionality in vibrational calculations of polyatomic molecules. The later has been tested in 3-D calculations on KCN and ArHCl[18]. By adiabatically separating the motion of CN or HCl, whose coupling with the other degrees of freedom is introduced through the kinetic part of the hamiltonian, we obtain energy levels in excellent agreement with the fully coupled 3-D calculations.

4. VIBRATIONAL CHAOS IN CONVENTIONAL MOLECULES

In typical (more rigid) molecules the transition to chaotic regions of phase space occurs at relatively higher energies than in floppy molecules. Quantum mechanical calculations have not been performed at these very high energies and it is not obvious which coordinate can adiabatically be separated. Generally a normal coordinate description is valid at low energies but a zero order hamiltonian in local coordinates seems more appropriate at higher energies[47]. Classical mechanics is thus the only means available for dynamical studies of these fully coupled triatomic and tetraatomic molecules. Although the classical trajectory technique has extensively been used to study unimolecular dissociation and intramolecular energy transfer only a few molecules have been studied with the methods of non-linear mechanics. These are HCC[48], O_3[49], SO_2[49], HCN[12], OCS[50] and the tetraatomics H_2CO[51] and HCCH[52]. It would appear that triatomic molecules are still largely unknown species even in their ground electronic states. Only a few triatomics have been studied experimentally under high vibrational excitation. Among them are O_3 and HCN. Lehmann et al[12] recorded the spectrum of HCN in the range of 15000-28500 cm^{-1} in an attempt to observe chaos. All the levels could be assigned and well fitted by standard techniques for localized states. They were thus unable to identify any chaotic states in this region, despite classical calculations which predicted chaos above 12990 cm^{-1}. We note, however, that they failed to find any state with two, or more quanta in the bending mode. Similarly, Imre et al[53] measured the vibrational spectrum of O_3 to within 500 cm^{-1} of dissociation. Their results have been successfully fitted using both Darling-Dennison[54] and algebraic approach[55] hamiltonians, and all observed states classified as regular. However Imre et al[53] observed no states involving bending excitation. Again classical trajectory calculations predict an onset to stochasticity 5000 cm^{-1} below dissociation. Analysis by us[17] showed that this should not be considered as disagreement with the spectroscopy of the molecule since classical mechanics predicts quasiperiodic trajectories for the stretching modes excited at high energies. Of course one should not expect exact agreement with the experiment as the potentials are not accurate at these energies[12].

We have reexamined the classical dynamics of HCN by using the potential of Murrell, Carter and Halonen[27]. For this potential, the onset of stochasticity is at about 40000 cm^{-1} above the minimum of HCN when only the CH stretching mode is excited. This is 28000 cm^{-1} above the barrier to isomerization. In contrast, if the bending mode is excited and the other modes are given only their zero point energies, chaos is observed just above the barrier. Access to large regions of stochasticity is obtained with the destruction of the stable periodic orbit associated with the $\omega_{CH} : \omega_{CN} : \omega_{bend} = 6:4:1$ resonance (Fig.4). This occurs at energies about 16000 cm^{-1} above the minimum[56].

Most of the work comparing classical with quantum mechanics supports the argument that extented regions of phase space with quasiperiodic trajectories correspond to regular quantum states which in turn are responsible for a regular spectrum. We believe that the difficulty in observing highly excited bending states in triatomic molecules is associated with the chaotic character of these states. As found for LiCN chaotic states are responsible

for a large number of lines too low in intensity to be
resolved by direct excitation with present day techniques.
    In polyatomic molecules it is expected that other low
frequency modes will drive the molecule to chaotic motion
at even lower levels of excitation. On the other hand
extreme type of motions in the molecule will be responsible
for regular motion. For example, argon clusters attached
to excited chemically interesting species show highly
non-statistical unimolecular dissociation properties[57,58].
No fragmentation of argon clusters can be observed despite
an energy excess in the chemical species of two orders of
magnitude.

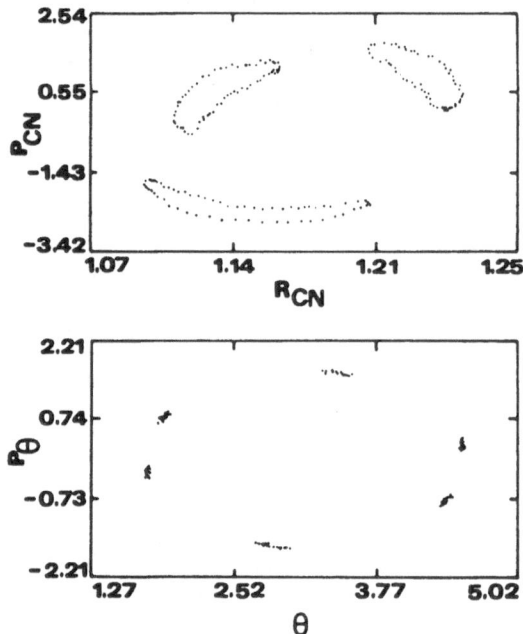

Figure   4.   A   trajectory   around   the   stable
$\omega_{CH} : \omega_{CN} : \omega_{bend}$ =6:4:1 periodic orbit of HCN.

    To our knowledge the vibrational spectroscopy of
acetylene studied by Abramson et al[11] is the only case
where irregular vibrational spectra have been investigated
in the light of present theories. With experimental
conditions, which favor the excitation of the trans-bending
mode, it was shown that above 9000 cm^{-1} the energy level
spacing distribution satisfies Wigner type statistics.
Below 9000 cm^{-1} the spectrum could be assigned by using
normal mode quantum numbers. However in the irregular
regime of the spectrum under low resolution, broad lines
called clumps are observed. These can be assigned

rotational quantum numbers and vibrational angular momentum. The regularity observed under low resolution, for a spectrum which is shown to be irregular under higher resolution, is a surprise. Classical trajectory studies[52] on a 6-D real potential energy surface of HCCH show an early onset of chaos by exciting the cis and trans bending mode. However at the energy of 28227 cm^{-1}, 11290 cm^{-1} above the barrier to isomerization to vinylidene, it was still possible to locate regular trajectories associated with large amplitude bending motion. These regular regions of phase space are created by stable periodic orbits embedded in the chaotic "sea". The islands of stability found in the phase space of acetylene at high energies could explain the regularity of the spectrum under low resolution. From the uncertainty principle

$$\Delta E \cdot \Delta t \approx \hbar \tag{3}$$

we deduce that high resolution (small ΔE) means that the spectrum is influenced by the long time behavior of the system. For low resolution (large ΔE) the short time evolution of the system influences the spectrum. Thus if the wave packet initially samples the regions around the periodic orbits, it will stay localized for a short time in these regions but over longer time it will spread in the chaotic regions of phase space[59]. Taylor[60], using the concept of adiabatic potentials and the theory of scattering resonances, offers another view which explains the regularity of the spectra under low resolution.

The uncertainty principle, eq.(3), has important consequencies for the properties of large polyatomic systems. For small molecules the states are sufficiently well separated and lifetimes are long enough for analysis of the time-independent Schrodinger equation to be valid. For polyatomic systems at intermediate or high energy, the density of states rapidly becomes very large. These states are then coupled in time and time-dependent approaches must be employed. This has led to the development of different approaches and terminology to treat this problem[61].

5. CONCLUSIONS

It is difficult not to accept that classical mechanics can play a vital role in understanding molecular spectroscopy and intramolecular dynamics. The division of phase space into regular and chaotic regions can at least qualitatively explain the behaviour of molecules like HCN and HCCH. Although a precise definition of quantum chaos is still missing, its manifestation through Wigner type distributions of the nearest neighbor spacing of energy levels has become increasingly accepted, GOE random matrix theory is valid at energies where classical mechanics is chaotic.

We believe that for a deeper understanding of quantum chaos more experimental work is needed. The ionization of hydrogen atom in a microwave field [62] is one promising experiment in this direction. Triatomic molecules are more complex systems but suitable for the investigation of quantum chaos. The type of experiments started by Lehmann et al [12] on HCN should be continued for non-rigid (floppy) molecular systems or van der Waals complexes. In order to compare theory with experiment we need potential energy surfaces accurate at the energies where the transition to chaos occurs. This is more likely to be achieved for light triatomic molecules than larger systems.

It was shown that excitation of the bending mode in rotationless triatomic molecules is a possible road to an early onset of chaos. The role of the rotational excitation in the appearance of chaos in molecules is also worth investigating. This is supported by the results obtained in model calculations [63] and from the spectrum of H_2CO [14]. Recent work on the highly rotationally excited states ($J \leqslant 20$) of H_2D^+ provides a first step in this direction [64].

REFERENCES

1. D.W. Noid, M.L. Koszykowski and R.A. Marcus, Ann. Rev. Rev. Phys. Chem. 32, 267, 1981.
2. G. Contopoulos, L. Galgani and A. Giorgilli, Phys. Rev. A18, 1183, 1978.
3. M. Pettini and A. Vulpiani, Phys. Lett. A106, 207, 1984.
4. V.I. Arnold and A. Avez, "Ergodic problems in classical mechanics" Benzamin, New York, 1968.
5. S.W. McDonald and A.N. Kaufman, Phys. Rev. Lett. 42, 1189, 1979.
6. D.W. Noid, M.L. Koszykowski, M. Tabor, and R.A. Marcus, J. Chem. Phys. 72, 6169, 1980.
7. G. Hose, and H.S. Taylor, J. Chem. Phys. 76, 5356, 1982; 78, 5845, 1983; G. Hose, and H.S. Taylor, Phys. Rev. Lett. 51, 947, 1983; G. Hose, H.S. Taylor, and A. Tip, J. Phys. A17, 1203, 1984.
8. M. Henon and C. Heiles, Astron. J. 69, 73, 1973.
9. M. Tabor, Adv. Chem. Phys. 46, 73, 1981.
10. I.C. Percival, Adv. Chem. Phys. 36, 1, 1977.
11. E. Abramson, R.W. Field, D. Imre, K.K. Innes and J.L. Kinsey, J. Chem. Phys. 83, 453, 1985; R.L. Sudberg, E. Abramson, J.L. Kinsey and R.W. Field,

J. Chem. Phys. 83, 466, 1985.
12. K.K. Lehmann, G.J. Scherer and W. Klemperer, J. Chem. Phys. 77, 2853, 1982; 78, 608, 1983.
13. J.L. Hardwick, J. Mol. Spectrosc. 109, 85, 1985.
14. H.L. Dai, R.W. Field and J.L Kinsey, J. Chem. Phys. 82, 2161, 1985.
15. J. Tennyson and S.C. Farantos, Chem. Phys. Lett. 109, 160, 1984.
16. S.C. Farantos, and J. Tennyson, J. Chem. Phys. 82, 800, 1985.
17. J. Tennyson, and S.C. Farantos, Chem. Phys. 93, 237, 1985.
18. S.C. Farantos and J. Tennyson, J. Chem. Phys. 85, 641, 1986.
19. J. Tennyson, G. Brocks, and S.C. Farantos, Chem. Phys. 104, 399, 1986.
20. J. Tennyson, Computer Phys. Reports 4, 1, 1986.
21. M.C. Gutzwiller, Physica D5, 183, 1982.
22. M.V. Berry and M. Tabor, Proc. Roy. Soc. A349, 101, 1976.
23. E.J. Heller, Phys. Rev. Lett. 53, 1515, 1984.
24. P.E.S. Wormer and J. Tennyson, J. Chem. Phys. 75, 1245, 1981.
25. R. Essers, J. Tennyson and P.E.S. Wormer, Chem. Phys. Lett. 89, 223, 1982.
26. J.N. Murrell, S. Carter, S.C. Farantos, P. Huxley and A.J.C. Varandas, "Molecular potential energy functions", John Wiley and Sons, 1984.
27. J.N. Murell, S. Carter, and L.O. Halonen, J. Mol. Spectrosc. 93, 307, 1982.
28. L.O. Halonen, M.S. Child and S. Carter, Mol. Phys. 47, 1094, 1982.
29. L.F. Shampine and M.K. Gordon, "Computer solution of ordinary differential equations", Freeman, San Francisco, 1975.
30. G. Benettin, L. Galgani and J.M. Strelcyn, Phys. Rev. A14, 2338, 1976.
31. H.D. Meyer, J. Chem. Phys. 84, 3147, 1986.
32. Ya.B. Pensin, Ups. Math. Nauk. 32, 55, 1977.
33. Ya.G. Sinai, "Introduction to ergodic theory", Princeton Univ. Press, Princeton, 1979.
34. A. Wolf, J.B. Swift, H.L. Swinney and J.A. Vartano, Physica D16, 285, 1985.
35. J. Tennyson and B.T. Sutcliffe, J. Chem. Phys. 77, 4061, 1982.
36. G. Brocks and J. Tennyson, J. Mol. Spectrosc. 99, 263, 1983.
37. J. Tennyson, Computer Phys. Commun. 29, 307, 1983.
38. R.M. Stratt, N.C. Handy and W.H. Miller, J. Chem. Phys. 71, 3311, 1979.
39. N. DeLeon, M.J. Davis, and E.J. Heller, J. Chem. Phys.

$\underline{80}$, 794, 1984.

40. D.W. Noid, M.L. Koszykowski, and R.A. Marcus, Chem. Phys. Lett. $\underline{73}$, 269, 1980; J. Chem. Phys. $\underline{78}$, 4018, 1983.
41. O. Bohigas, M.J. Giannoni and C. Schmit, Phys. Rev. Lett. $\underline{52}$, 1, 1984.
42. M.V. Berry and M. Tabor, Proc. Roy. Soc. London $\underline{A356}$, 375, 1977.
43. M.V. Berry and M. Robnik, J. Phys. $\underline{A19}$, 649, 1986.
44. V. Aquilanti, S. Cavalli and G. Crossi, in "Chaotic behavior in quantum systems. Theory and application", edited by G. Casati, Plenum, New York, 1985, p.299.
45. Y.Y. Bai, G. Hose, K. Stefanski, and H.S. Taylor, Phys. Rev. $\underline{A31}$, 2821, 1985.
46. J. Tennyson, Mol. Phys. $\underline{55}$, 463, 1985.
47. R.T. Lawton and M.S. Child, Mol. Phys. $\underline{37}$, 1799, 1979; $\underline{44}$, 799, 1981.
48. R.J. Wolf and W.L. Hase, J. Chem. Phys. $\underline{73}$, 3775, 1980.
49. S.C. Farantos and J.N. Murrell, Chem. Phys. $\underline{55}$, 205, 1981.
50. D. Carter and P. Brumer, J. Chem. Phys. $\underline{77}$, 4208, 1982; $\underline{78}$, 2104, 1983.
51. K.N. Swamy and W.L. Hase, Chem. Phys. Lett. $\underline{92}$, 371, 1982.
52. S.C. Farantos, J. Chem. Phys. $\underline{85}$, 641, 1986.
53. D.G. Imre, J.L. Kinsey, R.W. Field,and D.H. Katayama, J. Phys. Chem. $\underline{86}$, 2564, 1982.
54. K.K. Lehmann, J. Phys. Chem. $\underline{88}$, 1047, 1984.
55. I. Benjamin, R.D. Levine and J.L. Kinsey, J. Phys. Chem. $\underline{87}$, 727, 1983.
56. S.C. Farantos, M. Founargiotakis and J. Tennyson, work in progress.
57. A.J. Stace, J. Phys. Chem. $\underline{87}$, 2286, 1983.
58. S.C. Farantos, Chem. Phys. Lett. $\underline{92}$, 379, 1982; J. Phys. Chem. $\underline{87}$, 5061, 1983.
59. N. Moiseyev and A. Peres, J. Chem. Phys. $\underline{79}$, 5945, 1983.
60. H.S. Taylor, to be published.
61. E.B. Stechel and E.J. Heller, Ann. Rev. Phys. Chem. $\underline{35}$, 563, 1984.
62. R.V. Jensen, Phys. Rev. $\underline{A30}$, 386, 1984.
63. G.S. Ezra, Chem. Phys. Lett. $\underline{127}$, 492, 1986.
64. J. Tennyson and B.T. Sutcliffe, Mol. Phys., $\underline{58}$, 1065, 1986.

CORRELATION PROPERTIES OF SPARSE REAL SYMMETRIC RANDOM MATRIX

R. JOST, Service National des Champs Intenses CNRS, BP 166 X
38042 GRENOBLE CEDEX, France.

ABSTRACT. Sparse real symmetric random matrices are studied because
they model quantum mechanics problems like coupled oscillators. By
numerical calculations we determine the correlation properties of sets
of eigenvalues like Nearest Neighbor Distribution, NND and spectral
rigidity. We deal with the transition from Poisson statistics to Gaussian Orthogonal Ensemble (GOE) statistics as a function normalized
"size" of off diagonal matrix elements. The NND are fitted with one
parameter, Brody like, distributions. The transitions of NND from
Poisson to Wigner is governed by the normalized mean of absolute value
of off diagonal matrix elements. The long range correlation properties
are examined through the Fourier transform of the stick spectrum of
eigenvalues. Statistical analysis of eigenvectors shows that, for
intermediate coupling, i.e. between Poisson and GOE, each eigenvector
is a superposition of L_0 equally weighted (statistically) basis vectors.
L_0 is given by a dimensionless Fermi golden rule.

I. INTRODUCTION

Real symmetric random matrices are related with "quantum chaos" in the
sense that they model very complex hamiltonians. Only the statistical
correlation fluctuations (i.e. Nearest Neighbor Distribution NND and Δ_3
function (1)) of the eigenvalues for these random matrices are significant. These correlation properties have been found to be in good
agreement with the predictions of Gaussian orthogonal ensemble (GOE)
(1) when a system is "fully chaotic". We are interested in the transition from Poisson statistics, which holds for uncoupled systems, to GOE
statistics. Let us consider the hamiltonian of a multi-oscillator system as a model for the vibrations of a polyatomic molecule. We can write $H = H_0 + H_1$, where H_0 represents the hamiltonian of n uncoupled
oscillators and H_1 represents the coupling between them. In the basis
of H_0, the diagonal matrix elements are mainly related with the properties of H_0, and the off diagonal elements V_{ij} are related with H_1. The
large or infinite size of the H matrix usually prohibits direct
diagonalization.

31

R. Lefebvre and S. Mukamel (eds.), Stochasticity and Intramolecular Redistribution of Energy, 31–43.

The use of Direct Numerical Diagonalization (DND)(5) of truncated
matrix of Hamiltonian is rapidly growing because an array processor can
now perform diagonalization of full real symmetric (1000 x 1000)
matrix in less than 1 minute. The Cray II computer can diagonalize
10000 x 10000 matrices in a few hours ! Nevertheless the DND method
cannot give a large number of eigenvalues because few of the calculated
eigenvalues have converged (i.e. are independent of the truncation).

Typically only 10% of eigenvalues have converged for most of the
real calculations done for example on coupled oscillators (2-4) or for
molecular vibrations (5).

When we only look at the correlation properties of a set of eigenvalues of a hamiltonian, we don't necessarily need "exact" eigenvalues
but only a set of eigenvalues for which the correlation properties have
converged.

I think that this last requirement is much less "severe" than the
previous one. This idea is basically the idea of random matrix : the
statistical, correlation properties are independent of the details of
the hamiltonian and of the corresponding matrix elements but depend
only on global parameters of the distribution of the matrix elements.

Let us consider an infinite matrix which represents an hamiltonian
of coupled oscillators. The diagonal elements are ordered by increasing
values. In most cases, the non-zero off diagonal elements are located
in a band around the main diagonal. We can consider a sub-matrix of an
infinite matrix which corresponds to a subspace which includes all the
basis states for which diagonal energies are within Δ E around E.

The size of this submatrix can be small enough to have a roughly
constant density of basis state (it means that the mean spacing of diagonal elements is roughly constant over Δ E. The size of this submatrix
should be of the same order of magnitude as the width of the band of the
infinite matrix (this width is considered for basis states around E).

The diagonal part of the matrix which corresponds to uncoupled
oscillators is generated randomly with a uniform distribution (which
approximates the smooth variation of density of basis state).

Difficulties arise with off diagonal elements : Is there any correlation between the size and the location of off diagonal elements
(within the band structure)?

The following model supposes no such correlations i.e. the distributions of off-diagonal elements V_{ij} are independent of i and j.

2. THE MODEL

These real symmetric random matrices are constructed with a uniform
distribution of diagonal elements, corresponding to a constant density
of states.The mean spacing of diagonal values defines a unit of energy.
The off diagonal elements are uniformly distributed in the matrix (i.e.
the distribution of V_{ij} is independent of i and j). Their magnitudes are
determined by various distributions. When there is no coupling ($V_{ij}=0$),
the NND of the eigenvalues follow Poisson statistics. We want to define
how the correlation properties evolve with increasing coupling,i.e.for
increasing parameters of the distribution of V_{ij}.We compared the corre-

lation properties for many distributions of off-diagonal elements like gaussian,exponential,power laws (truncated), uniform, etc... The first moment V_{ij} can be zero or non zero.The parameter of sparseness α defines the ratio of zero and non-zero matrix elements.Figure 1 represents the shape of some used distributions and gives the corresponding parameters.

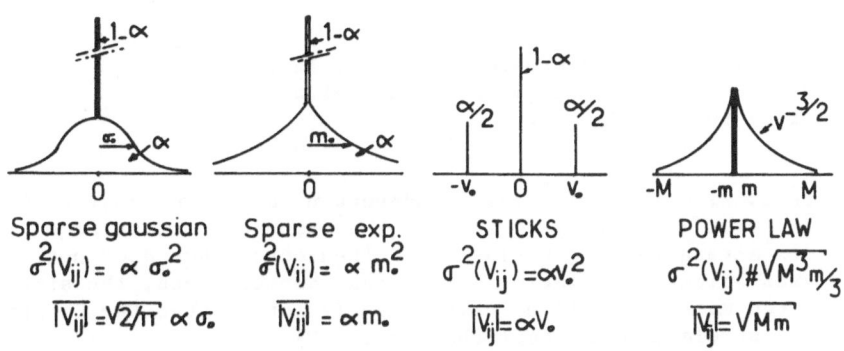

Figure 1: Used distributions of off diagonal matrix elements. The parameter of sparseness α range from 6/d to 1 (d is the matrix dimension).

3. METHOD OF ANALYSIS

For each type of distribution of V_{ij} for each set of parameters of these distributions and for each size of matrix, 18,000 eigenvalues have been obtained in order to get good statistics. The chosen distributions of V_{ij} give a NND intermediate between Poisson and Wigner. The Brody distribution (1) does not provide the best fit of the observed NND; however,the Brody parameter q remains a good description of the observed NND and we use it. This point will be discussed elsewhere (6). The mean spacing of eigenvalues in the central part of the (i.e. in region B of figure 2) remains very close to the mean spacing of diagonal elements (from H_0). Consequently, only the B part of the spectrum (11,600 eigenvalues) is used in the calculation of the statistical properties. As a result, the coupling H_1 can induce strong correlation properties (NND close to Wigner, Δ_3 close to GOE) without altering the initial density of states. Then, unfolding procedures are not necessary. This situation can be called "low coupling" in contrast with the GOE model for which the density of eigenvalues is much smaller than the density of diagonal values. Submatrices of dimension d = 25, 50, 100, 200, 400, 800 from a very large uniform matrix are diagonalized. We find that the statis-

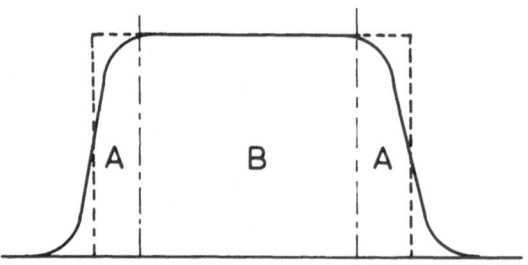

Figure 2: Typical shape of histogram of eigenvalues
    - dashed line for diagonal elements only, i.e. H_o only
    - solid line for $H = H_o + H_1$

tical properties like NND and Δ 3 are independent of the dimension of the submatrix for d $\geqslant$ 100 (figs.5 and 8). The truncation of a large uniform matrix alters the eigenvalues and eigenvectors but does not change the statistical correlation properties. Nevertheless, the size of the truncated matrix should be significantly larger than the length of long range correlations L_o (see chapter IV. D).

4. <u>RESULTS</u>

4.1. <u>Nearest neighbour distribution (NDD) versus "coupling parameter"</u>

A typical example of an NND histogram and of the fitted Brody function is given on figure 3.

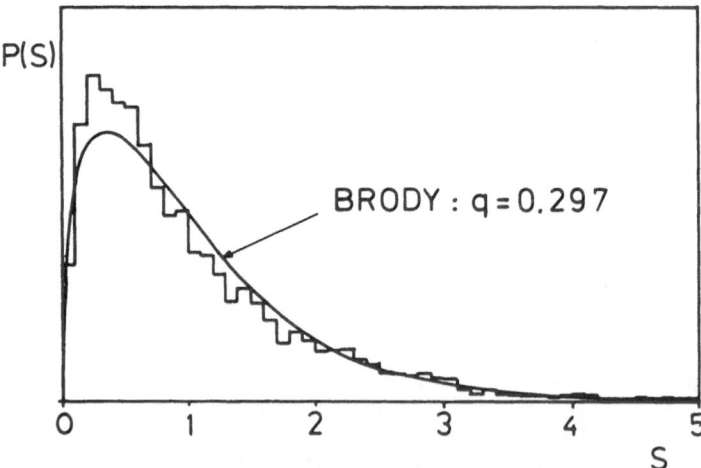

Figure 3: Typical NND histogram obtained with 11600 eigenvalues. The fitted Brody distribution deviates significantly from the histogram.

As displayed on figure 4, our numerical calculations show that the Brody parameter q which characterizes the NND is essentially a function of a reduced quantity, named "coupling parameter", which is the normalized mean value of the absolute value of off diagonal matrix elements: $|V_{ij}|$. In other words, q is a function of $\rho \overline{|V_{ij}|}$. Normalization is obtained by taking the mean spacing of ordered diagonal matrix elements as unit. As a result, the relationship between q and $\overline{|V_{ij}|}$ is independent of the shape of the distribution of off dia-gonal matrix elements, at least for the four distributions of figure 2. For intermediate distributions, i.e. for Brody parameter $0.3 \lesssim q \lesssim 0.8$ the fitted curves are significantly too far from the calculated histogram. This result means that the Brody distribution does not fit precisely our numerical histograms but, nevertheless the fitted Brody parameter q is a reasonable description of P (S).

Our histograms have been fitted with other distributions:the Berry and Robnik (7) distribution is numerically worse than the Brody distribution but when the corresponding parameter is plotted versus $\overline{|V_{ij}|}$ an analogous figure to figure 4 is obtained.

I have used a new one parameter distribution P (S) for which there is a linear dependence with S near the origin (6). With this new distribution the fit is much better than previously and again a figure analogous to figure 4 is obtained.

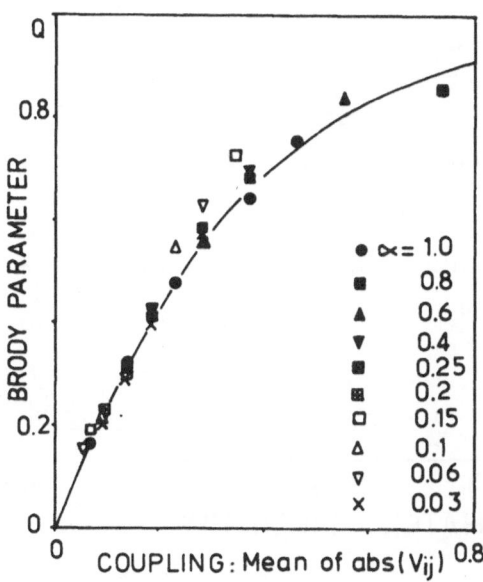

Figure 4 : Brody parameter versus coupling parameter: α defines the sparseness of the matrix (α = 1 corresponds to full matrix)

A keen fit procedure on a very large set of eigenvalues (up to 40000) and χ^2 test show that our numerical histograms of NND cannot be described by a universal one parameter distribution P (S). More precisely, two NND obtained with two distributions of off diagonal elements having the same "coupling parameter" $\overline{|V_{ij}|}$ but having different variances of V_{ij}, σ^2 (V_{ij}) are slightly different.

As a result NND cannot be described exactly by a universal one parameter distribution function; but at least a two parameter distribution is required.

Nevertheless, the largest deviations from a one parameter distribution remain small. Consequently, for practical purposes, i.e. to analyze a limited set of data (typically less than 1000 eigenvalues) a one parameter function is sufficient. Often, the experimental sets of data involve much less than 1000 eigenvalues.

The relationship between NND, characterized by q, and the "coupling parameter" $\overline{|V_{ij}|}$ is independent of the size of the matrix as shown on figure 5. A "coupling parameter" of the order of unity is sufficient to obtain a NND close to Wigner. This result has been found implicitly by J. Verbaarschot et al. (8).

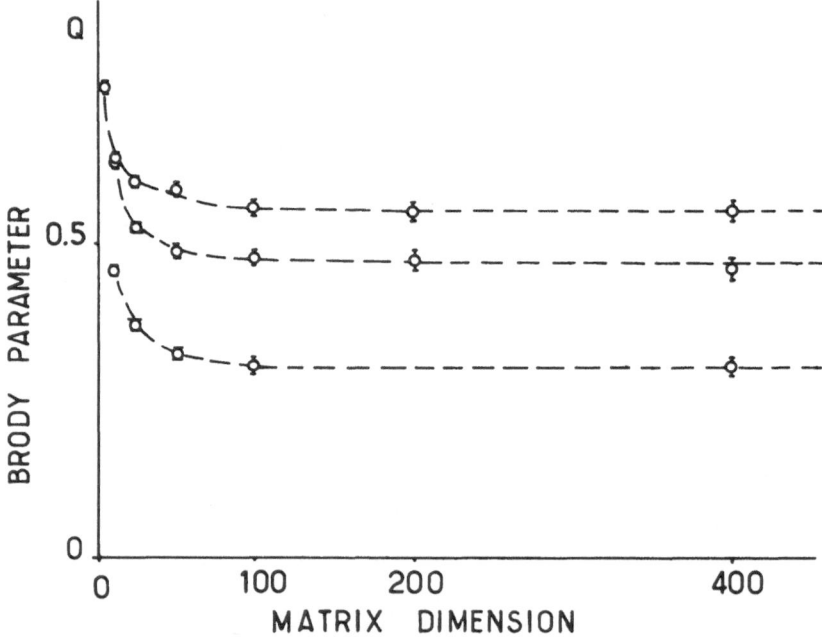

Figure 5: Brody parameter versus the matrix dimension. The distributions of diagonal and off diagonal matrix elements are identical within a dashed line

Figure 2b of reference 2 can be interpreted in the same way as figure 4. The horizontal axis kE is proportional to the normalized $|V_{ij}|$ (because E is proportional to the reverse of the mean spacing). As a result,we can predict the NND of eigenvalues of any real symmetric matrix without diagonalization and only by calculating the reduced coupling parameter $|V_{ij}|$. Again the main limitation of this model is related to the assumption of the random location of off diagonal matrix element.

4.2 Long-range correlation properties

Long range correlations (1)(or spectral rigidity) are studied with the 3 function and also with a new method based on the Fourier transform of stick spectrum. L. Leviandier et al. have shown (9) that the Fourier transform of a spectrum, $|FT|^2$, displays their correlations as well as $\Delta_{3(L)}$.

Examples of smooth $|FT|^2$ of stick spectrum of eigenvalues obtained at random matrix are displayed on figures 6, 7, 8.

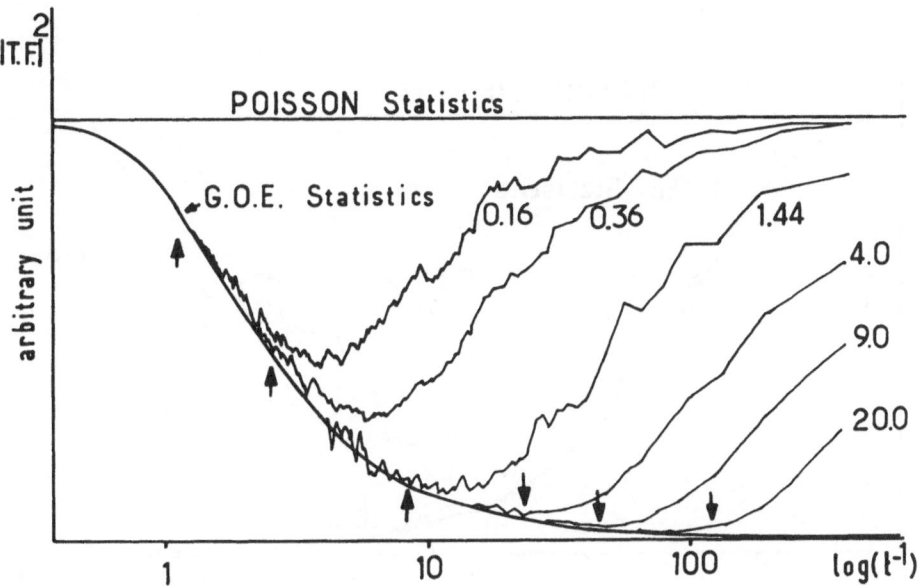

Figure 6: Square modulus of Fourier transform, $|FT|^2$ of stick spectrum of eigenvalues of 800 x 800 matrices.
The off diagonal matrix elements are given by a stick distribution (see figure 1) with $\alpha = 1$. The six curves correspond to different values of V_0. The corresponding variance, V_0^2 is given. The horizontal axis is log (t^{-1}), instead of t (which is the reciprocal of energy) in order to magnify the long range correlations. The numbers 1, 10, 100 are the corresponding L values, in order to compare with Δ_3 (L). Poisson and GOE lines are theoretical.

The horizontal axis corresponds to log (t^{-1}) instead of time t, in order
to magnify the correlation for short times. Time is considered as the
reciprocal of energy. The corresponding length of correlation in term
of L is given. Poisson statistics give no correlations and correspond
to a horizontal straight line. GOE statistics correspond to the smooth
line which decreases to zero for large L (i.e. for short times).

Figures 6, 7, 8 display long-range correlations of a set of eigenvalues obtained from matrices for which only one parameter varies.

From figure 8, we conclude that the long-range correlations are
roughly independent of the matrix. However, these long-range correlation properties cannot be determined for L larger than the size of the
matrix !

On figures 6 to 8 each arrow gives the quantity $L_0 = 2\pi \sigma^2 (v_{ij})$.
We observe that each calculated $|TF|^2$ follows the GOE curve up to the
corresponding arrow. Nevertheless, there is no universal one parameter
correlation function which evolves from Poisson to GOE curve.

The length of correlation L_0 is discussed in the next paragraph.

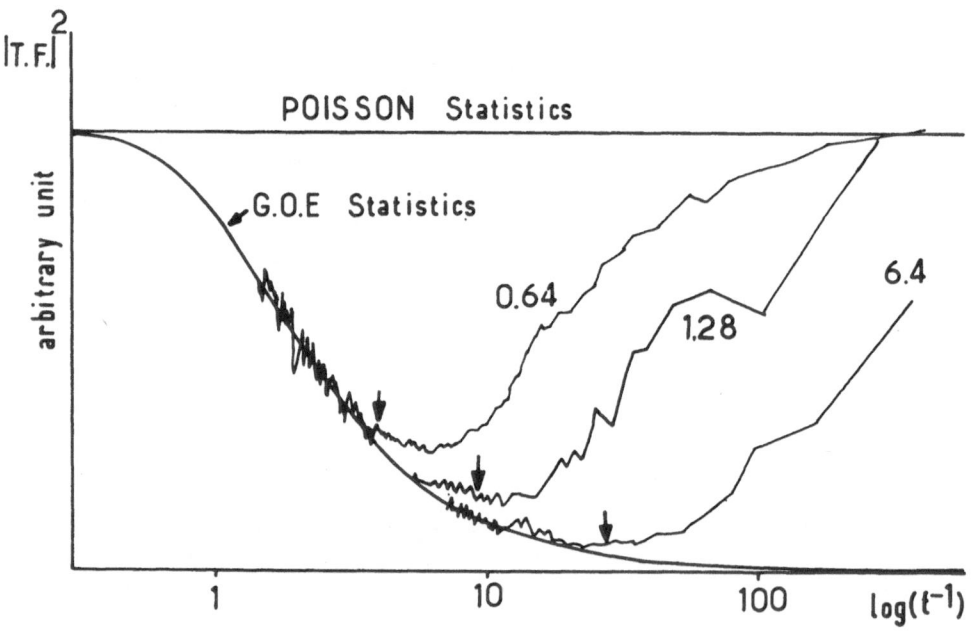

Figure 7: Same caption as figure 6. The three curves correspond to
three sets of eigenvalues having the same NND, i.e. the same q, i.e.
the same $|v_{ij}|$. In order to fix $|v_{ij}| = 0.8$, α is respectively 1,
0.5, 0.1 from left to right. The corresponding variance is given.

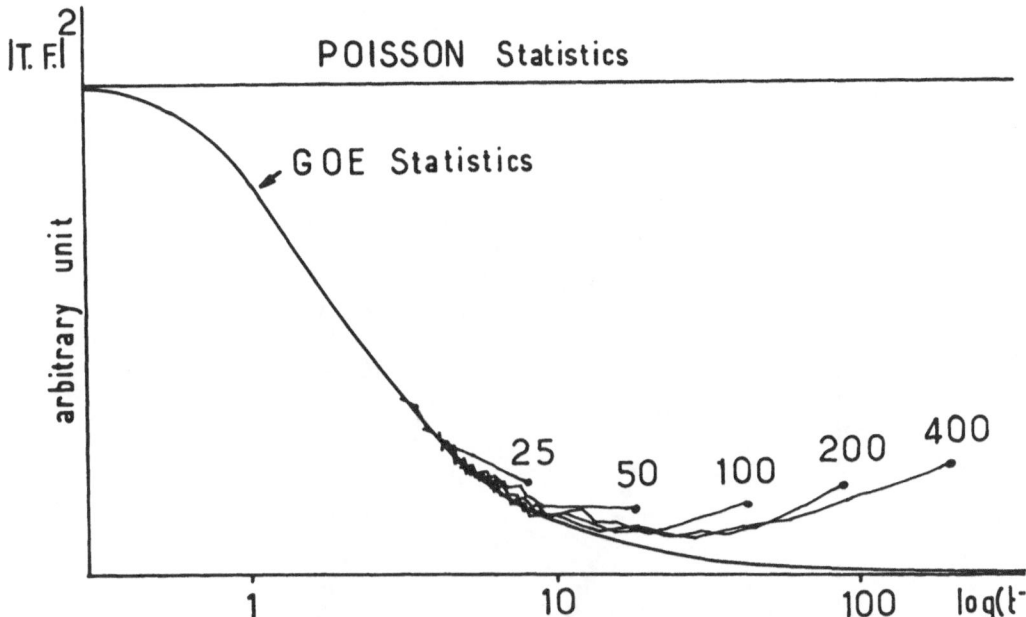

Figure 8 : Same caption as figure 6.
The distribution is sparse gaussian (see figure 2) with α = 0.2. Only
the dimension of the matrix changes from 25 to 400. The last point of
each $(FT)^2$ is slightly too high because a smoothing procedure is used.

4.3. Eigenvectors statistical properties

Each eigenvector is expressed as a sum of basis vectors

$$|\psi_i) = \sum_j c_{ij} |\varphi_j)$$

$|\varphi_j)$ and $|\psi_i)$ are respectively numbered according to increasing values of diagonal matrix elements and eigenvalues. Figure 9 shows two
examples ensemble averaging of c_{ij}^2, versus $|i-j|$.

a) For $H_1 \ll H_0$ we find the usual result of perturbation theory,
i.e. c_{ij}^2 decreases as $(i-j)^2$ because the zero order quantities
$(V_{ii}-V_{jj})^2$ are proportional to $(i-j)^2$ (diagonal matrix elements have
uniform distribution and then a uniform mean spacing)

b) When H_1 is large enough we obtain a typical shape of $\overline{c_{ij}^2}$
shown on figure 9.

For $|i-j| \lesssim L_0/2$, $\overline{c_{ij}^2}$ is constant. This means that any eigensta-
tes is mainly a combination of L_0 basis states. It means GOE properties
locally.

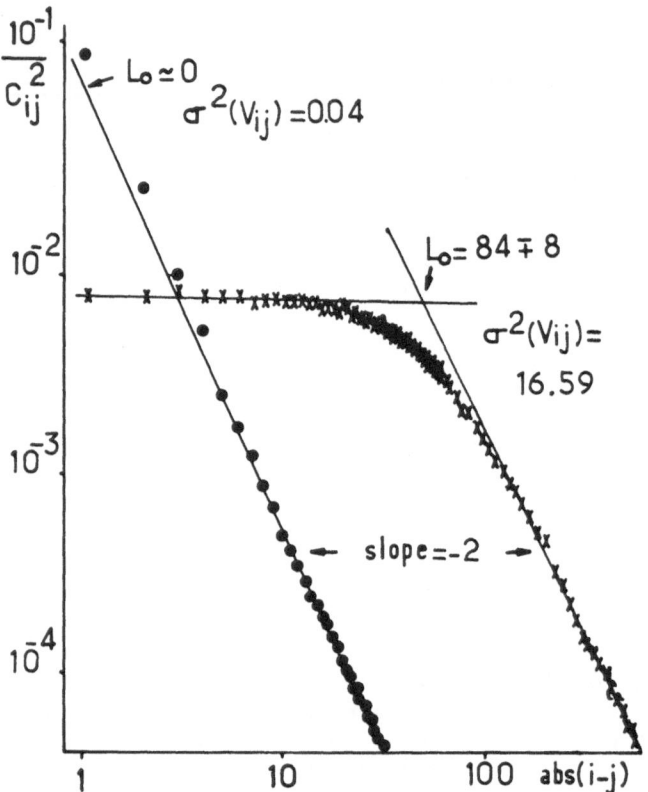

Figure 9: Two typical plots of C_{ij}^2 the average of squared eigenvectors coefficients versus the absolute value of (i-j). L_o is defined as the length of correlation.

On figure 10, the plot of L_o versus σ^2 (V_{ij}) shows a straight line with a slope of 2π. The comparison of the results obtained with matrices of size 160 and 800 shows the truncation effect on eigenvectors.

We concluded that for a very large uniform matrix (see Chapter II), $L_o = 2\pi \sigma^2$ (V_{ij}). As $\overline{V_{ij}} = 0$, σ^2 $(V_{ij}) = \overline{V_{ij}^2}$. Consequently, L_o is given by $L_o = 2\pi \rho^2 \overline{V_{ij}^2}$ which is a dimensionless Fermi golden rule (in our model $\rho = 1$).

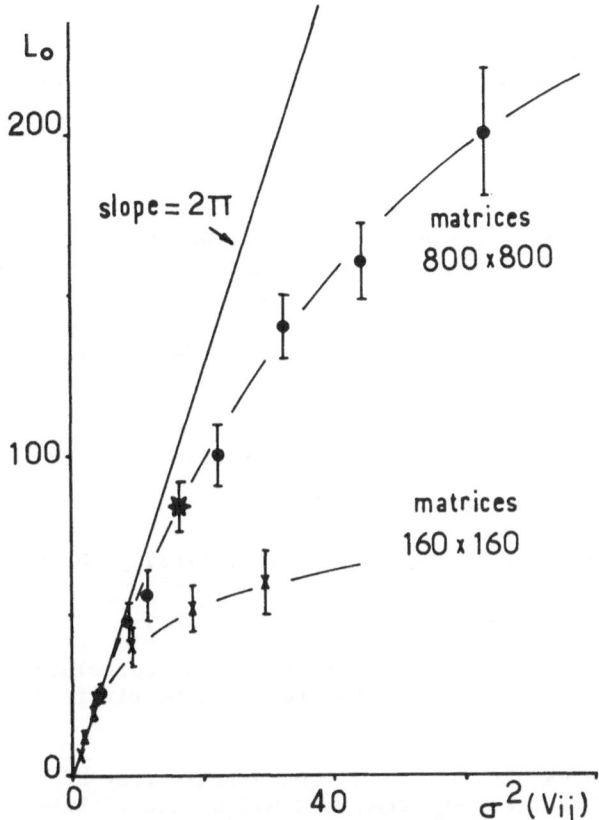

Figure 10 : Plot of L_o (see figure 9) versus the variance of V_{ij}. The star corresponds to figure 9 (L_o = 84 $\pm$ 8). The deviations from the full line of slope 2π are due to the truncation effect.

4.4. Truncation effects

A matrix which is very sparse may be decomposed into submatrices. Each submatrix may produce sets of strongly correlated eigenvalues but a superposition of several independent sets of eigenvalues gives much less correlations. This kind of decomposition is matrix size-dependent and therefore acts as a truncation effect. To avoid this effect the number N of non zero randomly distributed V_{ij} (for i > j) of d x d matrix, should be larger than approximately 3d. In other words, as the total number of off matrix elements is d(d−1)/2 (for i > j), the sparseness (α = 2N/(d−1)d) must be larger than 6/d. Conversely, consider a very large uniform random matrix, with a given degree of sparseness. Only a submatrix with d larger than 6/α will yield correct correlation properties of the very large initial matrix. The same effect qualitatively occurs even if the matrix has no strictly zero off diagonal matrix elements but only a lot of very small elements. As a result, the truncation of a large, uniformly filled matrix does not significantly change the correlation properties, as long as the matrix dimension remains larger than several times the normalized variance of off diagonal elements. On figure 8, 25 is roughly the smallest dimension which respects this criterion.

5. SUMMARY AND CONCLUSION

The correlation properties of random matrix for which the location of off diagonal matrix element is randomly distributed obeys very simple laws.

- NND is governed by our normalized coupling parameter $\overline{|V_{ij}|}$,

$$i.e. \quad \rho \; \overline{|V_{ij}|}$$

- long-range correlations are governed by our normalized $\overline{V_{ij}{}^2}$,

$$i.e. \quad \rho^2 \; \overline{V_{ij}{}^2}$$

- the eigenvectors are mainly an equally statistically weighted sum

  of about $2\eta \; \rho^2 \; \overline{V_{ij}{}^2}$ basis vectors.

- Truncation of a large matrix does not affect the correlation properties as long as the dimension of the truncated matrix is large enough.

    All these results may be used in the numerical study of vibrations of polyatomic molecules like SO_2 H_2CO and $H_2C_2O_2$ etc... Nevertheless, to apply these results to molecular systems, one must check that the assumption of uniform distribution of off diagonal elements holds.

6. ACKNOWLEDGEMENTS

I thank Drs T.H. Seligman, J.J.M. Verbaarschot, O. Bohigas, M. Lombardi and J.P. Pique for helpful discussions.

Computational facilities are provided by the scientific counsel of the "Centre de Calcul Vectoriel pour la Recherche", Ecole Polytechnique, France.

REFERENCES

1 - T.A. Brody, J. Flores, J.B. French, P.A. Mello, A. Pandey and S.S. Wong, Rev. Mod. Phys. 53, 385 (1981) and references cited within.

O. Bohigas and M. Gianoni, "Workshop on Mathematical and computational methods in Nuclear Physics" Lecture Notes in Physics 209 (1984) Springer Verlag.

2 - E. Haller, H. Köppel, L.S. Cederbaum, Phys. Rev. Lett. 52, n° 19 1665-68 (1984)

3 - T.H. Seligman, J.J.M. Verbaarschot, M.R. Zirnbauer, Phys. Rev. Letters 53 n° 3 215-217 (1984)

4 - D. Delande and J.C. Gay, Phys. Rev. Letters (accepted paper)

5 - R.B. Wattson and L.S. Rothman, J. of Mol. Spectros. 119 83-100 (1986)

6 - R. Jost, to be published

7 - M.V. Berry and M. Robnik, J. Phys. A : Math. Gen. 17, 2413 (1984)

8 - J. Verbaarshot, H.A. Weindenmuller and M. Zirnbauer, Ann. of Phys. 158, 78 (1984).

9 - L. Leviandier, M. Lombardi, R. Jost, J. P. Pique, Phys. Rev. Lett. 56, n° 23 2449-52 (1986) See also R. Jost and M. Lombardi, "Survey of correlation properties of polyatomic molecules vibrational energy levels using FT analysis" Proceedings of the Second International conference on quantum chaos Cuernavaca, Ed. T.H. Seligman (Springer Verlag) 1986.

10- J.M. DELORY and C. TRIC, Chem. Phys. 3, 54-69 (1974)

SPECTRAL FLUCTUATIONS : FROM ATOMIC NUCLEI TO MOLECULES

O. Bohigas
Division de Physique Théorique[*], Institut de Physique
Nucléaire, F-91406 Orsay Cédex

ABSTRACT. The existence of universality classes of spectral fluctuation
patterns is by now well established. For classically chaotic quantum
systems the fluctuations are those of eigenvalues of random matrices.
There is conclusive experimental evidence in atomic nuclei. The hydrogen
atom in a strong magnetic field deserves special mention. The search of
these fluctuation properties in molecular spectra is a challenging
problem

1. COMPOUND NUCLEUS AND RANDOM MATRICES

The whole development of nuclear physics has been decisively influenced
by the existence of a small window, in the region of the neutron binding
energy, within which the slow neutron reactions provide a probe of an
enormous resolving power. In this way it has been possible to see
individual resonances over some energy range. Since the earliest experiments Niels Bohr recognized that it is the strong coupling between the
incoming neutron and the many degrees of freedom of the target which
gives rise to the formation of a compound system —the compound nucleus—
with a lifetime very long compared to one-particle periods. It is the
purpose of the present contribution to briefly sketch developments, some
of them old, some other quite new, to which the study of the compound
nucleus resonances has given rise. Emphasis will be put on resonance
energies, whereas properties related to wave functions and strength
functions will remain untouched.

For our purpose the interesting experimental information is provided
by the positions of resonances of the compound nucleus. What is needed
are many *complete* and *pure* sequences of resonance energies corresponding
to states having the same quantum numbers J^π in a given energy range.
Indeed very characteristic spectral features are rapidly lost if the
sequence of levels is incomplete (missing levels) and/or polluted by
spurious levels due to erroneous spin-parity assignments. The relevant
information comes from neutron resonance spectroscopy and also from
high resolution proton scattering experiments, resulting from an impres-
sive long term collective effort performed mainly at Columbia University
(Liou et al 1975) and Duke University (Wilson et al 1975).

R. Lefebvre and S. Mukamel (eds.), Stochasticity and Intramolecular Redistribution of Energy. 45–55.
© 1987 by D. Reidel Publishing Company.

What is the situation from a theoretical point of view? At the region, for instance, of neutron threshold, the nuclear density is so high that one must give up a description of microscopic detail, a description of individual states. The aim of nuclear models at this and higher excitation energies is rather to describe average behaviours like level densities and special states like giant resonances, analogue states, etc., which have a peculiar structure. For the rest of the states, for the overwhelming and silent majority of anonymous states, what is needed is a statistical theory of energy levels which will not predict the detailed sequence of levels of each nucleus but which will describe the degree of irregularity of the level sequence that is expected to occur in any nucleus. In other words a statistical theory of level fluctuations (departures of the spectrum from its average behaviour). The random matrix theory initiated by Wigner and fully developped by Dyson, Mehta and others deserves this purpose (see Porter (1965), Mehta (1967), Brody et al (1981)). One considers the Hamiltonian H as an $N \times N$ stochastic matrix (the matrix elements are random variables) ; the random matrix ensemble is specified by the probability density $P(H) dH$ and one is interested in the eigenvalue fluctuations in the limit of large N. The underlying space-time symmetries obeyed by the system restrict the admissible matrix ensembles. If the Hamiltonian is time-reversal invariant and has integer angular momentum (or is rotation invariant with 1/2-odd-integer angular momentum), the Hamiltonian matrix can be chosen real and symmetric. When no other information except these general space-time symmetries is taken into account, one is led to the Gaussian Orthogonal Ensemble (GOE) of real symmetric matrices.

2. FLUCTUATION MEASURES. COMPARISON OF PREDICTIONS WITH EXPERIMENT

To discuss fluctuation measures, we shall refer to
  1) The *spacing distribution* $p(x)$ between adjacent levels
  2) Quantities directly related to the number statistic $n(L)$: given an interval of length L, it counts the number of levels contained in the interval. If one scales the spectra in such a way that the mean spacing is unity, the average value of $n(L)$ is L. We shall consider higher moments or cumulants of $n(L)$: *variance* $\Sigma^2(L)$, *skewness* $\gamma_1(L)$ and *excess* $\gamma_2(L)$.
  3) The Δ_3 *statistic of Dyson and Mehta*. Let $N(E)$ be the staircase function giving the number of levels up to energy E. Δ_3 measures, given an interval of length L, the least-square deviation of the staircase function from the best straight line fitting it. Its average value $\overline{\Delta_3}$ is related to the variance $\Sigma^2(L)$ of $n(L)$.

In general Σ^2 and $\Delta_3(L)$ can be expressed in terms of the 2-level correlation function, whereas $\gamma_1(L)$ ($\gamma_2(L)$) depends also on the 3-level (3- and 4- level) correlation functions(s).
  For the sake of comparison, let us first consider two limiting cases
  1) Take a random variable s whose probability density $p(x)$ is $\exp(-x)$. Construct a sequence $\{x_i\}$ as follows : $x_1 = 1/2$, $x_{i+1} = x_i + s_i$ $(i=1,2,...)$, where s_i are outcomes of independent trials of the variable s. The

resulting spectrum is what is called a *Poisson spectrum* (case of *maximum randomness*). One has $p(x)=e^{-x}$, $\Sigma^2(L)=L$, $\overline{\Delta}_3(L)=L/15$.

2) A *picket fence spectrum* or spectrum of the harmonic oscillator in one dimension $x_{i+1}=x_i+1$. It is the most ordered spectrum one can imagine, with *no randomness*. One has $p(x)=\delta(x-1)$, $\Sigma^2(L)=0$, $\overline{\Delta}_3(L)=1/12$.

The predictions of 'non trivial' models of level fluctuations, namely of GOE, are as follows. The spacing distribution is very well approximated by

$$p(x) = (\pi/2) \, x \, \exp(-(\pi/4)x^2)$$

It shows *level repulsion*, i.e. tendency to avoid level clustering or small probability of small spacings ($p(x)$ vanishes at the origin). The number variance for $L \gtrsim 1$ is given by

$$\Sigma^2(L) = (2/\pi^2)\ln L + 0.44$$

The logarithmic increase with L is to be compared to the linear increase for a Poisson spectrum. One can speak of *spectral rigidity* or of semi-crystalline nature of the spectrum, showing long range order. For large L, the average of $\Delta_3(L)$ is given by

$$\overline{\Delta}_3(L) \simeq (1/\pi^2)\ln L - 0.007$$

How well do GOE predictions compare with experimental data? Results are reproduced in Fig.1. As can be seen, all the fluctuation measures considered, which include a thorough study of 2-point measures and to some extent more than 2-point measures as well, are fully consistent with GOE predictions.

Now, from the theoretical point of view, are GOE-fluctuations specific of GOE or, on the contrary, are there other models that give GOE fluctuation patterns? And, from the experimental one, are the fluctuations of the nuclear resonances specific of nuclei or, on the contrary, are they also observed in other systems? Concerning the first question, one knows that there is a whole variety of models of random matrix ensembles that give GOE-fluctuations (some results are analytical, the other are obtained from Monte Carlo calculations). Concerning the second one, some atomic spectra have also been analyzed (Camarda et al 1983)[*]. And the outcome is that the fluctuations seem to be consistent with GOE predictions, although the statistical significance is much lower than for the nuclear case. We therefore see that, on the one hand, the spectra of very different systems (nuclei, atoms), when properly scaled, seem to have identical fluctuation patterns, even though they are governed by very different force laws (short range interactions and Coulomb long

---

[*] Some attempts to perform similar analyses of molecular spectra should be mentionned (Haller et al 1983, Abramson et al 1984, Mukamel et al 1984). However, although the spacing distribution seems to be consistent with the Wigner distribution, the results are inconclusive because of lack of spectral resolution resulting in a large position of missed and/ or spurious levels.

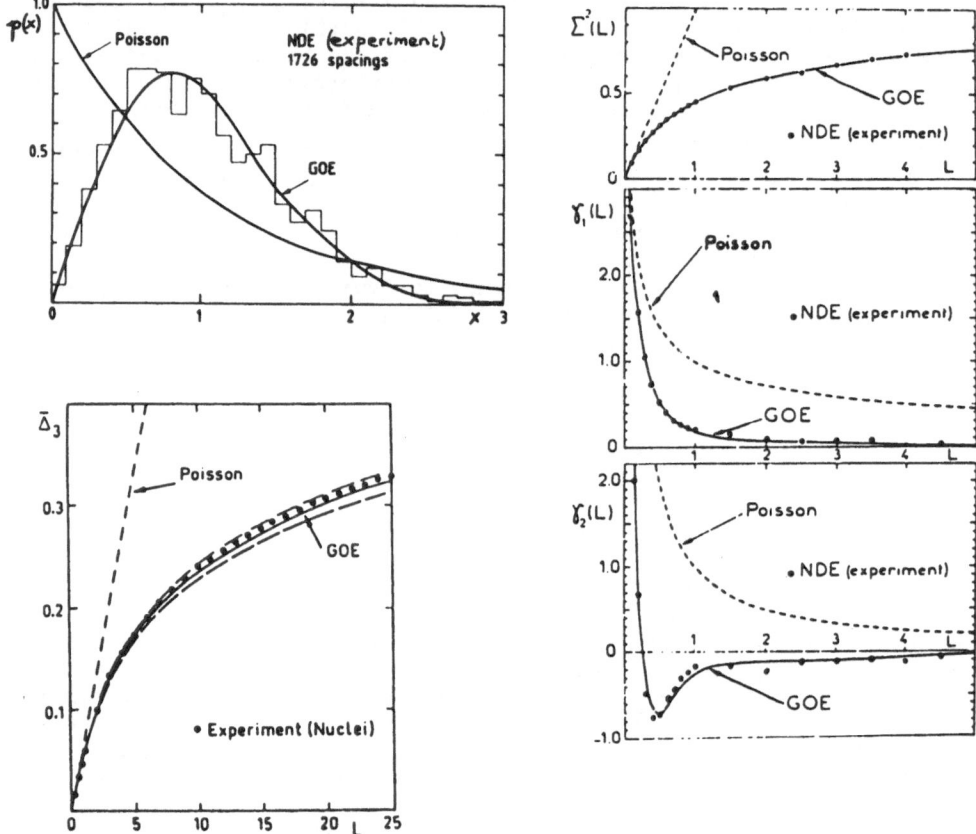

Figure 1. Results of fluctuation measures of nuclear resonances. Poisson and GOE predictions are given for comparison (taken from Haq et al (1982), Bohigas et al (1983), Bohigas et al (1985)).

range forces). On the other hand, these characteristic fluctuation patterns, although not specific of, are well reproduced by GOE, a *parameter-free theory*. Thus a simple picture emerges : the level fluctuation laws seem to be *universal*, as well from the experimental than from the theoretical point of view.

3. SPECTRAL FLUCTUATIONS OF CLASSICALLY CHAOTIC QUANTUM SYSTEMS

To obtain some clues on the origin of the universality of level fluctuation laws and also to investigate how 'complicated' a system must be in order to show GOE-fluctuations (Bohigas and Giannoni 1984a) let us very briefly remind the general scheme characterizing the route towards chaos for classically conservative Hamiltonian systems.

All conservative Hamiltonian systems with n degrees of freedom have in common three essential properties :
i) for a given set of initial conditions, the dimensionality of the accessible surface in phase space is less or equal to 2n-1 ; since the system is conservative, the energy is constant along this energy surface S_E.
ii) the volume element in phase space is conserved (Liouville's theorem); in other words, the Hamiltonian flow is incompressible.
iii) trajectories in phase space cannot cross.

Apart from these features the motion in phase space can exhibit a great variety of behaviours. For instance, one may ask how does a given volume element evolve with time. Does it tend to cover the whole energy surface S_E as time goes to infinity or does it remain in a restricted part of S_E? Does it conserve approximately its initial shape or does it display more or less dramatic deformations with time? According to the answers to such questions, one can define a hierarchy of regularity going from

integrable → ergodic → mixing → K-system

in the sense of regularity towards chaoticity. The integrable systems, which are the more regular ones and can be used as clocks, possess as many integrals of motion as number n of degrees of freedom. The motion in phase space of an integrable system is restricted to an n-dimensional torus, instead of a (2n-1)-dimensional energy surface for a generic system. For two-dimensional systems (n=2) one therefore has that a generic system will move on a 3-dimensional energy surface embedded in the 4-dimensional phase space, whereas an integrable system will move on a 2-dimensional torus.

In contrast with integrable systems, almost every trajectory of an ergodic system passes through almost every point of the energy surface, spending equal times in equal areas : the energy surface is asymptotically uniformly covered by a typical orbit.

Does ergodicity imply chaotic motion? For the motion to be irregular, erratic, a given volume element has to deform with time allowing for instability with respect to a perturbation. One therefore has to ask for a stronger property than simple ergodicity. Mixing systems are such that any volume element tends to 'dilute' uniformly in S_E as time goes to infinity, in the same way as a solute dilutes in a solvent if two liquids are miscible. Consequently, the distance between two points initially close to each other may become arbitrarily large as time in running.

The mixing property, however, tells nothing on the rate of separation of orbits. It only contains the concept of asymptotic equilibration. Ergodic systems which possess the strongest degree of irregularity (K-systems) have a further property besides mixing : their orbits separate exponentially with time. As a consequence of such a dramatic instability, long time predictions on the system are impossible, the memory of the initial state vanishing with time. Notice that such systems are deterministic in the sense that they are governed by causal equations : in principle they are predictable. However, due to the finite precision available for any practical purpose, one cannot follow their time evolution beyond some critical time. In the language of communication theory, K-systems are sources which continuously produce information (the so-called Kolmogorov entropy is positive), in contrast with integrable systems, whose motion is periodic or quasi-periodic, and for which knowing the history of any orbit during some given time interval is sufficient to determine with probability one its future evolution. Integrable systems considered as sources can be compared with records, which infinitely repeat the same message, whereas K-systems can be compared with a broadcast station, which is supposed to produce indefinitely new information.

There are very few explicit examples of dynamical systems for which it has been mathematically proved that they are K-systems. Among them, let us pay special attention to two-dimensional billiards. A billiard consists in the motion of a free point particle of mass m in a domain Γ of the plane of arbitrary shape. The particle is elastically reflected when it hits the boundary of Γ, according to the laws of specular reflection. There is at least one constant of the motion, namely the energy $E=(1/2)mv^2$. For particular shapes of the boundary, the system may be, at one extreme, integrable, or at the other extreme a K-system. For instance (see Fig.2) the circular billiard is integrable (the angular momentum is a second constant of the motion) whereas Sinai's and Bunimovich's stadium billiards are K-systems.

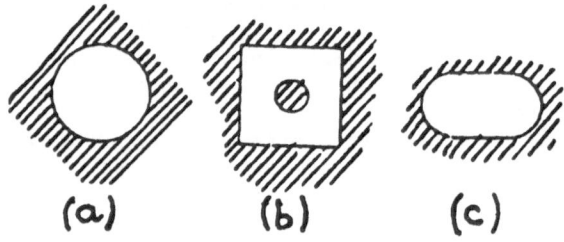

Figure 2. Different boundaries in which a
free particle is enclosed : (a) circular billiard ;
(b) Sinai's billiard ; (c) Bunimovich's stadium
billiard.

What are the spectral fluctuations of the corresponding quantum system? We have to solve the Schrödinger equation

$$(-\hbar^2/2m) \Delta \psi_n (\underset{\sim}{r}) = E_n \psi_n (\underset{\sim}{r})$$

with Dirichlet boundary conditions, namely vanishing of the wave function $\psi_n(\underset{\sim}{r})$ at the boundary of Γ. Results of the spectral fluctuations are reproduced in Fig. 3. The message seems fairly clear : the system whose classical analogue in integrable shows Poisson fluctuations (Berry and Tabor 1977) whereas the systems whose classical analogues are fully chaotic show GOE fluctuation patterns. By now, a considerable amount of

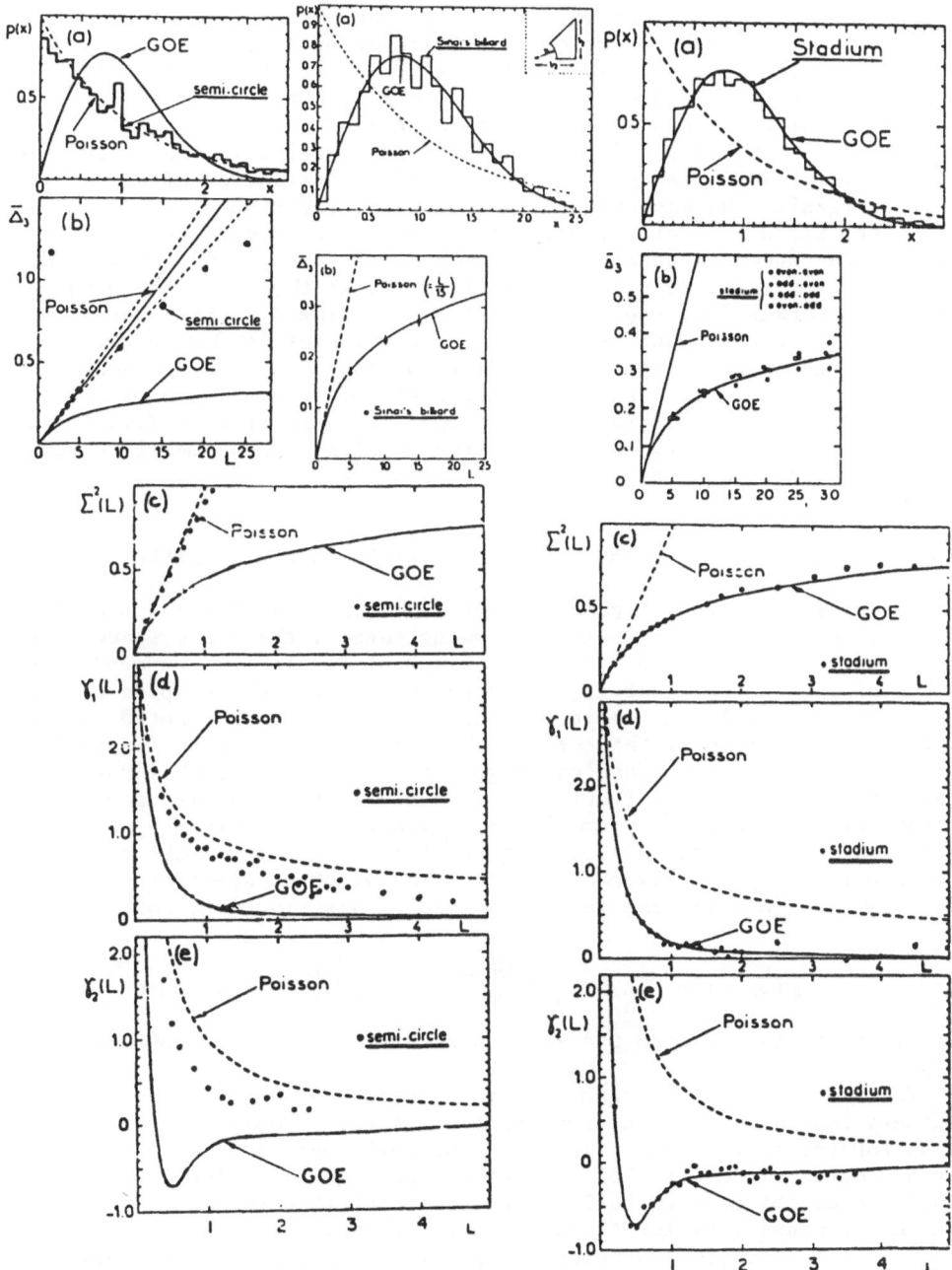

Figure 3. Spectral fluctuations of the systems shown on Fig.2 :
(a) spacing distributions p(x) ; (b) $\overline{\Delta}_3$(L) statistic ; (c) number
variance Σ^2(L) ; (d) number skewness γ_1(L) ; (e) number excess γ_2(L)
(taken from Bohigas et al 1984b, Bohigas et al 1984c).

numerical experience has accumulated and it supports the conclusions
drawn from the preceding examples (see Seligman et al 1986). To
summarize, these results establish the universality of the laws of
spectral fluctuations and explain the success of GOE in describing the
fluctuations of the compound nucleus resonances.

What is important now is to understand theoretically these findings.
Recently Berry (1985) has given a semiclassical derivation of the
spectral rigidity. He employs the semiclassical techniques introduced by
Gutzwiller, Bloch and Balian, which consists in an asymptotic representation of the spectral density as a sum over all the periodic orbits of
the classical system. He predicts universality in the regime $1 \ll L < L_{max}$,
with L_{max} determined by the shortest classical closed orbit ; for
$L \gg L_{max}$ there is a non-universal saturation value of $\overline{\Delta_3}(L)$. In the
universal regime, the long range predictions coincide with Poisson-,
(GOE-) predictions for classically integrable —, (chaotic and time-
reversal —) systems, in agreement with what is found numerically. It is
important to notice that, in deriving the results, random matrix theory
is not used and no statistical assumptions are made.

4. TRANSITION REGIMES. HYDROGEN ATOM IN A STRONG MAGNETIC FIELD

The classification of possible motions of classical systems mentioned above is not complete in the following sense : the phase space of a
generic system will show in general islands of regularity surrounded by
regions of chaoticity, that is, the structure of the phase space will be
neither purely regular nor purely chaotic. how is this reflected in the
spectral fluctuations of the corresponding quantum system (Seligman et
al 1984) ? Based on semiclassical arguments Berry and Robnik (1984)
suggest that the spectral fluctuations should result from independently
superposing a Poisson spectrum with relative weight μ, given by the
fraction of phase space of classical regular motion, and a GOE spectrum
with relative weight $\overline{\mu}$ given by the fraction of phase space of classical
chaotic motion ($\mu+\overline{\mu}=1$). The resulting spacing distribution, for instance,
is

$$p(x)=\mu^2 e^{-\mu x} \text{ erfc } (\frac{\sqrt{\pi}}{2}\overline{\mu}x)+(2\mu\overline{\mu}+\frac{\pi}{2}\overline{\mu}^3 x) \exp(-\mu x-\frac{\pi}{4}\overline{\mu}^2 x^2)$$

which interpolates between the Poisson result ($\overline{\mu}=0$) and the GOE result
($\overline{\mu}=1$).

The study of the hydrogen atom in a strong magnetic field, an
extremely interesting system for its fundamental 'simplicity' and for
its astrophysical relevance, is also especially well suites in the
present context (see Gay 1985 and references therein). The conceptual
importance of the problem follows from the fact that the two limiting
cases of zero and infinite field strength, namely the Coulomb and the
Landau (oscillator like) problems respectively, which have very different
symmetries, are the only three-dimensional problems which are exactly
soluble, as well classically than quantum mechanically. The motion of
the electron submitted to the action of the Coulomb and Lorentz forces
is governed, in the symmetrical gauge with vector potential $\vec{A}=(1/2)(\vec{r}\times\vec{B})$

(the magnetic field is in the z-direction), by the Hamiltonian (in a.u.)

$$H = \frac{p^2}{2} - \frac{1}{r} + \frac{\gamma}{2} \; L_z \; \frac{\gamma^2}{8} \; (x^2 + y^2)$$

γ is the reduced magnetic field strength B/B_c, $B_c = 2.35 \times 10^5$ T. The two last terms are the paramagnetic interaction associated with the normal Zeeman effect which is trivial in the present context and the diamagnetic interaction. L_z (and parity) are the only constants of the motion besides the energy. The importance of the diamagnetic effects are characterized by the ratio of the diamagnetic to Coulomb energy, which is proportional to $\gamma n^3 \propto |E|^{-3/2}$, where n is the principal quantum number and E is the energy. To magnify diamagnetic effects one can : i) increase the magnetic field (values of γ up to 10^{-4} can be achieved under laboratory conditions) ii) work with Rydberg atoms (one can achieve typically values of $n \approx 50$).

The phase space structure of this system (diamagnetic Kepler problem) has been recently studied in detail by Delande and Gay (1986). By increasing the value of $\gamma / |E|^{3/2}$, which is the significant parameter, one evolves from a motion which is fully regular, then a connected chaotic region appears and keeps increasing until it occupies the whole surface of energy, for a critical value of the parameter. How is this behaviour reflected in the spectral fluctuations of the corresponding quantum system? One expects a transition from the Poisson to the GOE regime (due to the particular spatial symmetry of the problem, although the system is not T-invariant one expects GOE-fluctuations, Robnik and Berry 1986). And this is indeed what is found when analyzing the computed spectrum. The results are illustrated in Fig. 4, where the fluctuation measures the states :

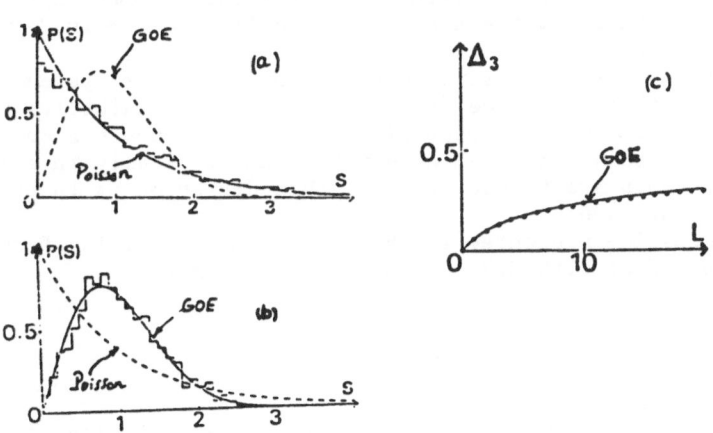

Figure 4. Hydrogen atom in a magnetic field :
(a) spacing distribution in the regular region ;
(b) id. in the fully chaotic region ; (c) spectral
rigidity in the fully chaotic region
(taken from Delande and Gay 1986).

having $L_z=0$ and even parity are reproduced (similar results have been obtained by Wintgen et al 1986 and Wunner et al 1986).

In summary, the Kepler diamagnetic problem offers a unique opport-unity for studying chaotic motion : i) it is an interesting problem from the point of view of symmetries, ii) the classical as well as the quantum properties can be computed accurately, iii) there is a transition from regular to chaotic motion when the field strength is increased, iv) it can be studied in the laboratory. For instance, the Bielefeld group (Holle et al 1986) performs measurements with magnetic field strengths B=6T and Rydberg atoms N=40, which correspond already to the chaotic regime.

5. REMARKS AND CONCLUSIONS

In the last few years, two areas —random matrix physics and the study of chaotic motion— that remained disconnected for a long time, have inter-penetrated each other. We know now that the fluctuations exhibited by the nuclear resonances, who gave rise to a large extent to the develop-ment of random matrix theories, are not specific of nuclei. On the contrary, they are generic of fully chaotic (time reversal invariant) systems as demonstrated by the study of many low-dimensional systems. It has also been possible to study in low-dimensional systems transitions from Poisson to GOE fluctuations corresponding to the regular to chaotic regime transition. The view that 'disorder' is a macroscopic concept, a property of systems made up of many particles, is incorrect as well classically then quantum-mechanically.

The hydrogen atom in a strong magnetic field deserves special mention : for its richness concerning symmetries, for its experimental feasibility, and because it can be theoretically treated in detail, it will probably become the chaotic system 'par excellence'.

The search of fluctuation properties of molecular systems is challenging. No clear evidence of molecular spectra exhibiting GOE fluctuation patterns has yet been given. To analyze spectra containing thousands of unresolved levels, as may be found typically in a molecular spectrum, new tools are needed. Work in this direction is in progress (Leviandier et al 1986).

The discussion presented in this contribution results from a collaborative effort with M.J. Giannoni and C. Schmit.

REFERENCES

Abramson E, Field R W, Imre D, Innes K K and Kinsey J L 1984 J. Chem. Phys. 80, 2298

Berry M V and Tabor M 1977 Proc. Roy. Soc. Lond. A356, 375

Berry M V and Robnik M 1984 J. Phys. A17, 2413

Berry M V 1985 Proc. Roy. Soc. Lond. A400, 229

Bohigas O, Haq R U, Pandey A 1983 in *Nuclear Data for Science and Technology*, Böckhoff K H (ed.), Reidel, Dordrecht

Bohigas O and Giannoni M J 1984 in *Mathematical and Computational Methods in Nuclear Physics,* Dehesa J S et al (eds.), Lecture Notes in Physics 209, Springer Verlag

Bohigas O, Giannoni M J and Schmit C 1984b Phys. Rev. Lett. 52, 1

Bohigas O, Giannoni M J and Schmit C 1984c J. Physique Lett. 45, L-1015

Bohigas O, Haq R U and Pandey A 1985 Phys. Rev. Lett. 54, 1645

Brody T A, Flores J, French J B, Mello P A, Pandey A and Wong S S M 1981 Rev. Mod. Phys. 53, 385

Camarda H S and Georgopulos P D 1983 Phys. Rev. Lett. 50, 492

Delande D and Gay J C 1986 Phys. Rev. Lett. 57, 2006

Gay J C 1985 in *Photophysics and Photochemistry in the Vacuum Ultraviolet,* Mc Glynn et al (eds.), Reidel

Haller E, Köppel H and Lederbaum L S 1983 Chem. Phys. Lett. 101, 215

Haq R U, Pandey A and Bohigas O 1982 Phys. Rev. Lett. 48, 1086

Holle A, Wiebusch G, Main J, Hager B, Rottke H and Welge K H 1986 Phys. Rev. Lett. 56, 2594

Leviandier L, Lombardi M, Jost R and Pique J P 1986 Phys. Rev. Lett. 56, 2449

Liou H I, Hacken G, Rainwater J and Singh U N 1975 Phys. Rev. C11, 462 and references therein

Mehta M L 1967 *Random Matrices and the Statistical Theory of Energy levels,* Academic Press

Mukamel S, Sue J and Pandey A 1984, Chem. Phys. Lett. 105, 134

Porter C E 1965 (ed.) *Statistical Theories of Spectra : Fluctuations,* Academic Press

Robnik M and Berry M V 1986 J. Phys. A19, 669

Seligman T H, Verbaarschot J J M and Zirnbauer M R 1984 Phys. Rev. Lett. 53, 215

Seligman T H and Nishioka H (eds.) 1986 *Quantum Chaos and Statistical Nuclear Physics,* Lecture Notes in Physics 263, Springer Verlag

Wilson W M, Bilpuch E G and Mitchell G E 1975 Nucl. Phys. A245, 285 and references therein

Wintgen D and Friedrich H 1986 preprint

Wunner G, Woelk U, Zech I, Zeller G, Ertl T, Geyer F, Schweitzer W and Ruder H 1986, preprint

EXCITED VIBRATIONAL STATES: SEMICLASSICAL SELF-CONSISTENT-FIELD AND
STATISTICAL CONSIDERATIONS

Mark A. Ratner
Department of Chemistry
Northwestern University
Evanston, IL 60201

R. B. Gerber
Fritz Haber Center for Molecular Dynamics and
Department of Physical Chemistry
Hebrew University
Jerusalem, Israel

V. Buch
Harvard College Observatory
Cambridge, MA   02138

ABSTRACT. An overview is presented of recent studies of highly-excited
vibrational states using both self-consistent-field (SCF) and statistical descriptions. The SCF scheme is the simplest possible extension
of the separable-modes picture, and provides quite adequate characterization of all vibrational levels at relatively low levels of excitation. At higher energies, above the classical threshold for chaos, most
SCF states are too strongly mixed for the description to remain valid.
These strongly-mixed situations are best described statistically, and we
present some results of statistical studies of molecular energy levels
and wavefunctions. Even at high energies, however, some SCF-type states
mix only very weakly. These states, for which the SCF-type description
remains valid, should show strong spectroscopic features in absorbtion
from lower states. They will generally be of "extremal-motion" type,
meaning that, because all vibrational energy greater than zero-point
energy is concentrated in a single mode (or possibly a very few modes),
the amplitudes for intramolecular vibrational energy transfer, and
therefore for mixing with other SCF states, are very small. It is
suggested that such high-energy SCF states have already been seen
spectroscopically in species such as ozone and HCN.

I.   INTRODUCTION

While excited vibrational states of molecules and of molecular models
are of interest for understanding such questions as the comparison of
classical chaos with quantum mixing, from a chemical viewpoint their
major importance is intrinsic: we wish to understand these vibrational

R. Lefebvre and S. Mukamel (eds.), Stochasticity and Intramolecular Redistribution of Energy, 57–80.
© 1987 by D. Reidel Publishing Company.

states, their energies, their spectra, and their dynamics. Figure 1 sketches the sorts of behavior expected in the vibrational states of a typical small molecule (H_2O, OCS, CH_2O...), as the vibrational energy content increases toward a value at which one of the bonds dissociates. In the low energy regime in which one or a few vibrational quanta are excited, the usual normal-mode picture will be valid, and vibrational energies will be given, to a fair approximation, by sums of normal-mode energies. For slightly higher levels of vibrational excitation, anharmonicities can no longer be ignored; the normal modes start to interact, harmonic selection rules start to fail as energy flow occurs among the harmonically approximated states. As energy continues to increase, such interactions become even more important, and as densities of states become even higher, many near-resonances occur and the states become strongly mixed. The work to be discussed in this paper is aimed at understanding and approximating, as well as possible, the vibrational states of typical three-to-six-atom molecules throughout this energy range.

Two types of state description will be used. At relatively lower energies (up to, say, one-third of the dissociation energy), we will employ a self-consistent-field (SCF) procedure[1-16] to correct, in an approximate but useful way, for the interactions among modes. The resulting SCF vibrational states are quite accurate in this energy range, but their accuracy depends substantially on the coordinate system chosen to describe the molecule. Single isolated resonances, for example of Fermi type, can be dealt with quite easily either by performing a small configuration interaction (CI)[7,17] calculation or by changing the choice of coordinates. As the energy continues to increase, overlapping resonances begin to occur and the classically chaotic regime is entered. In this energy range up to (and to some extent, even beyond) dissociation, the vibrational states are of two types: some SCF type states, which are largely unmixed and are responsible for strong spectral lines, will still be found, but most vibrational levels will be strongly mixed, and can best be described by a statistical picture[18-26] in which distributions of energy spacings and of wavefunction properties such as oscillator strength become relevant.

Section II is devoted to outlining the semiclassical self-consistent-field (SCF) technique for coupled vibrations, and to illustrations of its use for different molecules and in different coordinates. Section III describes SCF work on tunneling, in particular on the isomerization reaction HCN $\rightleftarrows$ HNC; it also includes an application of the time-dependent self-consistent-field (TDSCF) method to a model for libration in matrices, in which important energy transfer processes can occur among vibrational modes. Section IV discusses the regime of overlapping resonances, and presents both the SCF aspects and the statistical ones, briefly describing the sorts of states and energies involved. Some comments are given in Section V.

II. SCF VIBRATIONAL LEVELS, AND THE CHOICE OF COORDINATES IN SCF

A. SCF Formalism

The SCF approximation for coupled vibrations[1-17] is a precise analogue of the Hartree-Fock method for electronic systems: it attempts to find the best wavefunction of separable type for an interacting multimode system, and does so by replacing the true multimode potential acting on any given mode by its average over the other modes of the system. Formally, this means that, for the hamiltonian of an n-mode molecule

$$H = \sum_j (T_j + V_j) + \sum_{jk} W_{jk}^{(2)} +$$

$$+ \sum_{jkl} W_{jkl}^{(3)} + \ldots \tag{1}$$

$$= \sum_j T_j + V_{TOT} \tag{2}$$

where T_j and V_j are the kinetic energy and single-mode potential for the jth mode and $W^{(2)}, W^{(3)} \ldots$ are two-mode, three-mode, ... interactions, we approximate the wavefunction as a single product

$$\Psi(q_1 q_2 \ldots) = \prod_j \phi_j (q_j). \tag{3}$$

Then using the variational principle, the Schrodinger equation

$$H(q_1 q_2 \ldots q_n)\Psi(q_1 \ldots q_n) = E\Psi(q_1 \ldots q_n) \tag{4}$$

separates into n single-mode equations

$$\left[T_j + V_j + <_{i \neq j} \prod \phi_i | V_{TOT} |_{i \neq j} \prod \phi_i > \right] \phi_j(q_j) = \varepsilon_j \phi_j(q_j) \tag{5}$$

with the effective, averaged single-mode potentials

$$V_j^{(SCF)} = <_{i \neq j} \prod \phi_i | V_{TOT} |_{i \neq j} \prod \phi_i >. \tag{6}$$

The equation set (5) is solved self-consistently, and is the vibrational analogue of the Hartree equations. The densities $|\phi_j|^2$, energies ε_j, and effective potentials $V_j^{(SCF)}$ are actually found using semiclassical methods (Bohr-Sommerfeld quantization); details are given elsewhere.[6-14]

For coupled vibrations, just as for electronic-structure work, SCF approximations can be improved using the configuration interaction (CI) method.[4,7] Here the wavefunction is represented by a linear combination of SCF-like (product) states, with arbitrary coefficients determined using the variational principle. CI-type calculations on coupled vibrations have been given by several workers;[4,7] as the number of products increases, the results improve in quantitative accuracy, but become far less useful for physical interpretation.

B. Some Normal-Coordinate Results: CO_2

As a demonstration of the level of accuracy of SCF for relatively low-lying states, we present in table 1 some recent results of Romanowski[27] for the CO_2 molecule. For the states (01^10), (02^20), (00^01), and (01^11), which are quite far apart energetically, the SCF excitation energies are quite close to the experimental values (less than 0.5% off), and therefore CI correction is unnecessary. For the case of the (02^00) and (10^00) states, for which a standard Fermi resonance (stretch plus twice bend) occurs, the normal mode SCF data are nearly degenerate, and CI is necessary to yield the proper experimental energies. Thompson and Truhlar have shown[17] that a two-state CI, limited to the SCF states which are nearly degenerate, describes isolated pair resonances very well, and we have observed[27] that 3-state CI can deal very well with isolated 3-state resonances such as the (12^00), (20^00), and (04^00) states of CO_2, where an average error of 81 cm^{-1} for the SCF states is reduced to 12 cm^{-1} with a three-state CI.

The values in table 1 were obtained using a highly accurate potential surface derived from fitting observed vibrational data.[28] Some of the variation between SCF calculational results and experiment quoted in table 1 arises from the inaccuracy of the surface, some arises from SCF errors(lack of correlation among modes). The numbers show, however, that SCF techniques on proper surfaces yield quite good energies.

C. Normal and Local Coordinates

Extensive experimental and theoretical work on hydrocarbons and related molecules (SF_6, UF_6, H_2O, NH_3 ...) in which light atoms vibrate about heavy atoms with bond angles close to 90° has shown that, for vibrational states with quantum numbers greater than one or two, local modes prove a more accurate description of the spectrum than do normal modes.[29,30] A simple approximate local mode hamiltonian contains no coupling in the potential energy, and has only momentum coupling in the kinetic energy, and is written

$$H_{local} - \sum_{i,j} p_i p_j \cos\theta_{ij}/\mu_{ij} + \sum_i D_i \left\{1 - e^{-\beta_i(x_i - x_i^0)}\right\}2 \qquad (7)$$

where p_i is the momentum along local coordinate x_i, the potential is simply a set of Morse curves, and θ_{ij} is the angle between x_i and x_j. For saturated hydrocarbons, θ_{ij} is close to 109°, and the relevant mass μ_{ij} for coupling, say, two C-H stretches in methane is close to the mass of carbon, much larger than the reduced mass of the C-H bonds. Thus one expects, and indeed one finds, that in local coordinates there is only a very weak mode-mode interaction for such molecules, and local modes offer an attractive and accurate description.

Within the SCF, no corrections occur due to the $p_i p_j$ coupling, since the average of p_i in any bound state is zero. Therefore, one suspects that the SCF scheme might work better in normal than in local coordinates. Figure 2 presents some results obtained by Roth[8] in a two-mode model for HDO. Note that while the uncoupled local-mode energies indeed become better than the uncoupled normal-mode energies for quantum numbers greater than 5, the SCF corrections not only improve substan tially on the absolute energies of the states, but make the normal-mode choice most appropriate within the SCF.

The transformation to local coordinates from normal coordinates is a linear combination process. This transformation can be generalized, and a continuous set of coordinates can be generated by choosing arbitrary linear combinations. Thus, for two modes, one can write

$$\left. \begin{array}{l} Q_1 = x_1\cos\phi + x_2\sin\phi \\[6pt] Q_2 = x_1\sin\phi - x_2\cos\phi, \end{array} \right\} \tag{8}$$

where ϕ is an arbitrary value (not any physical variable), (x_1, x_2) are the local coordinates and (Q_1, Q_2) are the new, arbitrary coordinates. Eqn(8) is simply a generalized rotation, a continuous, linear transformation. Lefebvre,[31] Truhlar and Thompson,[17] and Moiseyev[32] have shown that the variational principle can be used to choose the mixing coefficient ϕ, and that, as the energy increases, ϕ changes so that the coordinates tend from normal to local.

D. Hyperspherical Coordinates

The SCF technique will yield different results when different coordinate systems are chosen--this is true in electronic structure work, as well as in coupled vibrations or nuclear structure theory, and is physically sensible, since certain coordinates seem appropriate for certain physical problems (spherical coordinates for rigid rotors, cartesian coordinates for hockey pucks). Table 2 outlines several possible schemes for selecting optimal coordinates for a given vibrational problem. The coordinates obtained from local or from normal coordinates via the generalized rotation of eq.(8) are just linear combinations of local modes, and as such may not be the most appropriate choices--for example, spherical coordinates cannot be obtained by such a rotation. This suggests that other sorts of coordinates, obtained by nonlinear transformation of the normal coordinates, might be of interest.

Self-consistent-field approximations are expected to be particularly useful if made in coordinates which are maximally collective, since collective coordinates lead to substantial dynamic screening of the localized motions.[12] Hyperspherical coordinates, which are of considerable current interest for scattering problems, are strongly collective, and would therefore seem very promising for SCF applications. One unpleasant feature of hyperspherical coordinates is more complexity: when more than two modes are involved, the hyperspherical kinetic energy operator is quite difficult to construct explicitly and the potential form can be quite difficult to deal with in SCF. Nevertheless, these modes have both formal and physical advantages and, from these preliminary studies, give quite good energies, so that these coordinates invite careful scrutiny.[14]

Gibson, et. al. have investigated the two-mode coupled Morse model of eqn(7) using hyperspherical coordinates: for this simple case, they are just the plane polar coordinates (r, θ) given by

$$x_1 = r\cos\theta \left.\begin{array}{c} \\ \\ \end{array}\right\} \qquad (9)$$
$$x_2 = r\sin\theta$$

where x_1, x_2 are the two local bond stretches. This yields a rather complicated potential, containing terms like $\exp\{2\beta r\cos\theta\}$. To perform the SCF calculations, these exponentials are expanded in a power series, so that averages such as those in eqn(6) can be evaluated. Some results are given in table 3; clearly, as expected from their collective nature, the hyperspherical energies do provide a significant improvement over local or normal coordinates for this CO_2 model. Similar calculations for H_2O again show some improvement,[14] though in that case it is marginal.

The hyperspherical coordinates are promising, and can produce quite good energy eigenvalues. Furthermore, at high energies the SCF eigenstates in spherical coordinates may be weakly mixed and therefore be responsible for observed[33-35] strong overtone spectra. However, hyperspherical coordinates are, as already suggested, clumsy to use; moreover, the transformation of eq(9), unlike the generalized rotation of eq(6), is not continuous and parametric--the coordinates have not been continuously optimized. More generally, one can define hyperspherical coordinates parametrically. Another coordinate system which is defined by a continuous, parametric nonlinear transformation of distance coordinates, not just by a rotation, is the set of spheroidal (or elliptical) coordinates, with interfocal distance 2a. They have been used recently in an SCF study of tunneling in the isomerization of HCN, to which we now turn.

III. SCF DESCRIPTIONS OF TUNNELING

A. HCN Isomerization: Static SCF

The SCF method can be employed in classical, quantum, or semiclassical mechanics.[12,13] It can deal easily with such special quantum phenomena as penetration into classically forbidden regions, zero-point phenomena and tunneling. Bacic et al. have recently[36] used semiclassical SCF methods to study the isomerization of HCN. Methodologically, this problem is particularly interesting, since it involves both use of a nonlinearly-determined coordinate set and a tunneling phenomenon.

The isomerization process in HCN involves the proton moving on a nearly elliptical path from the C to the N atom. Because the C≡N bond is so stiff, we freeze it at its equilibrium position. The minimum energy path on the potential surface is illustrated in figure 3, where its nearly-elliptical form is obvious. This strongly suggests that instead of discussing the motion of the proton in (x,y) coordinates, one might choose spheroidal coordinates, defined by

$$\left.\begin{array}{l} \xi = (r_1 + r_2)/2a \\[2mm] \eta = (r_1 - r_2)/2a \end{array}\right\} \tag{10}$$

where $2a$ is the interfocal distance between the defining foci and r_1, r_2 are the distances from the foci to the proton. The transformation (10) is linear, but since r_1 and r_2 are related to the protonic (x,y) coordinates by

$$\left.\begin{array}{l} r_1 = \{(x_1-x)^2 + y^2\}^{1/2} \\[2mm] r_2 = \{(x_2-x)^2 + y^2\}^{1/2} \end{array}\right\} . \tag{11}$$

The overall $(x,y) \rightarrow (\xi,\eta)$ transformation is both nonlinear and parametric--in this case the continuous parameter is $2a = x_1 - x_2$, the interfocal distance.

Choosing large values of $2a$ will move the (elliptical) contours of constant ξ farther out in space. Therefore one suspects that, as the total energy is increased, larger a values should be chosen. Note that $2a$ is not limited to be the C-N distance. In figure 3, contours of constant ξ with different choices of a are sketched.

There are two local energy minima on the HCN surface, corresponding to the two isomers HCN and HNC. Along the minimum energy path of figure 3, the energy contour is of the asymmetric double minimum type. Thus the rate of isomerization is simply the tunneling rate through this barrier. While a proper treatment of this rate involves a dynamical study, perhaps using time-dependent SCF (compare section IIIB), the rate can be estimated quite well, in cases such as this in which there is no substantial inter-mode energy transfer, using either static quantum or semiclassical techniques. Farrelly[37] has shown that uniform semiclassical quantization may be used to obtain good estimates of such rates. Here we use[36] an even simpler approximation to the tunneling rate

$$k_{m,n} = 2\omega_{m,n}P_{m,n} \tag{12}$$

$$P_{m,n} = \exp\left\{-2\int_{\eta_2}^{\eta_3} P_\eta \, d\eta\right\}, \tag{13}$$

where P_η is the momentum in the η coordinate, $\omega_{m,n}$ is the frequency along η in the SCF state with quantum numbers (m,n) for motion along (ξ,η) and η_2, η_3 are the two turning points. Verbally, eqns.(12,13) say that the tunneling rate is the attempt frequency times the barrier penetration probability per attempt; the factor of 2 arises from the bound character in the left and right wells.

Some results are shown in figures 4,5. In figure 4, the improvements due to the SCF procedure and to the choice of optimal coordinates are shown and clearly both the choice of proper coordinates and use of the SCF technique are needed to obtain good results. As expected the best choice of interfocal distance $2a$ is larger at larger total energies. Figure 5 shows interesting mechanistic behavior: as the quantum number m, corresponding to motion in the ξ coordinate (or to

C-H stretch in the reactant geometry), is increased, the isomerization
rate decreases; in contrast the rate increases as quantum number n along
the η coordinate (corresponding to H-CN bend in the reactant) is
increased. This is very sensible: larger m values take the system
point farther from the reaction coordinate, thus slowing the effective
reactant → product motion despite the increase in total energy.
Conversely, increase in n puts the energy into the elliptical/reaction
coordinate, just where it should be to promote reaction effectively.
This is a sort of mode-specificity effect: by using the physically-
motivated (ξ, η) coordinate set, we have been able to separate energy
deposition modes which stimulate or impede reaction.

These spheroidal coordinates, as suggested in table 2, are truly
optimal coordinates: they are physically motivated to correspond to a
real motion (the reaction path) in a molecule, but are defined by a
continuous, parametric transformation that can be optimized by choice of
the continuous parameter. We feel that choice of this sort of optimal
coordinates will, in general, render SCF studies both more accurate and
more useful in interpretation of experiments.

B. Tunneling in Matrices: Time-Dependent SCF

In the HCN isomerization, there is very little effective energy flow
from the ξ coordinate to the η coordinate, so that the static SCF
description may be employed to find the isomerization rate from the one-
dimensional result of eqn(12,13). In general, however, tunneling modes
can exchange energy with other motions such as stretches or rocks.
Under these conditions, the static SCF is not adequate to describe the
tunneling process, since the energy in any given mode is not constant.
The time-dependent SCF, which is a direct generalization of the SCF
idea, does describe energy flow quite effectively, and has been used to
do so in a number of situations ranging from dissociation of van der
Waals species to electronic excitation following atom/surface collisions
to vibrational relaxation in matrices.[38-42] We summarize here results
of a simple model study using TDSCF to examine tunneling rates for
reorientation in a matrix.[43]

Formally, the TDSCF equations are derived by solving the time-
dependent Schrödinger equation with the simple product ansatz

$$\Psi'(q_1 \ldots q_n, t) = \prod_j \phi'_j(q_j, t), \tag{14}$$

which means that at any time there is a single product of functions in
the different modes. The single mode TDSCF equations that can be
derived as

$$i\hbar\partial/\partial t \; \phi_j(q_j, t) = \left[T \; + < \prod_{i \neq j}\phi_i(q_i, t) \mid V_{TOT} \mid \prod_{i \neq j}\phi_i(q_i, t) > \right] \phi_j(q_j, t) \tag{15}$$

where Ψ, ϕ are related to Ψ', ϕ' of (16) by unimportant phase factors.
Just as was true for the static SCF equations (5), the TDSCF equations
(15) must be solved self-consistently. The time-dependent averaged
potential acting in mode q_j (last term of (15)) mediates energy flow
into and out of this mode.

The simple tunneling model consists of an oscillator coupled to a tunneling (libration) mode, and is written

$$H = H_{osc}(u) + T_0 + V_0(1 + 2u)(1 - \cos 2\theta),   \tag{16}$$

where H_{osc} is the harmonic oscillator hamiltonian, T_0 the kinetic energy of a librating mode, and V_0 the amplitude of the interaction, which is taken to first order in the oscillator displacement u. The parameters were chosen to correspond to libration of NH_3 in a matrix.

The calculation was performed[43] for several choices of the parameter $r_\omega = \hbar\omega/\Delta E$, where ω is the oscillator frequency and ΔE the energy gap from the lowest (split) level of the libration to the next lowest. Table 4 gives the relative tunneling times $r_r = r(n=1)/r(n=0)$ calculated from TDSCF and exactly for this model. Two features are of interest: first is the very good comparison between exact and TDSCF tunneling times, indicating the high accuracy of TDSCF for this problem. Second is the very much faster rate for tunneling in the resonance case ($\omega = \Delta E$), due to energy transfer from the phonon to the libron, thus permitting tunneling at higher energy (in the θ coordinate) corresponding to a thinner and lower tunneling barrier. This strong resonant enhancement is also well described in TDSCF. Thus we find here, as in previous work, that the TDSCF approach is a highly accurate approximation to the quantum dynamics of vibrational systems.

IV. HIGH VIBRATIONAL ENERGIES: SCF AND STATISTICAL BEHAVIOR.

Real molecular vibrations are anharmonic: as energy content increases, the density of states becomes very much larger, vibrational amplitudes (and therefore the sizes of anharmonic mixing amplitudes in the potential) increase sharply, and any simple separable approximation, such as the local-mode or normal-mode states, becomes untenable. Spectroscopically, direct absorption methods, using overtone spectroscopy or pump/-dump sequences, have been used to study states ranging quite close to, and even above, the dissociation threshold.[33-35] Figure 1 sketches, very schematically, the behavior of the vibrational eigenstates as energy is increased. Truly separable approximations are useful, in general, only for quite low energies. SCF states (or adiabatic states) are generally more accurate, since they include the average effect of the interactions. As was indicated in section II above, the SCF picture can be modified, by coordinate choice or by use of a small CI calculation, to deal with isolated few-state resonances. When the total energy exceeds several quanta, however, the density of states starts to climb very rapidly, and soon the spacing of levels becomes smaller than the average size of the mixing between zero order states. When this is true, we are in the regime of overlapping resonances, and the SCF or adiabatic states will, in general, be so strongly mixed by residual interactions that they represent a poor approximation to the true situation.

This regime of multiple overlapping resonances is, classically, chaotic. Quantum mechanically, the eigenvalue problem can still be

solved, but it becomes increasingly difficult and, perhaps, irrelevant
to do so. It has been suggested several times that, in this regime, the
statistical properties of the energy spectrum should be examined both to
indicate the existence of chaos and to characterize some of the
experimental observables.[21-26] We should like to extend these
statistical ideas to consider, on the one hand, the statistical
properties of wavefunctions and observables and, on the other hand, the
existence of SCF-type states which are nearly separable even at very
high energies.

A. Statistical Properties of Wavefunctions and Observables.

Even in the regime of overlapping resonances, one can still expand
eigenfunctions in basis states. Restricting ourselves, for notational
convenience, to a two-mode problem, one can then write

$$\Psi_\alpha(q_1, q_2) = \Sigma \ c^\alpha_{ij} \ u_i(q_1) \ u_j(q_2),$$ (17)

where the u_j are basis functions. Then the energy eigenvalues E_α are
still fixed by

$$H\Psi_\alpha = E_\alpha \Psi_\alpha.$$ (18)

The E_α will be spaced with respect to one another, and one can define
$g(\omega)d\omega$ as the number of nearest-neighbor energy spacings between $\hbar\omega$ and
$\hbar(\omega + d\omega)$. The quantity $g(\omega)$ has been studied extensively in nuclear
physics,[18] in metallic particles,[20] and in vibrational problems,[21-26]
where it is generally fit to a so-called Brody form[19]

$$g(\omega) = A\omega^\beta e^{-B\omega^{1+\beta}}$$ (19)

where A, B are constants. The parameter β takes the value o for a
Poisson distribution (appropriate for spacings of fully independent
states), 1 for a Wigner distribution (appropriate for the eigenvalues of
many types of random matrices and for vibrational eigenvalues far above
the chaotic threshold). The $g(\omega)$ distribution has been studied extensively for molecular vibrational problems; for models such as the
coupled Morse oscillators of eq(7),β is generally found to be close to
0.8 in value. The parameter β measures, in some way, the fraction of
energy space which is filled with overlapping resonances: for very low
energies, there are no overlapping resonances, β is nearly zero and
Poisson behavior is observed in $g(\omega)$, whereas for extremely high
energies all of energy space contains overlapping resonances, β
approaches unity and a Wigner distribution should be observed.
        Far less attention has been paid to the distributional properties
of wavefunctions and of wavefunction properties, though the results of a
numerical study of a two-mode coupled Morse system (eq(7)) have been
reported, and that study suggests some important conclusions about
highly excited vibrations.[21,22] One can discuss the three distributors
$P(c)dc$, $c^2(E_\alpha - \epsilon_i - \epsilon_j)$ and $\pi(\chi_{\alpha\beta})d\chi_{\alpha\beta}$ which are, respectively, the

probability distribution for observing a given coefficient value c in eq (17), the dependence of c^2 as a function of the energy gap between the eigenvalue and the energy of a basis state, and the probability distribution for observing a transition moment $\chi_{\alpha\beta} = \langle\Psi_\alpha|x|\Psi_\beta\rangle$. One expects, on prior grounds, that $P(c)$ should be a gaussian, that $c^2(\Delta E)$ should drop off with increasing ΔE (energy gap law) and that $\pi(\chi_{\alpha\beta})$ should decrease rapidly with $\chi_{\alpha\beta}$ (most transitions are to strongly mixed states, and are weak). Figures 6,7,8 show the results of calculation on the coupled-Morse model.[21,22] Both the gaussian behavior of $P(c)$ and the gap law behavior of $c^2(\Delta E)$ are very clearly seen, as is the dominance of very small terms in the $\pi(\chi)$ distribution.

The gaussian form of $P(c)$ and the preponderance of small values of $\chi_{\alpha\beta}$ both reflect the fact that the eigenstates are very highly mixed -- that most eigenfunctions Ψ_α contain contributions from many basis functions, each with a relatively small value of c. This in turn suggests that a <u>statistical wavefunction model</u> might be defined,[23] in which, for example, one approximates the correctly calculated value of the transition moment

$$\chi_{\alpha\beta} = \sum_{ijkl} c^\alpha_{ij}\, c^\beta_{kl} \langle u_i u_j\, |x|\, u_k u_l \rangle \tag{20}$$

by a statistical form

$$\chi_{\alpha\beta} \cong \sum_{ijkl} \hat{c}^\alpha_{ij}\, \hat{c}^\beta_{kl} \langle u_i u_j\, |x|\, u_k u_l \rangle \tag{21}$$

In both the correct form (20) and the approximate form (21) the basis set matrix elements $\langle u_i u_j\, |x|\, u_k u_l \rangle$ are calculated properly. The difference is that in the statistical approximation one replaces the correct coefficients c^α_{ij} by values $\hat{c}^\alpha_{ij}$ which are chosen randomly from a distribution of c's that satisfies the distributions determined by figures 6,7. Eq (21) then defines a statistical wavefunction method (SWM) for computing $\pi(\chi_{\alpha\beta})$. We have used SWM for several problems, including the calculation of vibrational overtone line widths.[45] For our purposes here, the SWM was used to compute an approximation to $\pi(\chi)$, which is given by the circles in figure 8.

There are three features to note in comparing the exact and SWM values for this $\pi(\chi)$ distribution.[23] The first is that, qualitatively the two agree fairly well. The second is that there is a discrepancy at $\chi_{\alpha\beta}=0$: these are the transitions which are forbidden by selection rules, which the SWM cannot describe. The third, and most interesting, point is that there exist several very strong transitions, with transition moment $|\chi_{\alpha\beta}| > 0.10$, which are observed in the correct $\pi(\chi)$ but not in the SWM approximation. The eigenstates responsible for these transitions are far more weakly mixed than are the great majority of eigenstates, which yield the (far more probable) small values of $\chi_{\alpha\beta}$. Indeed, these weakly-mixed eigenstates are dominated by a few large values of c^α_{ij} --they are not statistically mixed. One then wonders what the nature of these states is, and what role they might play in experiment.

Information about these states may be obtained by examining mixing
behavior of SCF states at high energies. In table 5 we present CI
results obtained by Romanowski for the water molecule,[27] using an SCF
basis. These calculations were done in normal coordinates. Note that
there are two sorts of states: the first group includes (220), (250),
(202) and (050), which are fairly strongly mixed; the second includes
(000), (002), and (300) and (400), which are determined by one single
basis function (with $c^2 > 0.80$), and contain many fewer large contributions. For the states in the second group, the SCF approximation is
quite good, as shown by the relatively small mixing. These states are
dynamically nearly decoupled from other SCF configurations, even ones
that lie quite close in energy. Similar results have been found by
Gibson in a study of OCS.[46]

We have also examined the state mixing more specifically for the
two-mode coupled Morse model (eq(7)). In table 6 we present some
results on normal-mode SCF states of that model.[47] Once again, we see
that there exist strongly mixed states and other states, at comparable
or even higher energies, which are only very weakly mixed (nearly SCF).
These strong SCF states are responsible for the nonstatistical (strong
transitions) behavior of figure 8; notice that they exist quite high up
in energy.

Thus we suggest, as indicated in figure 1, that there exist two
classes of states in the high-energy vibrationally excited region. One
class, which is by far the larger, is statistically distributed over the
basis functions in accordance with the P(c) distribution of figure 6,
exhibits energy spacings characterized by a Brody distribution for $g(\omega)$
with $\beta \geq 0.8$ and contains only relatively weak transition moments from
the ground state. The other, far smaller class contains SCF-like states
which are relatively weakly mixed, may contain fairly large progression-
like spacings and may often have very large transition moments from the
ground state. These SCF-like states, though small in number, may well
dominate the spectroscopic properties in this energy region, and we
suggest that the quite strong transitions seen at high vibrational
energies in molecules such as CO_2, HCN, C_2H_2, O_3, and NO_2 may arise from
transitions to these strong, SCF-type states.

V. COMMENTS

The self-consistent field model is the simplest correction to the
picture of separable vibrational modes. It assumes that the effect of
the true, many-mode potential energy on a given coordinate can be
replaced by its instantaneous average over all other modes, resulting in
single-mode equations containing either a static (eqn(5)) or dynamic
(eqn(15)) average potential. Clearly there is an error in this SCF
approximation -- this error is simply the correlation effect, arising
from the difference between the correct potential and the average
potential. This can be corrected for static problems by considering a
CI wavefunction (a weighted sum of many products) or for dynamical
problems by use of multiconfiguration TDSCF. For nearly all vibrational
states at low energies, and for many vibrational states at high ener

gies, the simplest SCF approximation is quite accurate both for energies
and for other properties such as geometries and transition moments.
Thus the SCF technique itself is of real value in understanding vibrational states and processes.

The semiclassical SCF has been used to study both model systems and
real molecules; it has also been used for a generalized Rydberg-Klein-
Rees inversion scheme to obtain potential surfaces[9,10,48] and in
analysis of geometric stability and dynamical geometric distortion.[11]
Its accuracy for any given energy range in any given molecule will
depend upon the coordinate system in which the SCF equations are
written; optimal coordinates should be chosen to correspond, if at all
possible, to the physically appropriate motions in the selected molecule
(perhaps based on preliminary classical trajectory studies).

The taxonomy of just which SCF states remain valuable (i.e. weakly
mixed) at higher energies is very incomplete. There are suggestions
that states with very unbalanced energy distributions will mix less
effectively than those in which the energy is already partly spread;[49]
examples are given in table 5, where the (400) state is far less mixed
than is the (202). This is expected on the basis of simple classical
ideas of interatomic collision: any modes in which the quantum number of
excitation is zero will have (except for zero-point) no excursions away
from equilibrium, and therefore will be far less prone to collisional
intramolecular energy transfer (resulting in a mixing of static states
in the time-dependent wavefunction) than will a comparable species with
nonzero quantum number. This argument may be slightly modified in
differing coordinates and, indeed, the best choice of SCF coordinates
for a given energy level is that in which the amplitude for this
intramolecular mixing is smallest; nevertheless, this point about the
nature of intramolecular transfer processes easily rationalizes why
particular states are far less mixed than others. These SCF-like,
weakly-mixed states will generally have all of the excitation energy
grouped in one or possibly a few modes. Examples include the (400)
state of table 5, the strong line states of O_3 which have been observed
spectroscopically,[50] and the states of the Henon-Heiles model that have
all their energy (other than zero point) either in the radial or the
angular coordinate.[49] States of this type are good SCF states even at
quite high energy, might exhibit strong spectral lines, and (classically) are quasiperiodic even well above the chaotic threshold.

There remains the challenge of characterizing completely, both
experimentally and theoretically, the nature of vibrationally-excited
states. In classical mechanics, one can distinguish quasiperiodic from
chaotic trajectories; it is tempting to identify the SCF-type states
with the quasiperiodic trajectories, and the statistically-defined
states with the chaotic behavior. That the regular states are nearly
separable (SCF-like) is an unanticipated, but useful, result. The
statistical characterization in terms of distributions of wavefunction
coefficients, matrix elements and energy spacings seems appropriate for
most states (above the classical chaotic threshold), while an SCF-type
description seems to characterize the (rarer) unmixed states. A
combination of the two theoretical descriptions should be possible, and

we have suggested the "statistical wavefunction model" as a first step
in that direction. Much more needs to be done, both to characterize
these states and to understand the validity limits of the characterization. The SCF picture appears to provide a powerful, adaptable and
flexible starting point for such characterization.

ACKNOWLEDGMENTS

The Fritz Haber Center is supported by the Minerva Gesellschaft, MBH,
München, West Germany. We are grateful to the Chemistry Division of the
NSF and to the donors of the Petroleum Research Fund, administered by
the American Chemical Society, for partial support of this research. We
thank Z. Bacic, B. Barboy, L. Gibson, H. Romanowski, R. M. Roth, and G.
C. Schatz for active and stimulating collaboration. M.R. thanks H. S.
Taylor for some valuable insights on mode mixing.

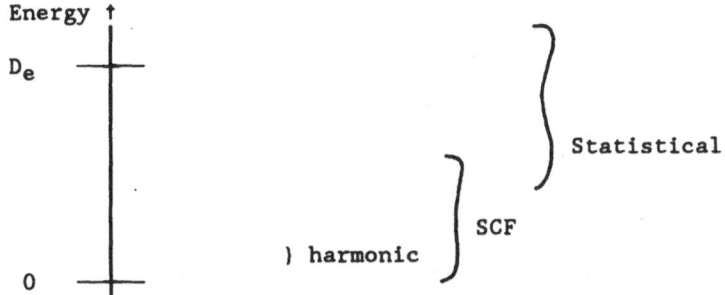

Figure 1. Highly schematic representation of the vibrational energy
behavior of a small polyatomic. The lowest-energy vibrational states
can be well described by harmonic or SCF treatments; at higher energies,
the harmonic approximation becomes quite inaccurate. Above the classical onset of chaos, most levels are statistically mixed, but some retain
SCF-type character up to, and even beyond, the dissociation energy.

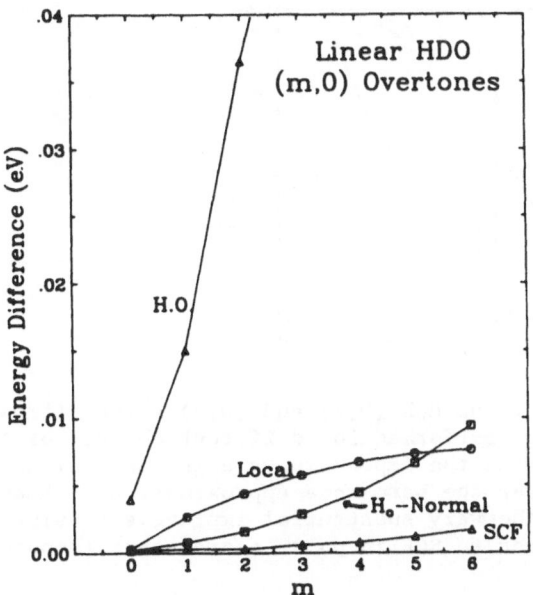

Figure 2. Semiclassical calculation of the (m,o) overtones for a linear
model of HDO (eqn (7)). The ordinate is the error between the exact and
approximate energies. Note that the uncoupled harmonic oscillator
(H.O.) is always quite inaccurate, that the local-mode approximation is
better than normal modes above m - 6, and that the SCF, in normal modes,
is quite accurate. From ref 7.

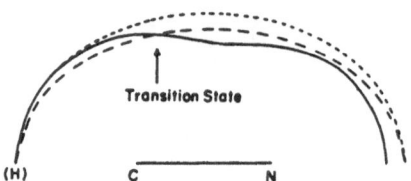

Figure 3. The minimum-energy path for HCN isomerization. The C ≡ N
distance is assumed fixed, and the curves show the exact mep (----), the
ellipse approximation with a = 1.2237Å(---) and the ellipse with a =
1.0631Å(---). The potential is that of J.N. Murrell et. al. (*J. Mol.
Spectr. 93*, 307 (1982)); 2a is the interfocal distance.

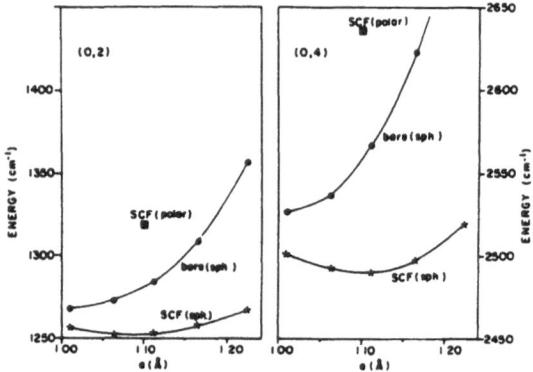

Figure 4. Energy levels of the HCN (0,2) and (0,4) states (from ref
36). THe calculations were performed for different choices of the
interfocal distance 2a. Note the improvements using optimal coordinate
SCF (denoted SCF (sph)) over the bare-mode approximation in these
optical coordinates, and the very substantial improvement using SCF in
optimal coordinates compared to SCF in polar coordinates (denoted SCF
(polar)).

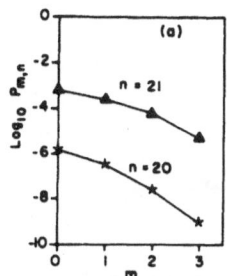

Figure 5. Tunneling probability for isomerization of HCN (from ref 36, 43). As the stretch quantum number m increases, or as the hyperspherical quantum number n decreases, the tunneling probability drops.

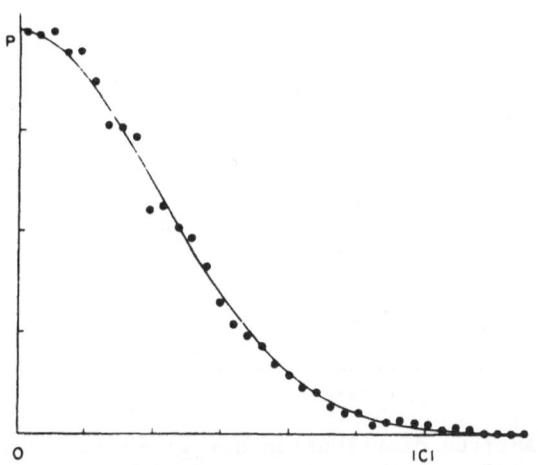

Figure 6. The probability distribution P(c) for wavefunction expansion coefficients c_{ij}^{α} of eq (17). The results (from ref 21, 22) are for a two-mode coupled Morse oscillator system as in eq (7).

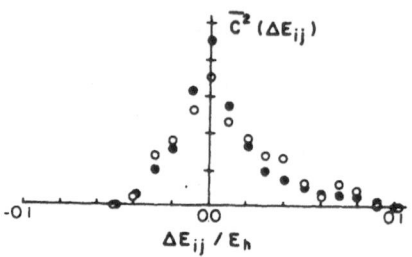

Figure 7. The variation of wavefunction expansion coefficients with energy gap $\Delta E = E_\alpha - \epsilon_{ij}$, where E_α is the eigenvalue and ϵ_{ij} is the energy of the basis function. Data as in figure 6.

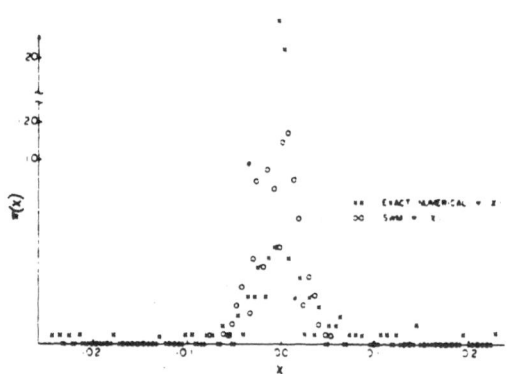

Figure 8. The probability distribution $\Pi(\chi_{\alpha\beta})$ for transition moments $\chi_{\alpha\beta}$. The crosses are for the exact computed distribution (eq (20), and the circles from the statistical wavefunction approximation (eq (21)). From ref 23.

Table I.   Some SCF Excitation Energies[a]

SCF State	$\Delta E_{SCF(cm^{-1})}$	ΔE_{CI}	ΔE_{expt}
01^10	671.1	668.1	667.4
02^20	1339.9	1337.7	1335.1
00^01	2356.1	2351.1	2349.2
01^11	3008.6	3006.6	3004.0
$\left\{ \begin{array}{l} 02^00 \\ \\ 10^00 \end{array} \right.$	1343.8 1345.9	1285.9 1390.1	1285.4 1388.2

(a)   Computed values from H. Romanowski and M. A. Ratner, to be published.   The potential surface used is that of A. Chedin, *J. Mol.
Spect.* **90**, 74 (1981).

Table II.   Choice of Optimal Coordinates Used for Vibrational Semi
classical SCF Work.

Coordinate Type	Transformation	Molecules	Reference
normal	linear, parametric	H_2O, CO_2	7
local	linear, parametric	H_2O, CO_2	7
hyperspherical	nonlinear, nonparametric	H_2O, CO_2	14
spheroidal	nonlinear, parametric	HCN	36

Table III.   Computed Energy Levels of CO_2.[a]

Level[b]	$E_{SCF}^{hyperspherical}$	E_{exact}	E_{SCF}^{normal}
(0,0)	.2303	.2309	.2309
(1,0)	.3849	.3849	.3862
(0,1)	.5305	.5306	.5316
(2,0)	.5390	.5377	.5404
(1,1)	.6824	.6808	.6858
(0,2)	.8259	.8268	.8264

(a)   From L. Gibson, M. A. Ratner, R. M. Roth and R. B. Gerber, *J. Chem. Phys.*, *85*, 3425 (1986).   Values are given in eV.

(b)   The quantum numbers refer to the two modes (lower, then higher frequency).

Table IV.   Calculated TDSCF Tunneling Times for the librator/vibrator Model[a]

$\hbar\omega_{osc}/\Delta E_r$	τ_1/τ_0 (SCF)	τ_1/τ_0 (EXACT)
.333	0.880	0.866
.667	0.675	0.642
1.000	0.074	0.069

(a)   Results for Model Hamiltonian of eq (16).   From R. B. Gerber and M. A. Ratner, *Proc. Seventeenth Jerusalem Symposium* (J. Jortner, B. Pullman, eds., Riedel, Dordrecht, in press).

Notation: τ_1 and τ_0 are tunneling times in $v = 1$, $v = 0$ states of oscillation; ΔE_r is the energy spacing of the lowest two libration tunneling states.

Table V.   SCF Studies of H_2O: Extent of CI mixing[a]

<u>H_2O Normal Modes</u>

(n_s	n_b	n_a)	E_{CI}[b]	c^2_{max}	#($|c| > .15$)[c]
0	0	0	4679	.99	1
0	5	0	12214	.64	4
0	0	2	12310	.85	1
2	2	0	15300	.64	5
3	0	0	15865	.86	1
2	5	0	19560	.25	11
4	0	0	19579	.81	2
2	0	2	19854	.38	9

(a)   From H. Romanowski and M.A. Ratner, to be published.   The three
      quantum numbers are, in order, symmetric stretch, bend, and
      asymmetric stretch. The coefficients c are those in the CI
      expansion of the exact state in virtual SCF states.

(b)   In cm^{-1}

(c)   The number of coefficients of absolute value exceeding 0.15.

Table VI.   Extent of mixing in a Linear Model of ABA[a]

MODEL: Linear ABA

$$\delta^2 = \left\{ \frac{H_{ij} - H_{ii} S_{ij}}{H_{ii} - H_{jj}} \right\}^2$$

n_a	n_{symm}	δ^2_{max}	E/D_e
2	3	.025	0.2
5	10	.26	0.35
7	20	1.51	0.65
0	40	.065	0.75
18	0	.23	0.55

(a)   From V. Buch, unpublished.   The model consists of two degenerate
      local modes mixed by the kinetic energy term $p_1 p_2/\mu$, as in eqn
      (7).

REFERENCES

1   J.M. Bowman, *J. Chem. Phys.* **68**, 608 (1978).
2   G.D. Carney, L.I. Sprandel and C.W. Kern, *Adv. Chem. Phys.* **37**, 305 (1978).
3   M. Cohen, S. Greita and R.P. McEachran, *Chem. Phys. Lett.* **60**, 445 (1979).
4   J.M. Bowman, K. Christoffel and F. Tobin, *J. Phys. Chem.* **83**, 905 (1979).
5   H. Romanowski, J.M. Bowman and L.B. Harding, *J. Chem. Phys.* **82**, 4155 (1985).
6   R.B. Gerber and M.A. Ratner, *Chem. Phys. Lett.* **68**, 195 (1979).
7   M.A. Ratner, V. Buch and R.B. Gerber, *Chem. Phys.* **53**, 345 (1980).
8   R.M. Roth, M.A. Ratner and R.B. Gerber, *J. Phys. Chem.* **87**, 2376 (1983).
9   R.B. Gerber, R.M. Roth and M.A. Ratner, *Mol. Phys.* **44**, 1335 (1981).
10 R.M. Roth, M.A. Ratner and R.B. Gerber, *Phys. Rev. Lett.* **52**, 1288 (1984).
11 B. Barboy, G.C. Schatz, M.A. Ratner and R.B. Gerber, *Mol. Phys.* **50**, 353 (1983).
12 R.B. Gerber and M.A. Ratner, *Adv. Chem. Phys.*, submitted.
13 M.A. Ratner and R.B. Gerber, *J. Phys. Chem.* **90**, 20 (1986).
14 L.L. Gibson, R.M. Roth, M.A. Ratner and R.B. Gerber, *J. Chem. Phys.* **85**, 3425 (1986).
15 A.D. Smith, W.-K. Liu and D.W. Noid, *Chem. Phys.* **89**, 345 (1984).
16 P. Schatzberger, E.A. Halevi and N. Moiseyev, *J. Phys. Chem.* **89**, 4691 (1985).
17 T.C. Thompson and D.G. Truhlar, *J. Chem. Phys.* **77**, 3031 (1982).
18 C. E. Porter, *Statistical Theories of Spectra: Fluctuations* (Academic, New York, 1965).
19 T.A. Brody et. al., *Revs. Mod. Phys.* **53**, 385 (1981).
20 W.P. Halperin, *Revs. Mod. Phys.* **58**, 533 (1986).
21 V. Buch, R.B. Gerber and M.A. Ratner, *J. Chem. Phys.* **76**, 5397 (1982).
22 V. Buch, M.A. Ratner and R.B. Gerber, *Mol. Phys.* **46**, 1129 (1982).
23 R.B. Gerber, V. Buch and M.A. Ratner, *Chem. Phys. Lett.* **89**, 171 (1982).
24 P. Pechukas, *Phys. Rev. Lett.* **51**, 943 (1983).
25 E.B. Stechel and E.J. Heller, *Ann. Revs. Phys. Chem.* **35**, 563 (1985).
26 E. Haller, H. Köppel and L.S. Cederbaum, *Phys. Rev. Lett.* **52**, 1665 (1984).
27 H. Romanowski, M.A. Ratner and R.B. Gerber, unpublished.
28 A. Chedin, *J. Mol. Spectros* **76**, 430 (1979).
29 e.g. M.S. Child and L. Halonen, *Adv. Chem. Phys.* **57**, 1 (1984).
30 B.R. Henry, *Acc. Chem. Res.* **10**, 207 (1977).
31 R. Lefebvre, *Int. J. Quant. Chem.* **23**, 543 (1983).
32 N. Moiseyev, *Chem. Phys. Lett.* **98**, 223 (1983).
33 K.K. Lehmann, G. Scherer and W.A. Klemperer, *J. Chem. Phys.* **77**, 2853 (1982); G.J. Scherer, K.K. Lehmann and W.A. Klemperer, *J. Chem. Phys.* **78**, 2817 (1983).

34   D.E. Reisner, P.H. Vaccaro, C. Kittrell, R.W. Field, J.L. Kinsey
     and H-L Dai, *J. Chem. Phys.* **77**, 575 (1982); D.E. Reisner, R.W.
     Field, J.L. Kinsey and H-L Dai, *J. Chem. Phys.* **78**, 2817 (1983);
     E. Abramson, R.W. Field, D. Imre, K.K. Innes and J.L. Kinsey, *J.
     Chem. Phys.* **83**, 453 (1985).

35   D. Bailly, R. Farrenq, G. Guelachivili and C. Rossetti, *J. Mol.
     Spectros.* **90**, 74 (1981).

36   Z. Bacic, R.B. Gerber and M.A. Ratner, *J. Phys. Chem.* **90**, 3606
     (1986).

37   D.H. Farrelly and A.D. Smith, *J. Phys. Chem.* **90**, 1599 (1986).

38   R.B. Gerber, V. Buch and M.A. Ratner, *J. Chem. Phys.* **77**, 3022
     (1982).

39   G.C. Schatz, V. Buch, M.A. Ratner and R.B. Gerber, *J. Chem. Phys.*
     79, 1808 (1983).

40   V. Buch, R.B. Gerber and M.A. Ratner, *Chem. Phys. Lett.* **101**, 44
     (1983).

41   Z. Kirson, R.B. Gerber, A. Nitzan, and M.A. Ratner, *Surface Sci.*
     137, 527 (1984).

42   L.A. Eslava, R.B. Gerber and M.A. Ratner, *Mol. Phys.* **56**, 47 (1985).

43   R.B. Gerber, Z. Bacic and M.A. Ratner, Proc. 19 Jerusalem Confer-
     ence, Ed. B. Pullman and J. Jortner (Reidel, Dordrecht, 1986).

44   e.g. R.L. Sundberg, E. Abramson, J.L. Kinsey and R.W. Field, *J.
     Chem. Phys.* **83**, 466 (1985); S. Mukamel, J. Sue and A. Pandey, *Chem.
     Phys. Lett.* **105**, 134 (1984).

45   V. Buch, R.B. Gerber and M.A. Ratner, *J. Chem. Phys.* **81**, 3393
     (1984).

46   L. Gibson, unpublished.

47   V. Buch, unpublished.

48   H. Romanowski, M.A. Ratner, R.B. Gerber, to be published.

49   K. Stefanski and H.S. Taylor, *Phys. Rev. A.* **31**, 2810 (1985);
     H.S. Taylor, preprint.

50   D. Imre, Ph.D. Thesis, Mass Inst. of Technology, 1985.

SEMICLASSICAL QUANTIZATION BY USING THE METHOD OF ADIABATIC SWITCHING OF THE PERTURBATION

H.S. Taylor and T.P. Grozdanov*
Department of Chemistry
University of Southern California
Los Angeles, California 90089

ABSTRACT. The semiclassical method based on the Einstein-Brillouin-Keller quantization of invariant phase-space tori and the hypothesis of the adiabatic invariance of classical action variables for near-integrable systems is discussed. The method is applied to calculation of energy spectra of two-dimensional coupled oscillator systems. Results of calculations for both, non-resonant and Fermi-resonant systems are in good agreement with quantum-mechanical predictions even when the corresponding classical dynamics is characterized by mild chaos. The general limitations of the method are discussed.

1. INTRODUCTION

The fully quantum-mechanical description of excited vibrational motion in polyatomic molecules rapidly increases in difficulty as the level of excitation and/or number of degrees of freedom increases. In the traditional variational approach this is manifested as a rapid increase in the size of the basis set required to achieve convergence. On the other hand, the numerical generation of classical trajectories for these systems is more feasible and many efforts have been made to address the problem by resorting to semiclassical methods of quantization[1]. From the more fundamental point of view, the development of various semiclassical methods of quantization contributes to a better understanding of some problems related to correspondence between the classical and quantum mechanics, such as, for example, the repercussions of the onset of classical chaos on the dynamics of the corresponding quantal system.

The semiclassical quantization of near-integrable (i.e. not strongly perturbed integrable) systems has for the most part been based on the well known Einstein-Brillouin-Keller (EBK) quantization[2] of the invariant phase-space tori, i.e. classical action variables[1a]. By adding the hypothesis of adiabatic invariance of classical action variables for near-integrable systems, Solov'ev[3] has proposed a method of semiclassical quantization to which we hereafter refer to as adiabatic

---

*On leave of absence from: Institute of Physics, 11000 Belgrade, Yugoslavia

R. Lefebvre and S. Mukamel (eds.), Stochasticity and Intramolecular Redistribution of Energy, 81–94.

switching (AS) method. The previous studies and numerical experiments[3-10] have revealed the possibilities and limitations of the method which will be discussed throughout the present work. We start by outlining below the general scheme of the method.

In order to quantize a near-integrable classical Hamiltonian H one starts by chosing an integrable reference Hamiltonian $\tilde{H}$ whose corresponding dynamics should be qualitatively as close as possible to that of H. The basic idea of the AS method is to perturb the motion on a quantized (by means of EBK rules) torus of $\tilde{H}$ by adiabatically switching on the perturbation $H-\tilde{H}$. Importantly, $H-\tilde{H}$ need not be small. If, in addition, one assumes that the action variables are adiabatic invariants (i.e. remain unchanged during the switching time-interval) then the trajectory obtained at the end of the switching procedure will sweep out a quantizing torus of Hamiltonian H which can be labeled by the same set of quantum numbers as the starting one. The quantized energy of H is simply given by the energy of the trajectory at the end of the switching time-interval. The classical adiabatic theorem[11,12] under certain conditions, guarantees that this scheme is well founded for the systems which remain separable during the switching procedure. There is no rigorous result of this king for non-integrable systems. Nevertheless, as previous studies[3-10] have shown, one can expect that the adiabatic invariance of the action variables is approximately (in the sense discussed in Section 2) maintained for near-integrable systems. The practical realization of the above scheme consists in construction of the Hamiltonian

$$H(s) = \tilde{H} + s(t/T)(H-\tilde{H}) \tag{1}$$

which is time dependent via the slowly-varying, smooth switching function $s(t/T)$, satisfying $s(t=0)=0$, $s(t=T)=1$, so that $H(t=0)=\tilde{H}$ and $H(t=T)=H$. Various forms of switching functions have been discussed in ref. 7. The most commonly used one is given by:

$$s(t/T) = t/T - Sin(2\pi t/T)/2\pi, \tag{2}$$

The switching time-interval T(proportional to the inverse of the switching rate) should be chosen in such a way as to ensure the approximate adiabatic invariance of action variables. One then takes a phase space point on a quantizing torus of $\tilde{H}$ as an initial condition, solves numerically from t=0 to t=T the Hamilton's equations of motion corresponding to Hamiltonian given by eq. (1) and obtains the quantized energy level of H. Note that in principle only one trajectory is needed to obtain a semiclassical approximation to the corresponding quantized level. In practice, one can never ensure the exact adiabatic invariance and a weak dependence on the initial conditions (i.e. the initial values of angle variables) developes. Hence the method[3,6] can be better implemented by randomly choosing N sets of initial conditions on the particular quantizing tours of $\tilde{H}$ and by averaging the set of results obtained for quantized energies of H as

$$\bar{E} = \frac{1}{N} \sum_{i=1}^{N} E_i, \tag{3}$$

The corresponding standard deviation

$$\Delta E = \left[\frac{1}{N} \sum_{i=1}^{N} (E_i - \bar{E})^2 \right]^{\frac{1}{2}} \tag{4}$$

is useful for internal consistency checks but in addition turns out to be an important indicator of the violation of adiabatic invariance during the switching procedure[6].

In Section 2 we summarize the limitations of the AS method by referring to previous work[3-10]. It should be emphasized that at present no rigorous results exist concerning this subject and most of the conjectures rely only on the numerical experiments. The two examples presented in Section 3 demonstrate the power of the method and illustrate some of the difficulties and limitations which are encountered in the practical applications of the AS method. Finally, Section 4 contains concluding remarks.

2.   LIMITATIONS OF THE AS METHOD

When analysing the limitations of the AS method it is instructive to assume for the moment that H(s) and hence H is separable. Then for a particular choice of H the classical adiabatic theorem[11,12] ensures the adiabatic invariance of the action variables provided that the switching rate (~ 1/T) is much smaller than any instantenous frequency of the system. Thus the adiabatic invariance can be violated only if one of the frequencies in any of the separable degrees of freedom passes through zero. (The most familiar violation of adiabaticity occurs in one dimensional systems when energy "touches" (is equal to) the top of the potential barrier). In phase space this corresponds to the "passage" of an adiabatically followed torus through a multidimensional separatrix and therefore corresponds to a change in the topology of torus and eventually in the type of the motion (i.e. librational to rotational and vice versa) in some of the separable degrees of freedom. Therefore, for a particular choice of H a restricted family of invariant tori of H can be quantized by the AS method.

It is also important to note that in the course of adiabatic switching instantenous frequencies associated with the separable degrees of freedom change continuously and therefore pass through resonances (become commensurate) infinitely often. For separable systems this causes no non-adiabatic effects. Note also that during the switching interval resonances occur at set of points of measure zero along the time axis.

If H(s) in eq. (1) is near-integrable the situation becomes more complex. In this case, in addition to non-adiabatic effects of the type analogous to that described above for separable systems, the role of resonances become crucial[5,6]. Namely, as the non-separable interaction is turned on, in the vicinity of resonant tori (i.e. those with commensurate frequencies) resonances zones[13] develope. Regular dynamics in the vicinity of a particular resonance zone can be in principle approximated by an integrable Hamiltonian with a pendulum like Hamiltonian describing the motion in one of the degrees of freedom. Thus, one can distinguish between the tori interior and exterior to resonance zone which are separated by multidimensional separatrix and characterized by different topologies. Moreover, as the Kolmogorov-Arnold-Moser (KAM) theorem[12] states, some of the tori are destroyed and replaced by phase space regions of small measure, filled with chaotic orbits. As the coupling increases the latter develop into chaotic layers located near the separatrices[13].

It is clear now, that in an AS calculation the "adiabatically" followed torus, when passing through resonance zones will undergo change in topology and even be temporarily lost during the switching procedure. Rigorously speaking these effects clearly cause the violation of adiabatic invariance, but here, one should distinguish between the effects caused by the high-order (weak) and low-order (strong) resonances[13]. In the first case one expects[5,6] that the non-adiabatic effects are manifested only on the very short timescale, so that one can choose the switching rate sufficiently small to ensure the approximate adiabatic invariance on the larger timescale, but sufficiently large to avoid a thorough explorations of high-order resonance zones. In other words, the working hypothesis of the AS method is that as long as one deals with high-order resonances which do not occupy substantial regions in phase space, their nonadiabatic effects can be ignored. In a sense they are passed diabatically.

The situation changes drastically when in the course of adiabatic switching a low order strong resonance is encountered. Since in this case the resonance zone occupys a substantial volume of phase space it cannot be ignored nor meaningfully passed through in a diabatic manner. The passage through the corresponding separatix, in this case, causes a violation of adiabatic invariance even on a larger timescale. In order to overcome the problem[6,9] it is necessary to choose a new H and/or quantization conditions in order to quantize tori interior to resonance zone. The practical realization of this task is far from simple. It is also clear that one should expect to encounter chaotic dynamics in the stochastic layers located near the separatrix of the strong resonance. For the trajectories generated by the AS method which end up in these chaotic regions, meaningful results can still be obtained by chosing the switching rate so as to be fast enough so as not to allow the detailed exploration of the chaotic zone but still small enough to preserve adiabatic invariance of the actions on the large timescale. This is connected with the conjecture[6] that even in the chaotic regions the actions are conserved for short times. Hence one can quantize imperfect partially destroyed "vague" tori[14] if the chaos is not too great. Clearly, the method is expected to fail if the dynamics requires too long a propagation in the stochastic layer.

3. APPLICATION TO NONLINEARLY COUPLED OSCILLATORS

Here we consider the application of the AS method to the semiclassical quantization of dynamical systems described by the Hamiltonian

$$H = H_o + \alpha x(y^2 + \beta x^2) \tag{5a}$$

$$H_o = \frac{1}{2}(p_x^2 + p_y^2 + \omega_x^2 x^2 + \omega_y^2 y^2) \tag{5b}$$

with two different choices of system parameters[9,10].

3.1 Nondegenerate case

As an example of straight forward application of the AS method we consider the near-integrable Hamiltonian given by eq. (5) with parameters

$$\omega_x = 0.7, \ \omega_y = 1.3, \ \alpha = 0.1, \ \beta = 0.1 \tag{6}$$

As an obvious but clearly not unique choice we take the harmonic part in eq. (5.a) to be the reference Hamiltonian, i.e. $\tilde{H} = H_o$. The initial conditions sampling the quantizing torus of H_o are

$$x^o, \ y^o = \left[\frac{2 I_{x,y}}{\omega_{x,y}} \right]^{1/2} \quad \text{Sin} \ (\theta_{x,y}^o) \tag{7a}$$

$$p_{x,y}^o = [2I_{x,y} \omega_{x,y}]^{1/2} \cos(\theta_{x,y}^o) \tag{7b}$$

where the quantized action variables are given by (we use the system of units where $\hbar=1$):

$$I_{x,y} = n_{x,y} + 1/2, \ n_{x,y} = 0, 1, 2, \ldots \tag{7c}$$

and angle variables $0 \leq \theta_{x,y}^o < 2\pi$ can be chosen arbitrarily.

Table I shows the results for two sets of excited (high-lying) energy levels. The second set represents the last 12 levels closest to the "escape energy" at E=11.4601. Each eigenvalue is the result of averaging 10 final energies obtained by running trajectories with initial conditions given by eqs. (7a-c) and using a switching function given by eq. 2 with T=400. The corresponding standard deviations as calculated by using eq. (4) are also shown and can be seen, in principle, to increase with the increasing excitation energy. For comparison, Table I also contains the quantal results obtained by diagonalizing the Hamiltonian in the basis of 30x30 harmonic oscillator eigenfunctions[15] along with the results of semiclassical quantization of fourth order Birkhoff-Gustavson normal form[16]. The present results are in better agreement with the quantal calculations than the perturbational results of ref. 16 which is to be expected for highly excited states.

Table I: Energy levels of highly excited states of Hamiltonian defined by eqs. (5) and (6).

State	(n_x,n_y)	BGNF[a]	QM[b]	AS[c]	$10^3 \Delta E$
36	(10, 0)	7.7445	7.7423	7.7484	1.4
37	(1, 5)	7.9019	7.8996	7.8983	0.7
38	(3, 4)	7.9564	7.9524	7.9505	0.9
39	(5, 3)	8.0135	8.0259	8.0224	1.3
40	(7, 2)	8.1288	8.1220	8.1194	0.9
41	(9, 1)	8.2500	8.2435	8.2405	0.8
42	(11, 0)	8.3973	8.3939	8.3902	1.0
74	(4, 6)	10.8933	10.8652	10.8626	2.5
75	(6, 5)	10.9434	10.9053	10.9022	1.7
76	(15, 0)	10.9627	10.9439	10.9336	1.8
77	(8, 4)	11.0172	10.9700	10.9658	2.2
78	(10, 3)	11.1166	11.0612	11.0588	1.6
79	(12, 2)	11.2439	11.1856	11.1802	7.1
80	(14, 1)	11.4014	11.3484	11.3376	11.1
81	(1, 8)	11.4369	11.4129	11.4149	4.8
82	(3, 7)	11.4460	11.4158	11.4057	10.9
83	(5, 6)	11.4765	11.4325	11.4294	3.6
84	(7, 5)	11.5305	11.4703	11.4651	1.1
85	(16, 0)	11.5902	11.5324	11.5519	1.8

(a) Semiclassical quantization of fourth order Birkhoff-Gustavson normal form, ref. 16.
(b) Quantum mechanical (variational) calculations, ref. 15.
(c) Adiabatic switching results, ref. 9.
(d) Standard deviations as defined in eq. (4).

For each eigenvalue shown in Table I, a few (of 10 in this case) "final" quantizing trajectories have been run[9]. These have been obtained by solving the Hamilton's equations of motion of the full Hamiltonian (eqs. (5) and (6)) with the initial conditions corresponding to final (at t=T) positions and momenta as obtained from the AS calculations. The term "quantizing trajectories" should be understood here conditionally. Namely, due to the approximate adiabatic invariance, these are not exact quantizing trajectories but are rather close to them.

Most of the levels from the first set of data in Table I are characterized by quantizing trajectories with slightly deformed rectangular caustics, a typical example of which is shown in Fig. 1. Clearly, this trajectory has adiabatically evolved from an initial quantizing trajectory of H with exact rectangular caustics.

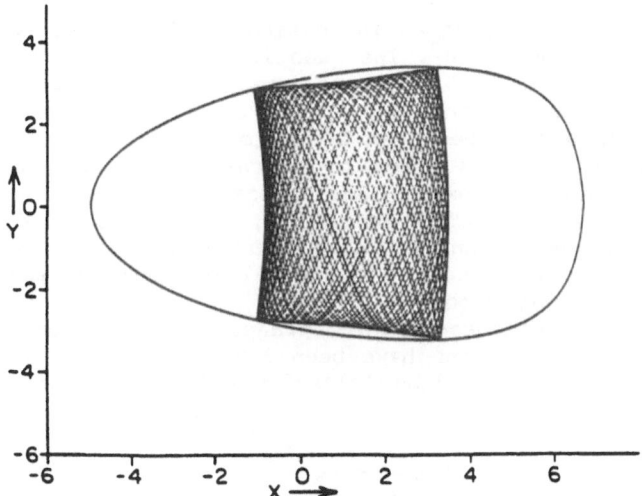

Figure 1: Typical quantizing trajectory with deformed rectangular caustics. Here $n_x=1$, $n_y=5$.

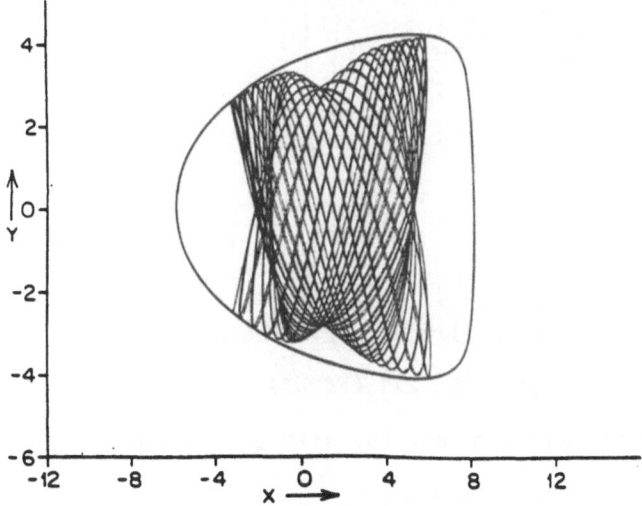

Figure 2: Quantizing trajectory with strongly deformed rectangular caustics. Here $n_x=4$, $n_7=6$.

Among the final quantizing trajectories of the highly excited levels from the second set of data in Table I, the prevailing type is shown in Fig. 2. It is characterized by strongly deformed "rectangular" caustics, which, in this case, is an indication of the proximity to the separatrix of 1:2 resonance zone[9]. The quantizing trajectories for the states labeled by (1,8) and (3,7) are found to lie within the 1:2 resonance zone and are characterized by parobolically shaped caustics[9]. This is an indication of the breakdown of the method.

Close to the separatrix of a low-order resonance zone one would expect to find the chaotic dynamics. Indeed, some of the final quantizing trajectories are found to be inrregular, as shown in Fig. 3. The corresponding standard deviations in Table I are larger than for quasiperiodic trajectories, but the semiclassical estimates of the eigenvalues are still reasonably good.

In addition to those discussed above, trajectories affected by the presence of high-order resonances have been found too[9], but the effect on the accuracy of the calculated semiclassical eigenvalues was minor.

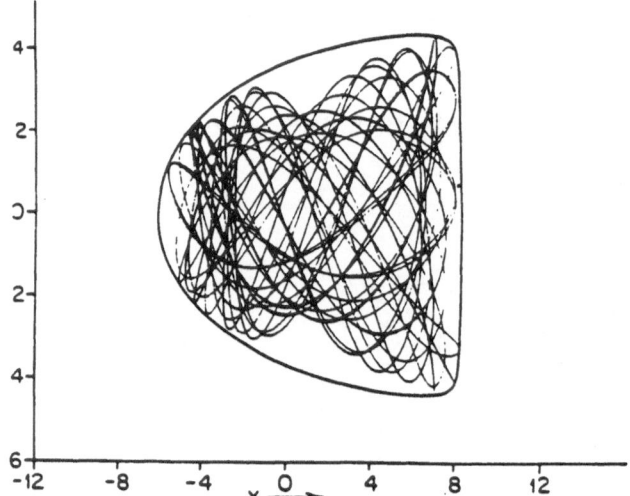

Figure 3: Irregular quantizing trajectory. Here $n_x=14$, $n_y=11$.

3.2 Oscillator with an intrinsic 2:1 (FERMI) resonance

The Hamiltonian is now given by eq. (5) with parameters:

$$\omega_x/2 = \omega_y = \omega = 0.7, \quad \alpha = -0.08, \quad \beta = -0.08, \tag{8}$$

so that the harmonic part H_0 is characterized by the 2:1 (Fermi) resonance, which is present in model Hamiltonians describing the vibrational motion of many molecules; for example CO_2[17]. In situations like this, when the zero-order Hamiltonian is degenerate, it is necessary[3,4,6] to apply classical perturbation theory, in order to lift the degeneracy, identify proper quantization conditions and eventually choose better reference Hamiltonian.

In our case we take as a reference Hamiltonian

$$\tilde{H} = H_0 + \frac{\alpha}{8\omega^2} K \tag{9}$$

where H_0 is given by (5.b) and

$$K = 2\omega^2 y^2 x + 2yp_x p_y - 2xp_y^2 \tag{10}$$

Hamiltonian (9) is an integrable approximation of (5.a) (H_0 and K are two integrals of motion in involution for $\tilde{H}$), obtained by applying the first-order classical perturbation theory[18,19]. "Part" of the interaction from (5.a) is effectively accounted for in (9) and the degeneracy of H_0 is lifted. We note also that K corresponds to a separation constant in the Hamilton-Jacobi equation of H_0 when solved in parabolic coordinates[20,21].

Exact quantized values of H_0 and K are known from the study of the Schrödinger equation of H_0 in parabolic coordinates[21]

$$H_0 = E_0^n = (n+3/2)\omega; \quad n = 0,1,2,\ldots \tag{11}$$

$$K = k_\lambda^n, \quad \lambda = -j, \, -j+1, \, \ldots, \, j \tag{12}$$

where $2j = [n/2]$ and $[\cdot]$ is the integer part of a number. The energy levels of the unperturbed oscillator are $2j+1$-fold degenerate. Eigenvalues k_λ^n with properties

$$k_{-j}^n < k_{-j}^n + 1 < \ldots < k_j^n; \quad k_{-\lambda}^n = -k_\lambda^n, \tag{13}$$

can be found[21], for example, by diagonalizing the quantum analogue of K in the Cartesian basis of harmonic-oscillator eigenfunctions of H_0 belonging to the same n-subspace. The corresponding eigenvalues of the reference Hamiltonian $\tilde{H}$ form multiplets of $2j+1$ non-degenerate levels. For large values of n (or α) multiplets can overlap (mix).

We define the quantizing trajectories of reference Hamiltonian $\tilde{H}$ as those which correspond to exact quantized values (11) and (12) of H_0 and K. The ensembles of such trajectories can be generated[10] by using the fact that the Hamiltonian-Jacobi equation of $\tilde{H}$ is separable in parabolic coordinates (ξ, η). The adiabatic invariants (action variables) are given by

$$I_\xi = \frac{1}{\pi} \int_{\xi_1}^{\xi_2} [2H_0\xi^2 - \omega^2\xi^6 + K]^{1/2} \, d\xi \tag{14.a}$$

$$I_\eta = \frac{1}{\pi} \int_{\eta_1}^{\eta_2} [2H_0\eta^2 - \omega^2\eta^6 - K]^{1/2} \, d\eta \tag{14.b}$$

where $\xi_{1,2}$ and $\eta_{1,2}$ are corresponding turning points[10].

For H_0 and K given by (11) and (12), I_ξ and I_η have well-defined values, but these are not in general equal to half-integers. By using the exact quantum-mechanical values for $K = k_\lambda^n$ in (14.a,b) we effectively

account for the barrier effects present in one-dimensional $\xi-$ (or $\eta-$)
dependent "potentials" in (14.a,b). Uniform semiclassical qantization
is discussed in ref. 21.

A typical initial quantizing trajectory of reference Hamiltonian
(9), corresponding to states with $\lambda>0$ (i.e., $K=k_\lambda^n>0$), is shown in
Fig. 4. Quantizing trajectories corresponding to states with $\lambda<0$
(i.e., $K=k_\lambda^n<0$) are essentially obtained by reflection with respect to
$X=0$ axis, whereas those with $\lambda=0$ (i.e., $K=k_0^n=0$) correspond to a
special type of separatrix motion[10].

The AS calculations have been performed[10] for all bound states up
to the classical escape energy at E=9.375. Two selected sets of the
results are presented in Table II. Each semiclassical eigenvalue is the
mean of 25 final energies obtained by randomly choosing initial conditions and solving Hamilton's equations of motion for (1) up to t=T=800.
For comparison, also shown are converged quantum-variational results.
These have been obtained[10] by diagonalizing Hamiltonian (5) in a truncated Cartesian harmonic-oscillator basis of 484 eigenfunctions.

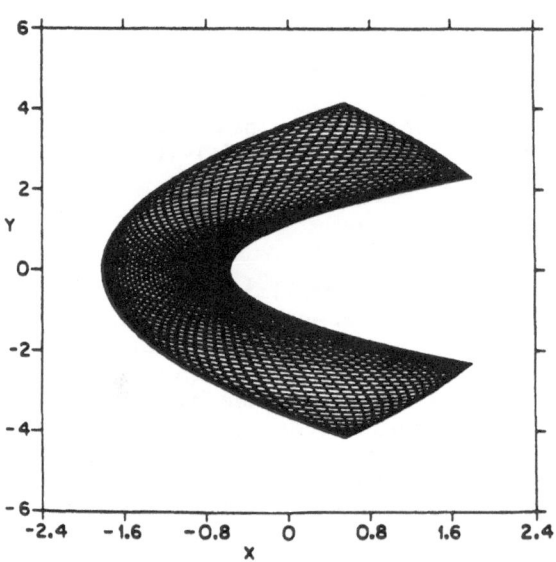

Figure 4: Quantizing trajectory of reference. Hamiltonian (9),
corresponding to n=5, $\lambda=1$.

As can be seen in Table II, for low-lying levels the AS results
agree well with variational calculations. The standard deviations are
small suggesting a rather high degree of invariance in the adiabatic
sequence. Final quantizing trajectories for these states differ from
those of reference Hamiltonian (see Fig. 4) only by having slightly
deformed caustics. In this, predominantely quasiperiodic, regime other
semiclassical methods provide the similar accuracy for the eigenvalues[18,22].

The highly excited states (see the second set of data in Table II) are generally characterized by large standard deviations. Nevertheless, for most of the levels the mean values of final energies provide a good semiclassical estimate. The large standard deviations are consequence of the violation of adiabatic invariance due to encounters with increasingly (with excitation) strong classical resonances and accompanying irregular dynamics. The final quantizing trajectories for these states are typically of the form shown in Fig. 5. They correspond to the motion on "vague torus"[14] rather than to a fully ergodic motion and therefore suggest the existence of the finite-time, approximate adiabatic invariants[6]. It has been demonstrated in ref. 6 that when the switching involves propagation through the chaotic layers on optimal value of the switching interval T can usually be found. Indeed, in a few cases marked in Table II it is possible to reduce the standard deviations by a factor of 2 by reducing the switching interval from T=800 to T=200. The corresponding mean values of final energies have been only slightly changed, but enough to resolve the ordering of some nearly degenerate levels.

TABLE II. Energy levels of Hamiltonian (5), (8)

State	n	λ	E_{AS}[a]	ΔE_{AS}[b]	E_Q[c]
1	0	0	1.0485	4.0(-5)	1.0485
2	1	0	1.7453	8.0(-5)	1.7404
3	2	1/2	2.3830	1.1(-4)	2.3859
4	2	-1/2	2.4840	1.8(-4)	2.4863
5	3	1/2	3.0323	1.8(-4)	3.0316
6	3	-1/2	3.2091	1.9(-4)	3.2070
7	4	1	3.6570	2.1(-4)	3.6584
8	4	0	3.8243	1.9(-4)	3.8275
9	4	-1	3.9485	1.5(-4)	3.9495
31	10	3/2	7.515	2.1(-2)	7.5176
32	11	5/2	7.676	3.0(-2)	7.6675
33	9	-2	7.6915	3.1(-4)	7.6914
34	10	1/2	7.814	1.6(-2)	7.8236
35	10	-1/2	7.999	1.0(-2)	8.0246
36	11	3/2	8.076	3.8(-2)	8.0780
37	12	3	8.178*	1.6(-2)	8.1760
38	10	-3/2	8.180	2.4(-3)	8.1874
39	11	1/2	8.415	2.8(-2)	8.4178
40	10	-5/2	8.4493	6.0(-4)	8.4485

(a) Results of AS calculations, ref. 10.
(b) Standard deviations, Eq. (4).
(c) Quantum variational results, ref. 10.
*AS calculations performed by using switching interval T=200

Finally, it should be noted that imbedded among the states characterized by irregular quantizing trajectories, are the states quantized
by quasiperiodic trajectories. As seen from Table II they are accompanied by small standard deviations and for them, the agreement with
quantal calculations is excellent. This is a generic property of model
Hamiltonians describing the molecular vibrational motion.

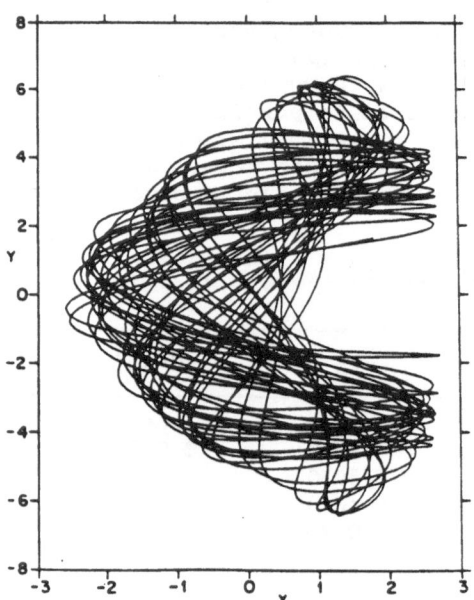

Figure 5: Irregular final quantizing trajectory of full Hamiltonian
(5). Here n=10, λ=3/2.

4.   CONCLUDING REMARKS

Strictly speaking, the nonintegrability destroys the adiabatic invariance of classical action variables. The main causes for this are attributed to effects of nonlinear resonances (i.e., change in topology
of the invariants) and accompanying chaotic dynamics (i.e., destruction
of the invariants). Still, under the condition that the encounters with
the strong (low-order) resonances can be avoided during the switching
interval, the AS method provides good semiclassical estimates of the
energy eigenvalues. Even some states with corresponding irregular trajectories can be approximately quantized.

The choice of a reference Hamiltonian and appropriate quantization
conditions is not a trivial task and often requires a considerable

insight into the dynamics of the system, such as previous application of the classical perturbation theory. In some cases more than one reference Hamiltonial is required in order to quantize states with topologically different quantizing trajectories. An example of that kind is provided by the well studied Henon-Heiles system[6,9].

Due to approximate adiabatic invariance one is led to the use of ensembles of trajectories and distributions in final energies with the corresponding standard deviations as indicators of the violation of adiabatic invariance. The monitoring of the change in global topology of the trajectories in the adiabatic sequence is also desirable.

The major advantage of the AS method is that it is the only method at present that can provide the semiclassical eigenvalues by using irregular quantizing trajectories. For the systems not strongly affected by resonances, straightforward application of the AS method takes less computer time as compared to other methods and, in principle, has no limitations on the number of degrees of freedom (the low-lying states of three-, four-, and five-dimensional coupled oscillators have been successfully treated[6]). In addition, when needed, the method automatically provides semiclassical eigenvalues in the whole range of the coupling constant.

ACKNOWLEDGMENT

This work was supported by NSF Grant No. CHE-8511496.

REFERENCES

1. For a review of standard semiclassical methods for calculating energy spectra, see, for example: (a) I.C. Percival, Adv. Chem. Phys. 36, 1 (1977); (b) D.W. Noid, M.L. Koszykowski and R.A. Marcus, Ann. Rev. Phys. Chem. 32, 267 (1981). For recently proposed methods see: (c) N. DeLeon and E.J. Heller, J. Chem. Phys. 78, 4005 (1983); J. Chem. Phys. 81, 5957 (1984); (d) W.H. Miller, J. Chem. Phys. 81, 3573 (1984); (e) C.W. Eaker, G.C. Schatz, N. DeLeon and E.J. Heller, J. Chem. Phys. 81, 5913 (1984); (f) C.C. Martens and G.S. Ezra, J. Chem. Phys. 83, 2990 (1985).
2. A. Einstein, Verh. Dtsch. Phys. Ges. (Berlin) 19, 82 (1917); L. Brillouin, J. Phys. Radium 7, 353 (1926); J.B. Keller, Ann. Phys. (New York) 4, 180 (1958); Also see ref. 1(a).
3. E.A. Solov'ev, Sov. Phys. JETP 48, 635 (1978).
4. T.P. Grozdanov and E.A. Solov'ev, J. Phys. B 15, 1195 (1982).
5. M.V. Berry, J. Phys. A. : Math. Gen. 17, 1225 (1984).
6. R.T. Skodje, F. Borondo and W.P. Reinhardt, J. Chem. Phys. 82, 4611 (1985).
7. B.R. Johnson, J. Chem. Phys. 83, 1204 (1985).
8. C.W. Patterson, J. Chem. Phys. 83, 4618 (1985).
9. T.P. Grozdanov, S. Saini and H.S. Taylor, Phys. Rev. A 33, 55 (1986).
10. T.P. Grozdanov, S. Saini and H.S. Taylor, J. Chem. Phys. 84, 3243 (1986).

11. L.D. Landau and E.M. Lifshitz, Mechanics, 3rd ed. (Pergamon, Oxford, 1976).
12. V.I. Arnold, Mathematical Methods of Classical Mechanics, (Springer-Verlag, New York, 1978).
13. (a) B.V. Chirikov, Phys. Rep. 52, 265 (1979); (b) A.J. Lichtenberg and M.A. Lieberman, Regular and Stochastic Motion, (Springer-Verlag, New York, 1983).
14. R.B. Shirts and W.P. Reinhardt, J. Chem. Phys. 77, 5204 (1982).
15. D.W. Noid - cited in ref. 16.
16. R.T. Swimm and J.B. Delos, J. Chem. Phys. 71, 1706 (1979).
17. E. Fermi, Z. Phys. 71, 250 (1931).
18. J.A. Sanders, J. Chem. Phys. 74, 5733 (1981).
19. T. Uzer and R.A. Marcus, J. Chem. Phys. 81, 5013 (1984).
20. D.W. Noid, M.L. Koszykowski and R.A. Marcus, J. Chem. Phys. 71, 2864 (1979).
21. T.P. Grozdanov, S. Saini and H.S. Taylor, J. Phys. A 19, 691 (1986).
22. C.W. Eaker and G.C. Schatz, J. Chem. Phys. 81, 2394 (1984).

EFFECT OF DIAGONAL AND OFF-DIAGONAL DISORDER IN THE MULTIPHOTON
EXCITATION OF AN ACTIVE MODE COUPLED TO A BATH: USE OF THE IRREDUCIBLE
RECURSIVE RESIDUE GENERATION METHOD

Israel Schek and Robert E. Wyatt
Department of Chemistry and
Institute for Theoretical Chemistry
University of Texas at Austin
Austin, Texas 78712

ABSTRACT. Passage from the quasiperiodic into the chaotic regime in a quantum system (a radiative
active mode coupled to a laser field and in turn, intramolecularly coupled to radiatively inert background
modes) is studied. Three main parameters are considered to be crucial in this passage: the
radiative/intramolecular relative coupling strength, the effectiveness of the background as a bath, and the
nature of the background detuning and coupling to the radiative pump mode.

I. INTRODUCTION

Features of chaotic behavior within various fields of chemical dynamics, such as the energy distribution in
internal modes and degrees of freedom [1], unimolecular reactions [2] and multiphoton excitation of
molecules [3], has attracted much attention in the last decade [e.g., see 4].
        Since the transition from a quasiperiodic regime (QPR) to a chaotic one in classical dynamics is
becoming well understood, attempts have been made to bridge the corresponding gap in the quantum
mechanical (QM) domain, by analogy to the classical characteristics [4,5]. However, in this context a
preliminary question is the very definition of the chaotic regime in quantum mechanics.
        Let us first briefly review some properties of the classical chaotic regime: an exponential
separation with respect to time of phase space trajectories which are initially slightly separated (linear
separation of the trajectories is observed in the QPR); a continuous frequency spectrum (sharp and discrete
in the QPR) of a dynamical variable; correspondingly, a decaying correlation function (oscillatory in the
QPR) of a dynamical variable; the tendency of chaotic trajectories to fill all of the energetically accessible
phase space cells (a limited number in the QPR); and related to this property is ergodicity in the chaotic
regime.
        Nordholm and Rice [6] were among the first to study the ergodic property in the QM domain by
computing the contributions in the eigenvectors of different components of a basis set. Termed "global"
are eigenfunctions which result from the mixing of very many basis functions. Although this is only a
limited condition [7], it may still serve as an indication of breakdown in the "regularity" of an otherwise
QPR wavefunction.
        It has also been asserted by Percival [8] that the chaotic QM spectrum would be irregular (in that
it has a broad series of lines and a continuous band, contrary to a series of sharp lines) and that the
eigenvalue spectrum would be very sensitive to a perturbation (which is plausibly equivalent to the
exponential divergence of the chaotic trajectories).

R. Lefebvre and S. Mukamel (eds.), Stochasticity and Intramolecular Redistribution of Energy, 95–107.
© *1987 by D. Reidel Publishing Company.*

Prevalence of chaotic features in a molecular system is determined either by (i) initial preparation, that is, the initial state is far from any eigenstate of the Hamiltonian [9], so that the expansion coefficients are randomly distributed, or (ii) by the very nature of the system, including high order non-linear coupling terms in the Hamiltonian or the absence of Fermi resonances (i.e., the separable part of the Hamiltonian consists of incommensurate harmonic oscillators) [10]. To this end, one should also consider the nature of the external force operating on the system; an appropriate example would be a non-ideal laser field with a typical band width, operating on an otherwise coherent molecular system.

It is pertinent to study the conditions for chaotic behavior in systems which are seemingly regular and integrable. It is expected that such behavior is enhanced in systems with a large number of internal degrees of freedom or a multitude of states. Presently, we are looking particularly for irregular, nonrecurrent phenomena in the multiphoton excitation (MPE) by an intense laser field and the role played by the coupling of a single active mode to optically inert modes.

Due to the immense number of relevant states which are to be considered in a realistic QM model, powerful techniques capable of solving the dynamics determined by the full Hamiltonian are necessary. One such approach, in which thousands of dressed molecule-field states can be considered, is the recursive residue generation method (RRGM) [11,12]. This approach employs the Lanczos algorithm [13,14] where the very large sparse Hamiltonian matrix is recursively converted into a much more compact tridiagonal Jacobi matrix. The RRGM calculates transition amplitudes between any two dressed states, by recursive generation of poles and residues of the Green operator, which is associated with the relevant time-propagator. The calculation of a myriad of eigenvectors of the huge Hamiltonian matrix is thus avoided. Development of the RRGM was mainly inspired by studies by Haydock et al [15-17] on the spectra of disordered solids, where the local density of electronic or vibrational states about a particular site was calculated and by the studies of Whitehead and Watt [18] on nuclear shell structure.

One defines a source state |0>, considered as the primary subsystem for transition probabilities, as a linear combination of the initial and final states. In the recursive procedure, additional states are always born successively one-by-one out of the previous two states. The dimension of the dynamically pertinent space is gradually extended. Previously [19], we developed a procedure for finding generating operators of the recursive states and used them to construct the self energies and linking elements, (the diagonal β_j and off-diagonal $\int_j$ elements of the tridiagonal representation, respectively) from successive power moments of the Hamiltonian with respect to the source vector, $<0|H^k|0>$. These elements can be interpreted as masses (diagonal) and stiffnesses (off-diagonal) of quasiparticle members of a mechanical chain, by which the dynamics of the system can be developed.

2. METHOD OF LINKED AND IRREDUCIBLE MOMENTS

The survival probability in the initial state |0> under the operation of the Hamiltonian H during the time t can be written in terms of the eigenvalues $\{E_j\}$ and the residues $\{R_j\}$

$$P_0(t) = |<0|\exp(-iHt)|0>|^2 = |\sum_{j=1}^{M} R_j \exp(-iE_jt)|^2 \tag{1}$$

where M is the number of recursion steps in the tridiagonalization procedure. It was shown [19,20] that since the tridiagonal representation (J) is equivalent to the original one (H), then the equivalent moments of any degree, k, with respect to the source state are equal:

$$\mu_k^H = <0|H^k|0> = \mu_k^J = <0|J^k|0> \tag{2}$$

This equality serves to express the tridiagonal elements $\{\alpha_j, \beta_j\}$ in terms of the original elements $\{H_{kj}\}$. But the numerical computation of the power moments, even for moderate values of k, is usually undesirable due to numerical problems resulting, in part, from the propagation of roundoff errors when very large or very small numerical magnitudes of the moments are required. Therefore power moments do not provide a viable route to computing transition or survival amplitudes. To bypass these numerical instabilities [e.g., 21,22] a concise form of the linked moments was described [23]. All paths in the linked moment diagram which involve intermediate hops back to the source state are avoided before returning to I0>. Consequently, the number of products and additions needed to calculate the relevant contribution to the tridiagonal chain elements decreases enormously as one proceeds to higher order moments. This procedure is equivalent [24] to elimination of the first row and column in the two equivalent matrix representations.

In certain cases by (when the initial state is coupled only to one single state), it can be shown [25] by the level scheme that the chain particle is definitely equivalent to the first excited state of the zero-order Hamiltonian, $|1>^H$ [25]. Consequently, arguments concerning the "linking procedure" can be repeated to eliminate intermediate "visiting" of |1> (once it is reached from I0>), before finally returning to I0> at the end of the process. A new difference moment which is linked also with respect to |1> (i.e., intermediate hops back to |1> are eliminated) is obtained by subtracting the corresponding moments. It contains the relevant contributions of the next zero-order states to the higher chain particles (|2>,|3>,···). All of the redundant paths are thus eliminated and need not be numerically calculated. Since there are no further redundant contributions that can be extracted (reduced) from these moments, they are termed irreducible moments.

In order to compare the convergence of calculations based on the full, linked and irreducible moments to the reference method (direct diagonalization of the Hamiltonian), we show in Fig. 1 the time development of a ground state survival probability obtained for this hierarchy of reduction the irrelevant information. The model system consists of a radiatively active mode (creation operator $a_1^\dagger$) coupled to a laser field (creation operator $b^\dagger$) with electric amplitude ε and frequency ω, via the dipole moment μ. In turn the active mode is intramolecularly coupled to a radiatively inert mode (creation operator $a_2^\dagger$) by the coupling term c. The system Hamiltonian is given by:

$$H = \omega_1 a_1^\dagger a_1 + \omega_2 a_2^\dagger a_2 + \omega b^\dagger b +$$
$$\mu\varepsilon (a_1^\dagger b + b^\dagger a_1) + c (a_1^\dagger a_2 + a_2^\dagger a_1) \tag{3}$$

where ω_1 and ω_2 are the corresponding mode frequencies. Obviously, the irreducible (and partly the linked) moment calculations do reproduce the results obtained from the reference calculations, particularly, for the longer time domain. In the full moment approach, the effect of round-off errors is very destructive. Conversely, due to conservation of the orthogonality of the recursive states with respect to the source state, in both the linked and irreducible moment approach, satisfactory convergence of the numerical results is obtained and the influence of numerical round-off error is suppressed (but not eliminated). As the electric amplitude ε increases (or in turn the intramolecular coupling c) the full moment approach increasingly fails to trace the reference results, due to the inherent inability to preserve credible orthogonality of the recursion vectors [26].

SURVIVAL PROBABILITY, HIERARCHY OF REDUCTIONS

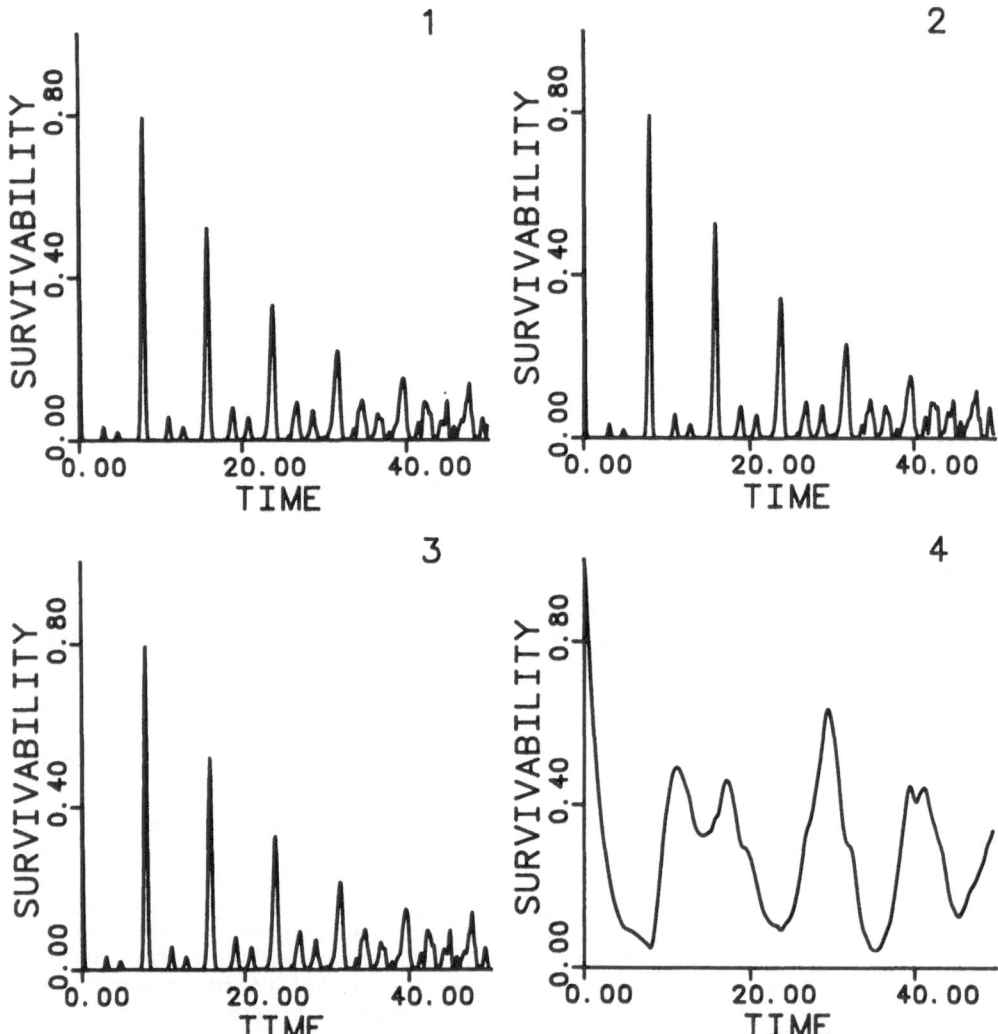

Figure 1. The ground state survival probability calculated for the two mode anharmonic system in Eq. (3) for various degrees of reduction of redundant information, compared to a direct diagonalization approach: (1) direct diagonalization; (2) irreducible moments; (3) linked moments; (4) full moments. Time units are in reciprocal energy units.

3. LASER EXCITATION OF A MULTI-MODE SYSTEM

The model system to be studied is a molecule with M modes, one of which is coupled via the term $\mu\varepsilon$ (as defined in Sec. 2) to the laser field. The other modes, coupled to the active one via the terms $\{c_j\}$, are detuned by $\{\Delta_j\}$ with respect to the frequency ω_1 of the active mode. The Hamiltonian is thus given by:

$$H = \omega b^\dagger b + \omega_1 a_1^\dagger a_1 + \sum_{j=2}^{M} (\omega_1 + \Delta_j)\, a_j^\dagger a_j +$$

$$+ \sum_{j=2}^{M} c_j\, (a_1^\dagger a_j + a_j^\dagger a_1) + \mu\varepsilon\, (b^\dagger a_1 + a_1^\dagger b) \tag{4}$$

To investigate the role played by the nature of the intramolecular coupling on the energy deposition into the active mode, we [27] vary the number of the background modes and the relative strength of their coupling with respect to the radiative (Rabi) coupling.

A medium sized molecule corresponds to M=4-15 modes, for which a quasicontinuum may be expected after the absorption of about 5 IR photons [28]. For such a molecule, it is reasonable to assume that indeed a single mode is radiatively active (e.g., υ_3 in SF_6), whereas at higher energies this assumption breaks down, and additional detailed "microscopic" information is required. A molecular system with 15 modes in which up to 5 photons have been absorbed is spanned by 15,504 product states and there are $>7.3\times10^4$ independent (diagonal and off-diagonal) elements in the corresponding Hamiltonian matrix. Consequently, the dynamics of systems with M>10 were studied on the Cray X-MP/24 supercomputer, whereas for smaller systems, the CPU time and storage requirements allowed use of the Cyber 170/750.

In order to explain the non-periodic nature of the dynamics of a medium sized molecule and the erosion of coherent effects accompanying multiphoton excitation, a random coupling model [28,29] will be introduced, either for the detunings $\{\Delta_j\}$ or the couplings $\{c_j\}$. In each case a random distribution was compared to the constant set with the same root mean square value: either $<\Delta_j^2> = \Delta_0^2$ or $<c_j^2> = c_0^2$, where $<\ >$ denotes the average over the random distribution.

Survival probabilities of the ground molecular state for a four mode system (M=4) are shown in Fig. 2 for the "channel ratios" $\mu\varepsilon/c$=1., 10., 100. In the left subfigures, time-development for constant detunings 0.5, 0.1, 0.05 are shown, and in the right ones -- random detunings. Likewise, in Fig. 3 results for the larger system are shown (M=15).

Under a weak field excitation ($\mu\varepsilon/c$<1.), only low levels are excited, and there is no essential difference between the M=4 and 15 cases, as shown by comparing Figs. 2 and 3. For a set of equally detuned modes, the background serves as one effective mode which is indirectly (via the active mode) coupled to the laser. Analysis of the corresponding eigenfrequencies (obtained by diagonalization of the Hamiltonian Eq. (4) in the mode representation) shows that two of them are essentially $\pm\mu\varepsilon$ (when Δ<<c, $\mu\varepsilon$). Basically, these eigenmodes are combinations of the laser and the active mode. Another eigenmode is a combination of the laser and the effective background mode, with a much smaller eigenfrequency. This nice behavior is clearly shown in Fig. 2. These three eigenmodes are well separated from the rest of the eigenmodes (all having an eigenfrequency equal to the constant background detuning).

When the background frequencies are randomly scattered, this picture is lost and there is no definite effective background mode and, consequently, no block factorization; there is a total mixing of the laser, active and background modes. Therefore, the population is scattered among M+1 (M molecular and laser) modes, as shown in the right subfigures. For a large number of molecular modes (M=15, Fig. 3), the probability of returning into the initial mode (and into the ground state) is consequently very low. As the frequency scatter of the background modes becomes larger, this chaotic effect is more pronounced, due to the more extensive frequency mismatch.

When the mode frequencies have an equal values, but <u>random coupling</u> terms $\{c_j\}$, it is again possible to define an effective combination mode with a proper lowering operator A, given by

$$A = \sum_{j=2}^{M} c_j a_j \left/ \left(\sum_{j=2}^{M} c_j^2 \right)^{1/2} \right. \tag{5}$$

Consequently, the dynamics is non-chaotic (in spite of the random nature of the coupling) and effectively only three modes are involved: the laser, the active mode and the collective mode A.

When the laser field is very high, then the background modes are screened by the radiative coupling and there is no effective intermode population exchange, which corresponds to the "extraction" of the active mode from the quasicontinuum [30].

One condition for chaotic behavior is the absence of quasiperiodic features in the observables such as the survival probability. In the model studied here, the conditions for this behavior are:

(i)   Random detuning of the background modes relative to the pump mode;

(ii) Large enough background to supply an effective sink;

(iii) Strong enough field to activate the dense molecular region, so that the power broadening smears the fine details of the background.

As is obvious for M=4, even if conditions (i) and (iii) are satisfied (Fig. 2), there is no obvious chaotic behavior.

ACKNOWLEDGEMENTS

This work was supported in part by research grants from Cray Research, Inc., the National Science Foundation, and the Robert A. Welch Foundation.

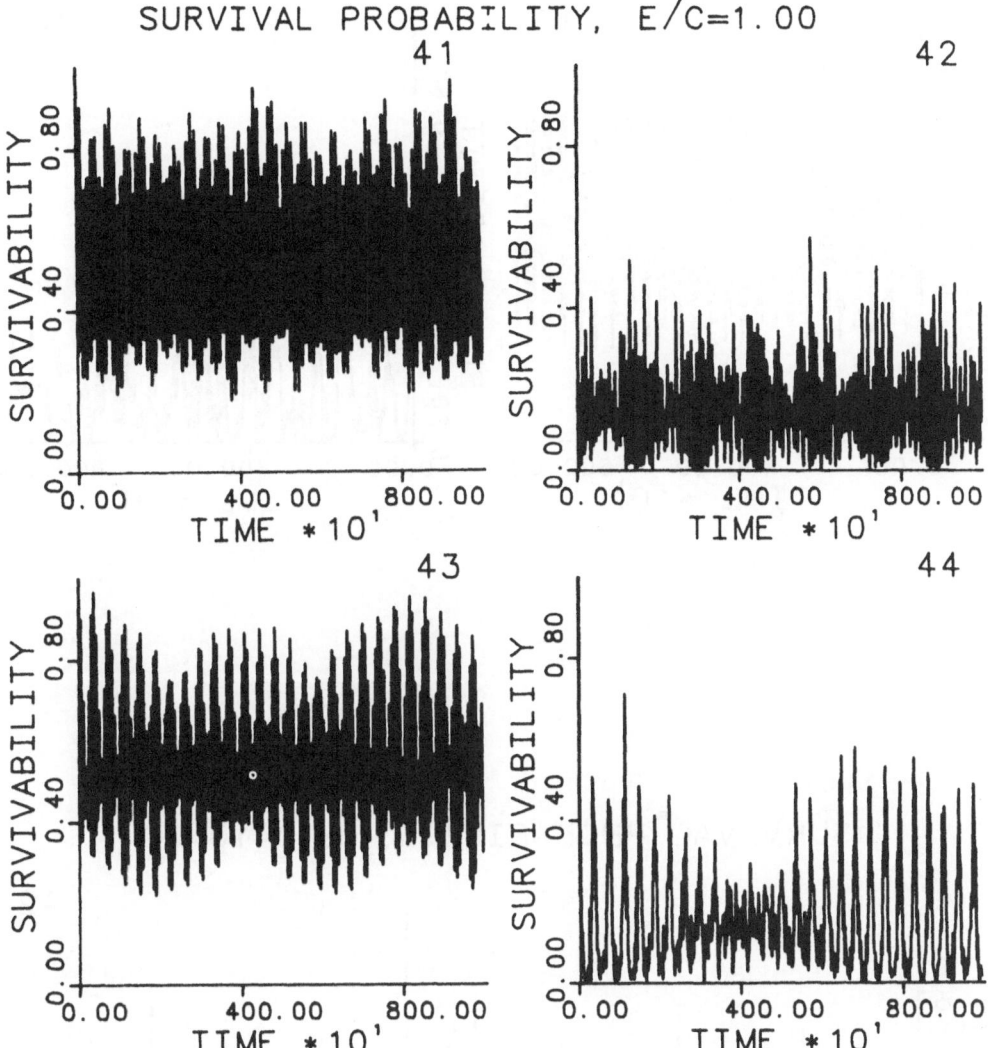

SURVIVAL PROBABILITY, E/C=1.00

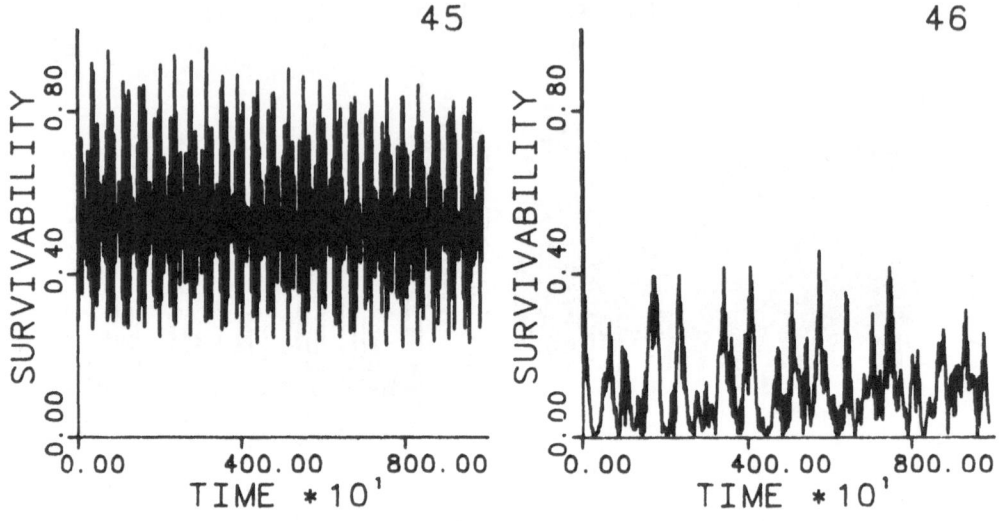

SURVIVAL PROBABILITY, E/C=10.0

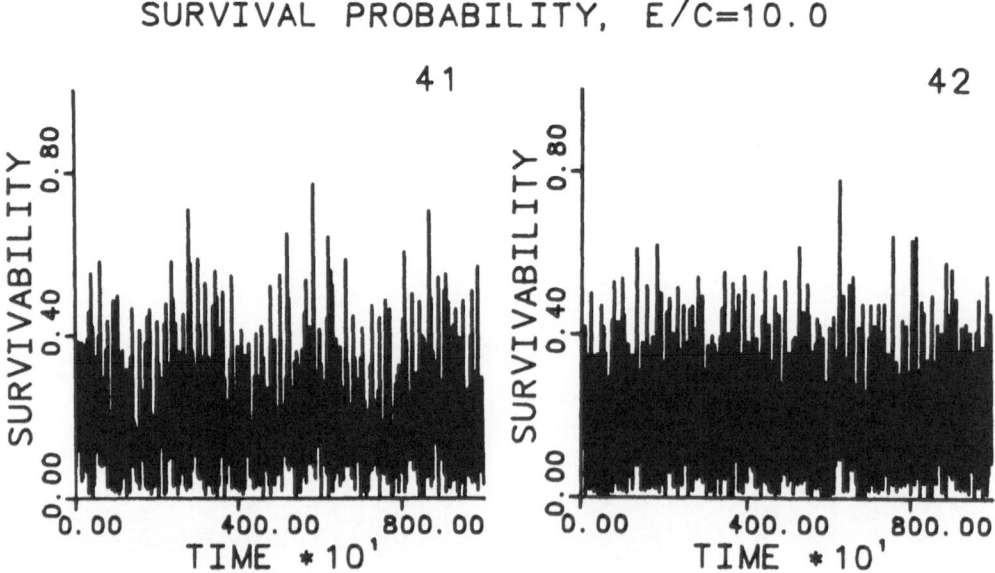

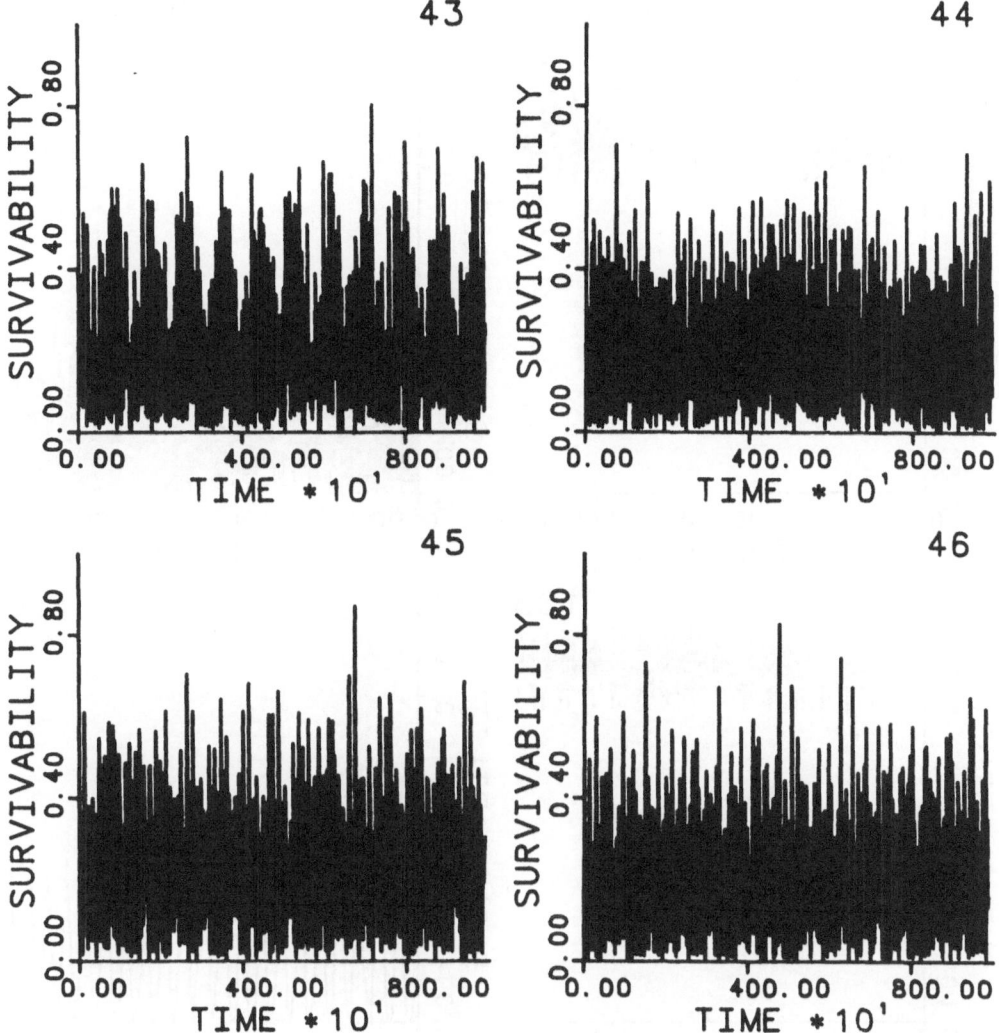

Figure 2. The survival probability for a four mode system
(M = 4) for the "channel ratios" μϵ/c = 1. (2a). 10. (3b). The
left subfigures: constant detunings 0.5, 0.1, 0.05; the right
ones: random detunings with the same corresponding r.m.s.
values.

SURVIVAL PROBABILITY, E/C=1.00

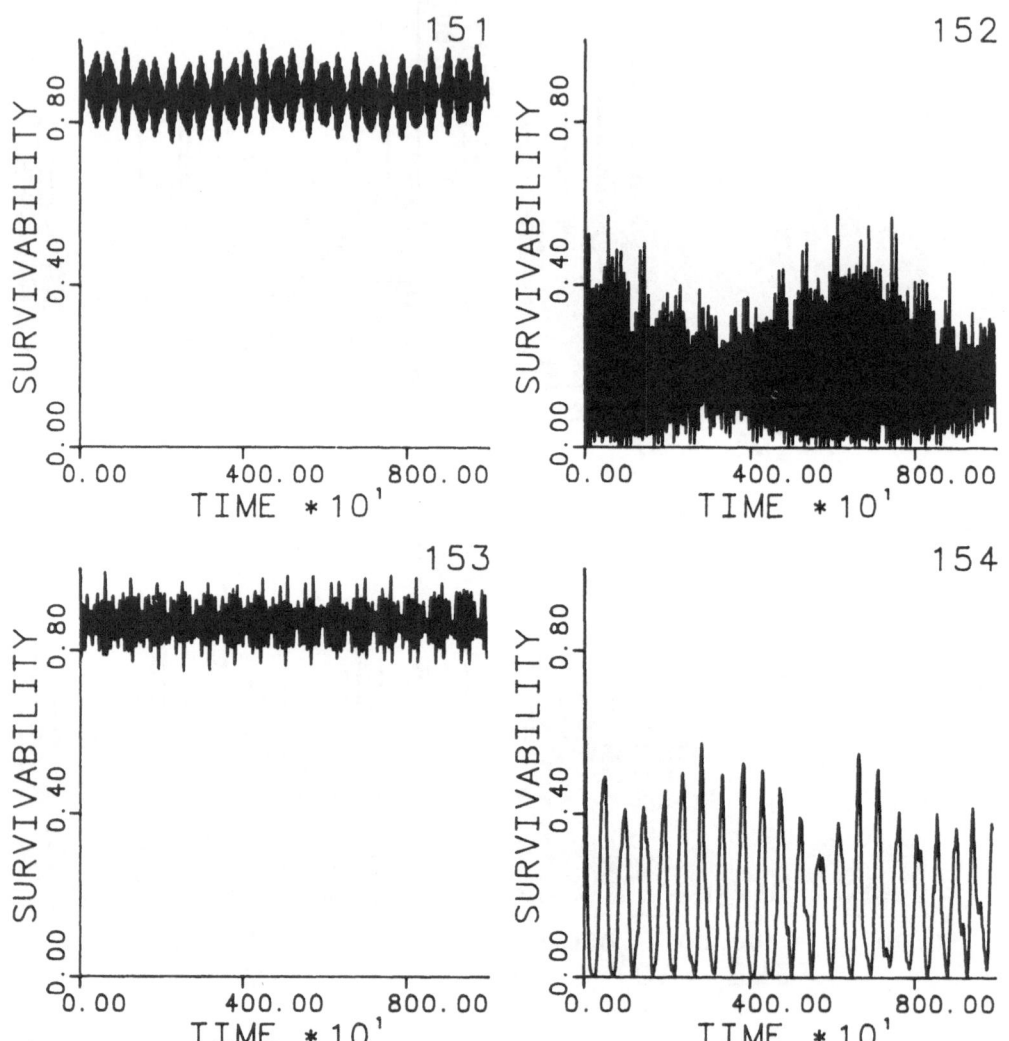

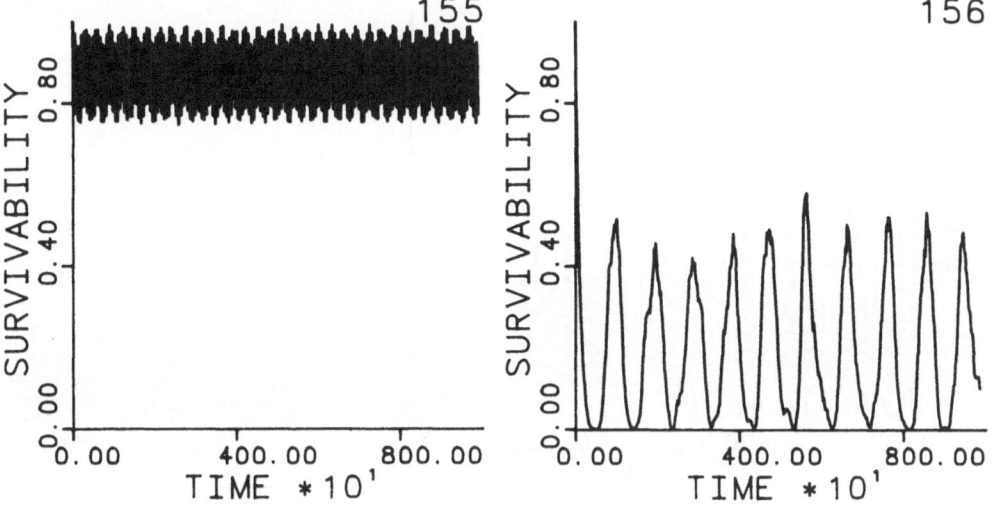

SURVIVAL PROBABILITY, E/C=10.0

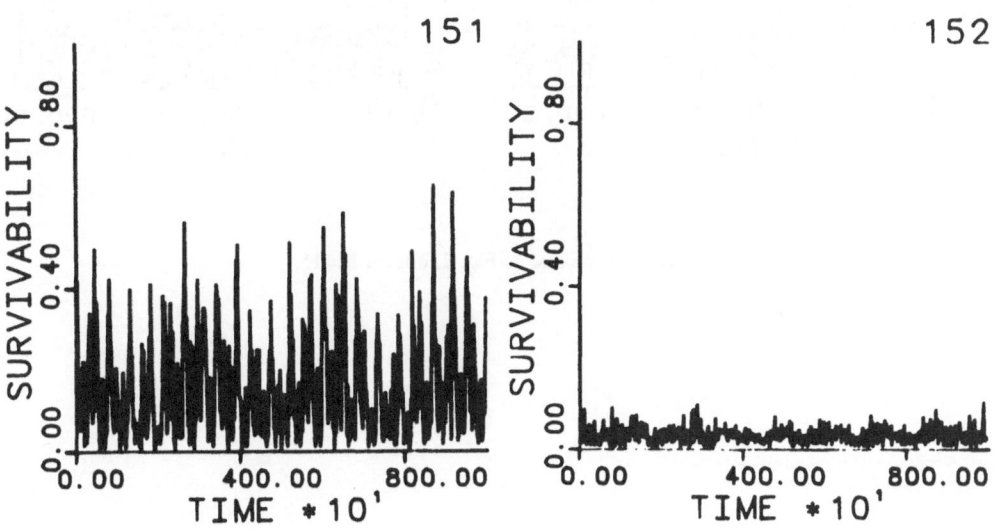

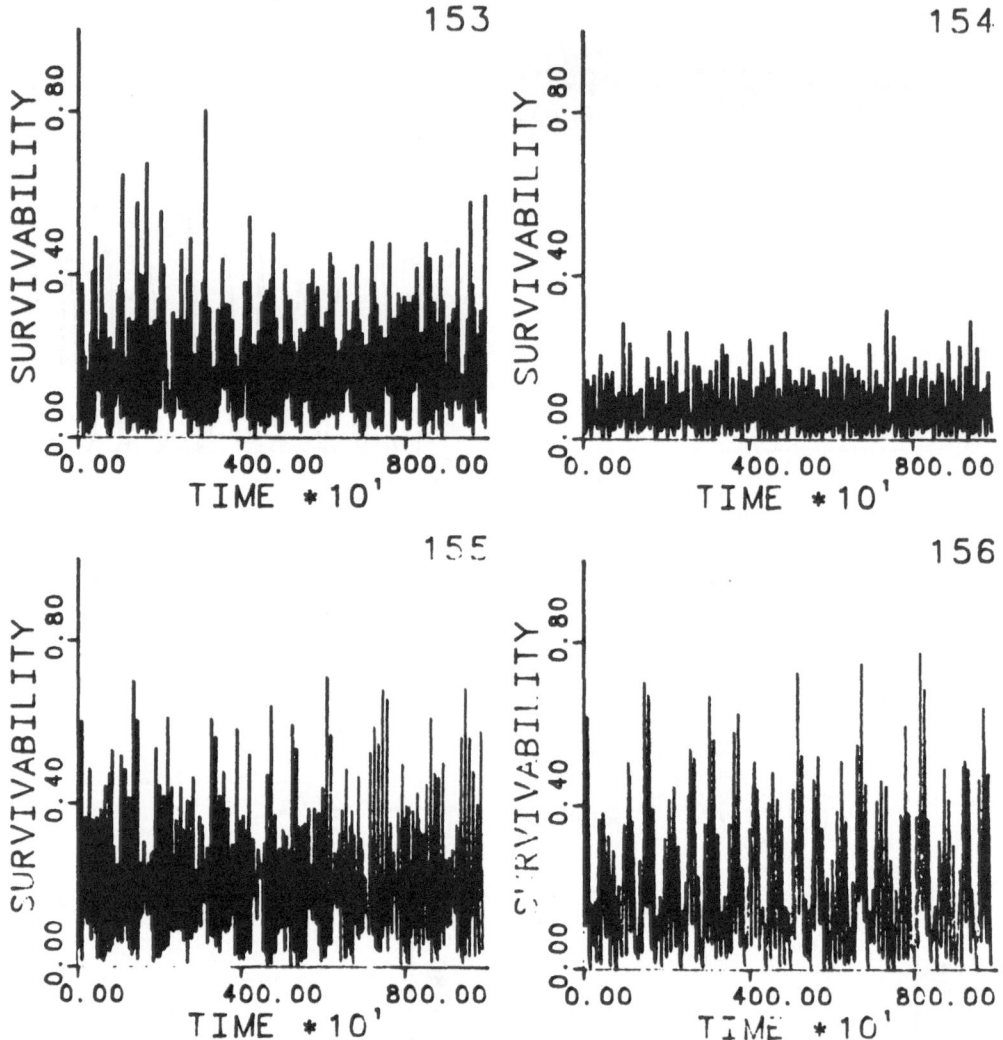

Figure 3.   Same as Fig. 2, but for M=15.

REFERENCES

1. W. B. Miller, S. A. Safron and D. R. Herschbach, J. Chem. Phys. 56, 358 (1972).
2. P. J. Robinson and K. A. Holbrook, "Unimolecular Reactions," Wiley, New York (1972).
3. P. A. Schultz, A. S. Sudbø, E. R. Grant, Y. R. Shen, and Y. T. Lee, J. Chem. Phys. 72, 4985 (1980).
4. D. W. Noid, M. L. Koszykowski and R. A. Marcus, in "Ann. Rev. of Phys. Chem.," Vol. 32, eds. B. S. Rabinovitch, J. M. Schurr and H. L. Strauss (1981).
5. S. A. Rice, Adv. Chem. Phys. 24, 155 (1973).
6. K. S. J. Nordholm and S. A. Rice, J. Chem. Phys. 61, 203 (1974); ibid., 61, 768 (1974); ibid., 62, 157 (1975).
7. D. W. Noid and R. A. Marcus, J. Chem. Phys. 67, 559 (1977).
8. I. C. Percival, J. Phys. B. 6, L229 (1973).
9. M. Feingold and A. Peres, Physica B9, 433 (1983).
10. C. D. Cantrell, H.W. Galbraith, and J. R. Ackerhalt, in "Multiphoton Processes," edited by J. H. Eberly and P. Lambropoulos, (Wiley, New York, 1978), p. 331.
11. A. Nauts and R. E. Wyatt, Phys. Rev. Lett, 51, 2238 (1983).
12. A. Nauts and R. E. Wyatt, Phys. Rev. A, 30, 872 (1984).
13. C. J. Lanczos, J. Res. Nat. Bur. Stand., 45, 255 (1950).
14. C. C. Paige, J. Ins. Math. Appl., 10, 373 (1972).
15. R. Haydock, Comp. Phys. Commun., 20, 11 (1980).
16. R. Haydock, in "Solid State Physics," H. Ehrenreich, F. Seitz, and D. Turnbull, eds. (Academic, New York, 1980).
17. R. Haydock, in "Excitations in Disordered Systems" M. F. Thorpe, ed., (Plenum, New York, 1981).
18. R. R. Whitehead, A. Watt, B. J. Cole and I. Morrison, in "Advances in Nuclear Physics," Vol. 9, ed. M. Baranger and E. Vogt (Plenum, New York, 1977), p. 123.
19. I. Schek and R. E. Wyatt, J. Chem. Phys. 83, 3028 (1985).
20. I. Schek and R. E. Wyatt, J. Chem. Phys. 83, 4650 (1985).
21. P. Lambin and J.-P. Gaspard, Phys. Rev. B., 26, 4356 (1982).
22. J.-P. Gaspard and P. Lambin, in "The Recursion Method and its Applications," D. G. Pettifor and D. L. Wearie, eds. (Springer, Berlin 1984), p. 72.
23. I. Schek and R. E. Wyatt, J. Chem. Phys., 84, 4497 (1986).
24. N. Moiseyev, I. Schek and R. E. Wyatt, Chem. Phys. Lett. (submitted).
25. I. Schek and R. E. Wyatt, Chem. Phys. Lett. 129, 99 (1986).
26. R. Haydock, V. Heine and M. J. Kelley, J. Phys. C., 8, 2591 (1975).
27. I. Schek, N. Moiseyev and R. E. Wyatt, J. Chem. Phys. (submitted).
28. I. Schek and J. Jortner, J. Chem. Phys. 70, 3016 (1979).
29. B. Carmeli, I. Schek, A. Nitzan and J. Jortner, J. Chem. Phys. 72, 1928 (1980).
30. I. Schek and J. Jortner, Theor. Chem. Acta 69, 323 (1986).

SEMICLASSICAL DYNAMICS IN PHASE SPACE;
TIME-DEPENDENT SELF-CONSISTENT FIELD APPROXIMATION

Shaul Mukamel, Yi Jing Yan, and Jonathan Grad
Department of Chemistry
University of Rochester
Rochester, N.Y.   14627

ABSTRACT. Gaussian wavepackets in phase space, which are constructed
to have the exact first and second moments with respect to the
coordinates and momenta, are used to develop self-consistent equations
of motion for the semiclassical time evolution of interacting
anharmonic systems. The equations apply to pure as well as to mixed
states and may, therefore, be particularly useful for molecular
dynamics in condensed phases. Numerical calculations of Raman
excitation profiles, using a Morse oscillator, demonstrate the accuracy
of the present equations.

I.   Introduction

    The development of efficient methods for calculating the time
evolution of interacting quantum systems is one of the major problems
in molecular dynamics. Considerable recent activity in this area has
focused on the usage of Gaussian wavepackets to represent the time-
dependent wavefunction. The propagation is usually made by taking the
interaction potential to be locally quadratic[1-6]. This method is
exact for harmonic systems. Its applicability to anharmonic systems
has been critically analyzed by several authors[4-7]. In this
article we develop a new type of self-consistent semiclassical
reduced equations of motion, which is based on projection operator
techniques of nonequilibrium statistical mechanics in phase
space[7,8]. To lowest order the equations provide a time-dependent
self-consistent field approximation (TDSCF) which can then be
systematically improved. It should be noted that the existing
Gaussian wavepacket formalism[1-6] treats the time evolution of pure
states using the Schrodinger equation. Our approach is capable of
handling mixed states as well, since it is based on the density
matrix[7-9]. It therefore has a multitude of potential applications
to molecular dynamics in condensed phases. In this article, we
evaluate explicitly the TDSCF equations of motion and perform
numerical calculations of the time evolution and the Raman lineshapes
of a diatomic molecule with a harmonic ground state and an excited
state given by a Morse oscillator[8]. The calculations demonstrate

R. Lefebvre and S. Mukamel (eds.), Stochasticity and Intramolecular Redistribution of Energy, 109–121.
© *1987 by D. Reidel Publishing Company.*

that the present equations are by far superior to the Thawed Gaussian procedure[1].

II. THE TDSCF EQUATIONS IN PHASE SPACE

We consider a quantum system characterized by N coordinates x_j, $j=1\ldots N$, their conjugate momenta $p_j = -i\hbar\,\partial/\partial x_j$, and masses m_j. Its Hamiltonian is

$$H = \sum_j p_j{}^2/2m_j + V(x_1,x_2\ldots x_N), \tag{1}$$

where V is the interaction potential. We wish to develop a semiclassical self-consistent procedure for the approximate solution of the Liouville equation

$$\frac{d\rho}{dt} = -i\,L\rho \equiv \frac{-i}{\hbar}\,[H,\rho]. \tag{2}$$

Here L is the Liouville operator and $\rho(\mathbf{x},\mathbf{x}',t)$ is the total density matrix where $\mathbf{x}$ is a vector with N components $x_1,\ldots x_N$. If the system is in a pure state with a wavefunction $\psi(\mathbf{x})$, then $\rho(\mathbf{x},\mathbf{x}',t) = \psi(\mathbf{x},t)\psi^*(\mathbf{x}',t)$ and Eq.(2) is identical to the Schrodinger equation. Equation (2) applies, however, to mixed states as well. The self-consistent procedure for solving Eq.(2) is obtained in the following steps[7,8]. We start with a set of dynamical operators, whose expectation values are believed to be relevant for the dynamics. In the present reduced description, we shall focus on the following set of 6N operators:

$$A_{0j} = 1,\ A_{1j}=x_j,\ A_{2j}=p_j,\ A_{3j}=x_j^2,\ A_{4j}=p_j^2,\ A_{5j}=x_jp_j+p_jx_j$$

$$j=1\ldots N. \tag{3}$$

The expectation values of these operators will be denoted $a_{\alpha j}$.

$$a_{\alpha j}(t) \equiv \langle A_{\alpha j}\rangle \equiv \mathrm{Tr}\,[A_{\alpha j}\,\rho(\mathbf{x},\mathbf{x}',t)] \qquad \alpha=0\ldots 5,\ j=1\ldots N. \tag{4}$$

Note that normalization requires that $a_{0j}=1$. We next construct an approximate density matrix $\sigma(\mathbf{x},\mathbf{x}',t)$ chosen to be in the form of a product of single particle density matrices

$$\sigma(\mathbf{x},\mathbf{x}',t) \equiv \prod_{j=1}^{N} \phi_j(x_j,x_j',t) \tag{5}$$

ϕ_j are taken to be Gaussian

$$\phi_j(x_j,x_j',t) = \exp[-b_{0j}-b_{1j}x_j-b_{1j}^*x_j' - b_{2j}x_j^2-b_{2j}^*x_j'^2- b_{3j}x_jx_j']. \tag{6}$$

The time-dependent parameters $b_{\alpha j}$ may be uniquely expressed in terms of $a_{\alpha j}$ by requiring that the expectation values of our operators $A_{\alpha j}$

evaluated using the exact (ρ) and the approximate (σ) density matrix
will be the same, i.e.,

$$a_{\alpha j}(t) \equiv \text{Tr} [A_{\alpha j}\rho(\mathbf{x},\mathbf{x}',t)] = \text{Tr} [A_{\alpha j} \sigma(\mathbf{x},\mathbf{x}'t)]. \qquad (7)$$

Making use of projection operator techniques[9,10], we can then
derive closed reduced equations of motion (REM) for $a_{\alpha j}$. The
equations are constructed to yield the exact values of $a_{\alpha j}$.
$\sigma(\mathbf{x},\mathbf{x}',t)$ may then provide a reasonable approximation for $\rho(\mathbf{x},\mathbf{x}',t)$.
The detailed derivation of the equations is given elsewhere[7,8]. For
each degree of freedom j we define a 6 x 6 matrix

$$S_{\alpha\beta}^{j}(t) = \text{Tr} [A_{\alpha j}^{+} \sigma(t) A_{\beta j}] \qquad \alpha,\beta=0,1...5 \qquad (8)$$

the time dependent projection operator,

$$P(t) = \sum_{j=1}^{N} \sum_{\alpha,\beta=0}^{5} |\sigma(t) A_{\alpha j}>> [S^{j}(t)]_{\alpha\beta}^{-1} <<A_{\beta j}| \qquad (9)$$

and the complementary projection $Q(t)=1-P(t)$. We are using here
Liouville space notation, whereby an ordinary operator A is
written as a ket $|A>>$ and $<<A|B>>=\text{Tr}(A^{+}B)$ is the scalar product of two
operators. Using the assumption that at some initial time t_0
$\rho(t_0)=\sigma(t_0)$, we can derive the following exact REM for $a_{\alpha j}(t)$:

$$\dot{a}_{\alpha j}(t) = -i <<A_{\alpha j}|L|\sigma(t)>> - \int_{t_0}^{t} ds <<A_{\alpha j}|LK(t,s)Q(s)L|\sigma(s)>>, \qquad (10)$$

where

$$K(t,s) = \exp_{+}(-i\int_{s}^{t} d\tau Q(\tau)L) \qquad (11)$$

Here the dot represents a derivative with respect to time, i.e.,

$\dot{a}_{\alpha j} \equiv da_{\alpha j}/dt$, $\exp_{+}$ denotes the time-ordered exponential, and

$$<<A_{\alpha j}|L|\sigma(t)>> \equiv \text{Tr}[A_{\alpha j} L \sigma(t)]$$

$$= \text{Tr}[A_{\alpha j} H \sigma(t) - A_{\alpha j} \sigma(t)H] \qquad (12)$$

Equations (10) are 5N equations for the 5N variables $a_{\alpha j}$. It should
be noted that Eqs.(10) are exact, provided K(t,s) is evaluated to
infinite order. The first term in the r.h.s. of Eq.(10) represents
the "mean field". If $\sigma(t)=\rho(t)$ at all times, then the second term
vanishes identically, and the first term represents the exact

evolution. The second term represents the effect of fluctuations (the fact that the actual density matrix $\rho(t)$ is different from $\sigma(t)$). The TDSCF equations, which will be developed here, are obtained by neglecting the second term, i.e., taking

$$\dot{a}_{\alpha j} = -i \ll A_{\alpha j}|L|\sigma(t)\gg \quad . \tag{13}$$

It will be convenient to make a minor change of variables and use instead of $a_{\alpha j}$ the following 5N variables $\sigma_{\alpha j}$.

$$\sigma_{1j} = \langle x_j \rangle; \ \sigma_{2j} = \langle p_j \rangle$$

$$\sigma_{3j} = \langle x_j^2 \rangle - \langle x_j \rangle^2; \ \sigma_{4j} = \langle p_j^2 \rangle - \langle p_j \rangle^2$$

$$\sigma_{5j} = \langle x_j p_j + p_j x_j \rangle - 2\langle x_j \rangle \langle p_j \rangle \quad . \tag{14}$$

Making use of Eqs.(7), we can express the parameters $b_{\alpha j}$ (Eq.(6)) in terms of the moments $\sigma_{\alpha j}$, resulting in[8]

$$b_{0j} = \frac{\sigma_{1j}^2 + \sigma_{3j} \ \log(2\pi \ \sigma_{3j})}{2 \ \sigma_{3j}} \tag{15a}$$

$$b_{1j} = \frac{- \hbar\sigma_{1j} + i \ [\sigma_{1j}\sigma_{5j} - 2\sigma_{2j}\sigma_{3j}]}{2\hbar\sigma_{3j}} \tag{15b}$$

$$b_{2j} = \frac{4\sigma_{3j}\sigma_{4j} + \hbar^2 - \sigma_{5j}^2 - 2i\hbar\sigma_{5j}}{8\hbar^2\sigma_{3j}} \tag{15c}$$

$$b_{3j} = \frac{\hbar^2 + \sigma_{5j}^2 - 4\sigma_{3j}\sigma_{4j}}{4\hbar^2\sigma_{3j}} \tag{15d}$$

We have evaluated Eq.(13) explicitly for the Hamiltonian Eq.(1). The final result, recast in terms of the new variables $\sigma_{\alpha j}$, is[8]

$$\dot{\sigma}_{1j} = \sigma_{2j}/m_j \tag{16a}$$

$$\dot{\sigma}_{2j} = - \langle V_j(\mathbf{x}) \rangle \tag{16b}$$

$$\dot{\sigma}_{3j} = \sigma_{5j}/m_j \tag{16c}$$

$$\dot{\sigma}_{4j} = - \langle V_{jj}(x) \rangle \, \sigma_{5j} \qquad\qquad (16d)$$

$$\dot{\sigma}_{5j} = \frac{2}{m_j} \, \sigma_{4j} - 2\langle V_{jj}(x) \rangle \, \sigma_{3j} \quad , \qquad (16e)$$

where

$$V_j(x) = \frac{\partial V}{\partial x_j} \qquad\qquad (17a)$$

$$V_{jj}(x) = \frac{\partial^2 V}{\partial x_j^2} \qquad\qquad (17b)$$

$$\langle V_j(x) \rangle = \int dx \, V_j(x) \, \sigma(x,x,t) \qquad\qquad (17c)$$

$$\sigma(x,x,t) = \prod_{j=1}^{N} \phi_j(x_j,x_j,t) \qquad\qquad (17d)$$

$$\phi_j(x_j,x_j,t) = \frac{1}{\sqrt{2\pi}\,\sigma_{3j}} \quad \exp[- \frac{(x_j - \sigma_{1j})^2}{2 \, \sigma_{3j}}] \; . \qquad (17e)$$

$\langle V_{jj}(x) \rangle$ is defined in an analogous way to Eq.(17c) by replacing $V_j(x)$ with $V_{jj}(x)$. Equations (16) are the phase space TDSCF equations, and they will be analyzed in the next section.

III. Discussion

We shall now discuss the significance of the phase space TDSCF procedure, (Eqs.16):

(1) We first note that Eqs.(16) do not contain $\hbar$, this suggests that they are completely classical. Indeed, the present procedure may be repeated for classical mechanics by replacing L in Eq.(2) with the classical Liouville operator. $\phi_j(x_j,x_j',t)$ should then be replaced by a phase space distribution function of coordinates and momenta $\bar{\phi}_j(x_j,p_j,t)$. Taking $\bar{\phi}_j$ to be Gaussian in x_j and p_j, we can repeat the present derivation step-by-step and derive Eqs.(12). The TDSCF for the moments using Gaussian wavepackets, is therefore completely classical.

(2) If $\phi_j(x_j,x_j',t)$ is to represent a pure state, it should be factorized in the form of a product of a function of x_j and a function

of x'_j. A necessary and sufficient condition for that is $b_{3j}=0$, i.e. $\sigma_{5j}^2 = 4\,\sigma_{3j}\sigma_{4j} - \hbar^2$. In this case, σ_{5j} (up to a sign) is uniquely determined by σ_{3j} and σ_{4j} and is not independent. The density matrix (Eq.(6)) may represent, however, mixed states as well, whenever the above condition is not satisfied.

(3) The Thawed Gaussian equations of motion[1] may be obtained from our TDSCF equations, (Eqs.(16)), if the following approximations are made:

(i) replacing $\langle V_j(x)\rangle$ by $V_j(\langle x\rangle)$ and $\langle V_{jj}(x)\rangle$ by $V_{jj}(\langle x\rangle)$.

(ii) Assuming that initially the system is in a pure state. If we then make the substitutions $\sigma_{1j}=x_t$, $\sigma_{2j}=p_t$, $\sigma_{3j}=\hbar/(4\alpha_1)$ and $\sigma_{4j}=\hbar|\alpha_t|^2/\alpha_1$, we obtain the Thawed Gaussian (TG) equations for x_t, p_t, and α_t. Here α_1 is the imaginary part of α_t. Another point to be noted is that in the present TDSCF equation all five moments are coupled, whereas in the TG procedure the first moments $\langle x_j\rangle$ and $\langle p_j\rangle$ obey the classical equations of motion and are decoupled from the second moments.

(4) It can be easily verified from Eqs.(16) that the solution of σ_{5j} is

$$\sigma_{5j}^2(t) = 4\,\sigma_{3j}(t)\,\sigma_{4j}(t) + \gamma,$$

where γ is a constant determined by the initial conditions. If $\gamma=-\hbar^2$, then the system will be in a pure state at all times. This shows that within the TDSCF, if the initial density matrices ϕ_j represent a pure state, they will represent a pure state at all times. This is no longer the case, however, once the fluctuation terms in Eq.(10) are included. The fluctuation terms allow a pure single particle state to evolve into a mixed state. This is a necessary requirement for a reduced description which should show, e.g., how a system relaxes to thermal equilibrium with a thermal bath.

(5) TDSCF equations using pure states were shown to provide useful approximations for a variety of molecular dynamical problems including molecular scattering, electronic spectra, the dissociation of clusters, and thermal desorption from surfaces[2-6,11,12]. They are also extremely useful for a mixed description in which some degrees of freedom are treated quantum mechanically and the others are treated classically. The present phase space TDSCF approach enjoys all these advantages. In addition, it is particularly suitable for dynamics in condensed phases since it may eliminate the necessity of performing tedious thermal averagings. The "bundle of trajectories" used by Gerber, Ratner, and coworkers[11] applies naturally within the phase space TDSCF.

(6) The present equations may be extended by various ways[7,8]. One possibility is to expand the fluctuation term perturbatively. Note that for harmonic systems the TDSCF formulation is exact, provided we take x_j to be the normal modes. This suggests that an expansion of the fluctuation kernel $K(t,s)$ in anharmonicities may be appropriate. Alternatively, we may add more dynamical variables to our chosen set $A_{\alpha j}$ (e.g., x_j^3, p_j^3, etc.) and construct a more elaborate wavepacket with more parameters. The TDSCF will then provide an improved reduced description of the system. The methodology of the present approach is analogous to the development of reduced equations of motion in nonequilibrium statistical mechanics[13,14].

IV. Raman Excitation Profiles of Anharmonic Molecules

In this section, we apply the TDSCF equations toward the calculation of molecular Raman lineshapes. Consider a diatomic molecule with two electronic states, a ground state $|g\rangle$ and an excited state $|e\rangle$. Its Hamiltonian is

$$H = |g\rangle H_g \langle g| + |e\rangle (\omega_{eg} + H_e) \langle e| \qquad (18)$$

with

$$H_g = p^2/2m + m\omega''^2 x^2/2 \qquad (19a)$$

$$H_e = p^2/2m + D\{1 - \exp(-a(x-x_0))\}^2. \qquad (19b)$$

We denote the eigenstates of H_g by $|\psi_n\rangle$

$$H_g |\psi_n\rangle = \hbar\omega''(n+1/2)|\psi_n\rangle \qquad n = 0,1,\ldots. \qquad (20)$$

The Raman excitation profiles were calculated within the Condon approximation using the Kramers-Heisenberg formula:

$$Q_{n0}(\omega_L) = |\int G_{n0}(t)\exp[-i(\omega_L - \omega_{eg})t - \Gamma t/2]|^2, \qquad (21a)$$

Γ^{-1} being the lifetime of $|e\rangle$. The Green function is

$$G_{nm}(t) = \langle\psi_n|\exp(-i H_e t)|\psi_m\rangle = \langle\psi_n|\psi_m(t)\rangle \qquad (21b)$$

and

$$|\psi_m(t)\rangle = \exp(-i H_e t)|\psi_m\rangle . \qquad (21c)$$

$Q_{n0}(\omega_L)$ denotes the intensity of the Raman transition in which the molecule changes its state from $|\psi_0\rangle$ to $|\psi_n\rangle$ as a function of the incident photon frequency ω_L. In order to calculate the spectra, the wavefunction corresponding to the reduced density matrix must be used. Thus, we must specialize our reduced density matrix to a pure state. Setting $\sigma_{s_j}{}^2 = 4\sigma_{3_j}$ $\sigma_{4_j} - M^2$, we have

$$\sigma(x,x',t) = \psi(x,t)\psi^*(x',t) \tag{22a}$$

$$\psi(x,t) = (2\pi\sigma_3)^{-1/4}\exp\left\{\left(-\frac{1}{4\sigma_3} + \frac{i\sigma_5}{4\hbar\sigma_3}\right)(x-\sigma_1)^2 + \frac{i\sigma_2}{\hbar}(x-\sigma_1) + \frac{i\gamma}{\hbar}\right\} \tag{22b}$$

The equation of motion for the phase factor $\gamma(t)$ (which does not appear in the density matrix σ) is derived by demanding that the quantum energy is conserved,

$$\langle E\rangle = \langle\psi|p^2/2m + V(x)|\psi\rangle = i\hbar\langle\psi|\dot{\psi}\rangle \tag{23}$$

We thus get:

$$\dot{\gamma} = \frac{-\hbar}{4m\sigma_3} + \frac{V_2\sigma_3'}{2} + \frac{\sigma_2{}^2}{2m} - V_0 \tag{24}$$

We have performed detailed numerical calculations of the time evolution, and the Raman excitation profiles of the model system of Eq.(18). In all calculations the molecule is assumed to be initially in the ground vibrational state of H_g, i.e., $|\psi_0\rangle$, and it evolves in time according to the excited state Hamiltonian H_e. Each calculation was performed with both the SCF and TG methods. For comparison, the exact calculations were made by expanding the wavefunction $|\psi(t)\rangle$ in the basis set of the discrete spectrum of the Morse oscillator. In all calculations, we took the ground-state frequency $\omega''=1.05$ and the Morse oscillator frequency $\omega = (2Da^2/m)^{1/2}=1$.

We further changed variables to dimensionless units in which the coordinate x is given in units of $(\hbar/m\omega)^{1/2}$, and the momentum p gives in units of $(\hbar m\omega)^{1/2}$. The calculations were done for two values of the dimensionless displacement $x_0 = -0.5$ and -1.5, and for the anharmonicity $\omega x_e \equiv \hbar a^2/2m = .05$. Figures 1 and 2 display the moments $\langle x^2\rangle - \langle x\rangle^2$, and $\langle p^2\rangle - \langle p\rangle^2$ for $x_0 = -0.5$ and $\omega x_e = 0.05$. It is clear that the SCF procedure yields reasonably accurate results. The second moments calculated in the TG procedure monotonically increase and are qualitatively incorrect. In Fig. 3, we display the fundamental Raman profiles Q_{10} using the same parameters of Figs. 1 and 2 and $\Gamma = 0.16$. The calculation is shown also on a logarithmic scale in Fig. 4. The SCF provides a reasonable approximation near the center. In the far wings it converges to the exact result. The TG procedure, which does not have the correct short time dynamics, does not converge, even at large detunings. The same observations hold also for Fig. 5, where we repeat these calculations using a larger displacement, $x_0 = -1.5$.

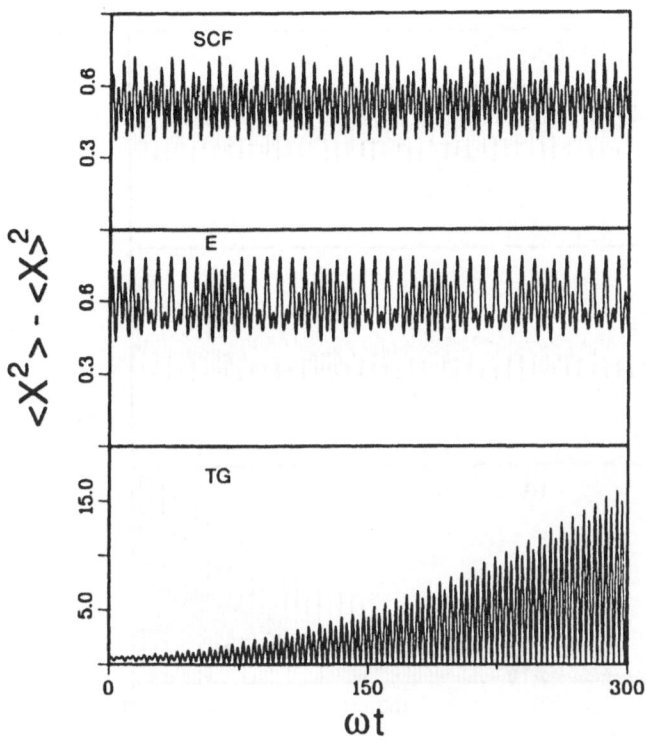

Figure 1
    The dimensionless variance of the displacement $\sigma_3 = \langle x^2 \rangle - \langle x \rangle^2$ vs.
time for the Gaussian wavepacket propagated on a Morse potential. The
Morse oscillator frequency is $\omega \equiv (2Da^2/m)^{.5} = 1$, $D/\hbar\omega = 5$, and $\hbar a^2/m\omega = .01$.
These parameters correspond to an anharmonicity of $\omega x_e = .05$. We have
calculated the dimensionless coordinate x in units of $(\hbar/m\omega)^{\frac{1}{2}}$ and
the dimensionless momentum p in units of $(m\omega\hbar)^{\frac{1}{2}}$. In these units,
we took $x_0 = -0.5$, and the initial conditions $\sigma_1(0) = 0$, $\sigma_2(0) = 0$, $\sigma_3(0) =$
0.48, $\sigma_4(0) = 0.53$. Since this is a pure state, $\sigma_5(0)$ is uniquely
determined by $\sigma_3(0)$ and $\sigma_4(0)$. Shown is the present calculation
(Eq.(16)) (SCF), the exact calculation (E), and the Thawed Gaussian
calculation (TG).

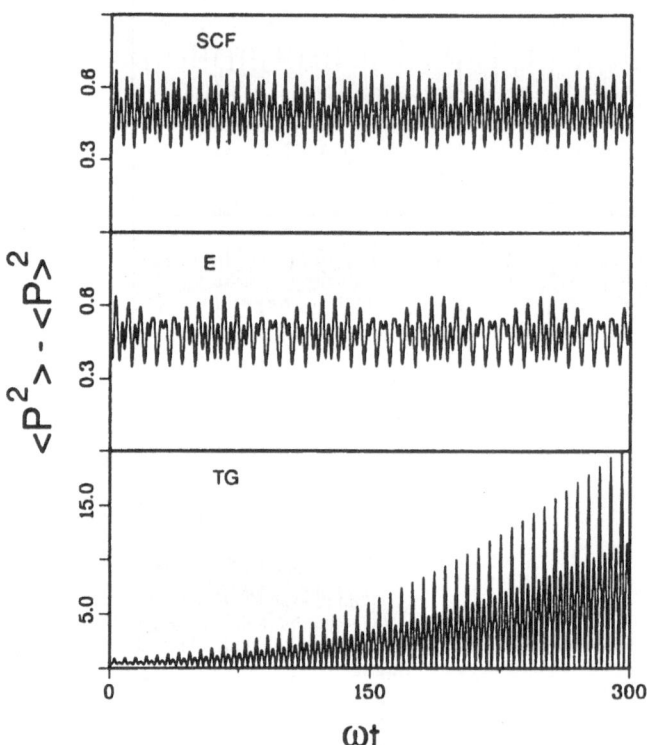

Figure 2
        The dimensionless variance of momentum $\sigma_4 = \langle p^2 \rangle - \langle p \rangle^2$ for the system of Fig. 1.

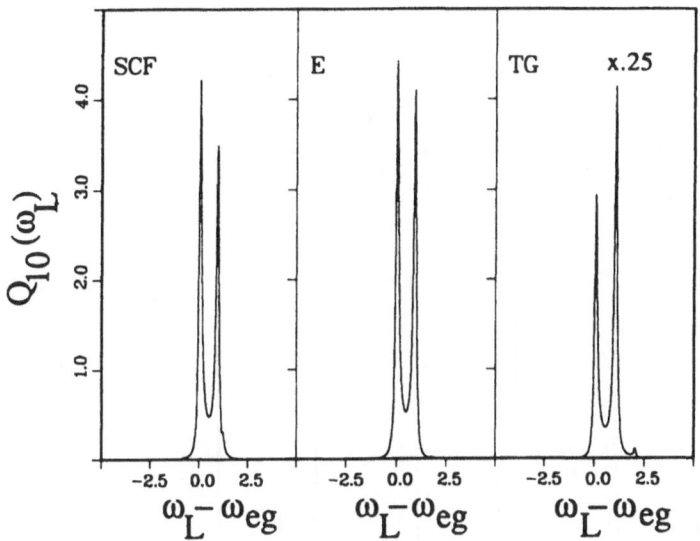

FIGURE 3

    The fundamental Raman excitation profile Q_{10} for the system of Fig. 1. Note that we had to multiply the calculated TG profile by 0.25 in order to put it on the same scale of the E and SCF calculations.

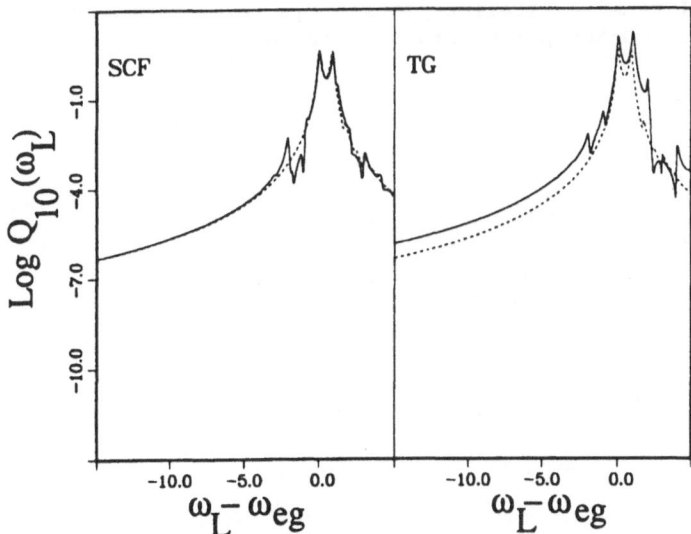

FIGURE 4

    Same as Fig. 3, shown on a logarithmic scale base (10). The dashed line is the exact calculation.

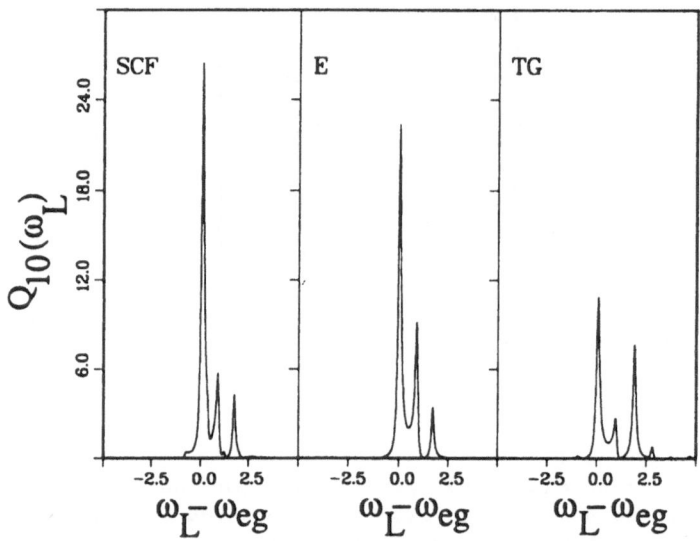

Figure 5
    The fundamental Raman excitation profile Q_{10} for a dimensionless displacement $x_0 = -1.5$. Other parameters same as in Fig. 3.

ACKNOWLEDGMENTS

The support of the National Science Foundation, the Office of Naval Research, the U.S. Army Research Office, the donors of the Petroleum Research Fund, administered by the American Chemical Society, and the Camille and Henry Dreyfus Foundation is gratefully acknowledged.

REFERENCES

1. E. J. Heller, J. Chem. Phys. 62, 1544 (1975); Acct. Chem. Res. 14, 368 (1981).

2. R. D. Coalson and M. Karplus, Chem. Phys. Lett. 90, 301 (1982).

3. R. F. Grote and A. E. DePristo, Surface Science 131, 491 (1983); D. C. Clary and A. E. DePristo, J. Chem. Phys. 81, 5167 (1984).

4. R. T. Skodje and D. G. Truhlar, J. Chem. Phys. 80, 3123 (1984).

5. D. Thirumalai, E. J. Bruskin, and B. J. Berne, J. Chem. Phys. 83, 230 (1985).

6. R. Heather and H. Metiu, Chem. Phys. Lett. 118, 558 (1985); S. Swada, R. Heather, B. Jackson, and H. Metiu, J. Chem. Phys. 83, 3009 (1985).

7. S. Mukamel, J. Phys. Chem. 88, 3185 (1984).

8. J. Grad, Y. Yan, and S. Mukamel, Chem. Phys. Lett. (submitted); J. Chem. Phys. (submitted).

9. S. Mukamel, Phys. Rep. 93, 1 (1982).

10. R. Zwanzig, Supp. Prog. Theo. Phys. 64, 74 (1978).

11. R. B. Gerber, V. Buch, and M. Ratner, J. Chem. Phys. 77, 3022 (1982); V. Buch, R. B. Gerber, and M. A. Ratner, Chem. Phys. Lett. 101, 44 (1983); G. C. Schatz, V. Buch, M. A. Ratner, and R. B. Gerber, J. Chem. Phys. 79, 1808 (1983).

12. R. Kosloff and C. Cerjan, J. Chem. Phys. 81, 3722 (1984).

13. L. Onsager and S. Machlup, Phys. Rev. 91, 1505 (1953).

14. B. Robertson, Phys. Rev. 144, 151 (1966); 160, 175 (1967).

VARIOUS ASPECTS OF THE RESONANT STATE

R. Lefebvre and M. Garcia-Sucre[*]
Laboratoire de Photophysique Moléculaire[+]
91405 Orsay
France

ABSTRACT. The analogy between quantum resonance states and bound states can be emphasized with the use of various devices which transform the normally divergent resonance wave functions into localized functions.

1. INTRODUCTION

Quantum resonance states have many points in common with quantum bound states. This is reflected in the classical picture which is associated to them in the simple case of a particle in a one-dimensional potential. While the bound state corresponds to a back and forth oscillation in a well, the resonance state is seen as a similar oscillation with leakage through or above a barrier. In multidimensional situations it is expected that at a resonance energy many classical trajectories will remain trapped for a long time in the internal region of the system before escaping. Such trajectories have been detected in several situations involving three particles. In the case of the collision of an atom with a diatomic molecule they were used by Noid and Koszykowski [1] to obtain with success the resonance energies through semi-classical quantization. Another example is the explanation by Segev and Shapiro [2] of some features observed and calculated in the photodissociation spectrum of H_2O as being associated to long lived trapped trajectories.

In view of this analogy all the questions examined within the context of classical or quantum stochasticity can be discussed [3] for resonance states which represent a kind of continuation of the discrete spectrum. Most criterions used for characterization of quantum chaos in bound states are immediately applicable to resonance states. We may quote the sensitivity of eigenvalues to external perturbations [4] or to a change in some internal coupling parameter [5], the examination of the nodal patterns [6] or the localization of their wave functions within dynamical potentials and the possibility to define associated quantum numbers [3].

---

*Permanent address : Instituto Venezolano de Investigaciones Cientificas (IVIC) and Universidad Central de Venezuela (UCV), Caracas, Venezuela
+Laboratoire du CNRS associé à l'Université Paris-Sud.

R. Lefebvre and S. Mukamel (eds.), Stochasticity and Intramolecular Redistribution of Energy, 123–132.
© 1987 by D. Reidel Publishing Company.

In this paper we use the analogies between bound and resonance states to develop a number of efficient devices which give the resonance energies and eventually the wave functions. Although illustrated for one channel model problems they can be extended to multidimensionnal situations since the combination of propagation and matching techniques which are used can be also implemented for coupled channel descriptions [7]. We will make reference below to some of these extensions. Systematic methods for the search and calculation of resonance energies are obviously a preliminary to any study of their characteristics.

The procedures we examine fall under the following categories :
- The integration of the wave equation along the real axis, without the enforcement of any asymptotic boundary conditions.
- The use of an optical potential in the asymptotic region.
- The change of a parameter in the potential into a complex number, followed by a Padé procedure to go back to the initial problem.
- The addition of a complex perturbation followed again by a Padé procedure to remove the perturbation.

2. RESONANCE CALCULATIONS WITH INTEGRATION ALONG THE REAL AXIS WITHOUT THE ENFORCEMENT OF ASYMPTOTIC BOUNDARY CONDITIONS

The definition of a resonance state is that its wave function should behave as an outgoing wave in the asymptotic region, say $\exp[ikr]$ for the one channel case if the potential goes to zero sufficiently rapidly. Since the associated wave number is of the form $k = k_0 - ik_1$, with k_0 and k_1 positive, the function contains a factor $\exp[k_1 r]$ which makes it divergent as $r \to \infty$. It is possible to calculate the resonance energies with integration along the real axis if this asymptotic behavior is properly taken into acount when starting the inward integration of the Schrödinger equation [8—9]. Another well known possibility is to make a rotation of the axis into the complex plane. This corresponds to changing r into $\rho\exp[i\theta]$, with ρ real. The wave number k can also be written $K\exp(-i\beta)$ with $\beta>0$. The wave function behaves now as $\exp[iK\rho\exp(\theta-\beta)]$ and if $\theta>\beta$ it contains a damping factor. We describe now a procedure which uses the real axis as the integration path, <u>without the enforcement of any boundary condition in the asymptotic region</u>. To state it more explicitely, the fact is that the final result is not affected by the initial choice of the quantity to be propagated. This circumstance is well known to occur in bound state calculations [10]. The procedure will be illustrated for an anlytically soluble model studied by Ginocchio [11].

The integration technique is derived from the Numerov procedure [12,13]. If $\tilde{h}$ is the step size of a grid of points, one considers the ratio of values of the wave function $\psi(r)$ at two adjacent points :

$$P(r) = \psi(r+\tilde{h}) \, [\psi(r)]^{-1}. \tag{1}$$

This quantity is propagated inward with the relation :

$$P(r-\tilde{h}) = [2\beta(r) - \alpha(r+\tilde{h}) \, P(r)]^{-1} \, \alpha(r) \tag{2}$$

α and β are two quantities related to the energy E and the potential V(r) according to

$$\alpha(r) = \tilde{h} \, (1 + \tilde{h}^2/12 \, (E-V(r))) \; ; \; \beta(r) = \tilde{h} \, [1 - 5\tilde{h}^2/12 \, (E-V(r))] \quad (3)$$

It is instructive to look at the propagation equation (2) in the asymptotic region where the potential practically vanishes. Writing :

$$\bar{\alpha} = \tilde{h} + \tilde{h}^3 k^2/12 \qquad \text{and} \qquad \bar{\beta} = \tilde{h} - 5\tilde{h}^3 k^2/12 \quad (4)$$

we have

$$P(r-\tilde{h}) = \cfrac{1}{\dfrac{2\bar{\beta}}{\bar{\alpha}} - P(r)} = \cfrac{1}{A - P(r)} \quad (5)$$

Itération of this equation leads to the construction of a continued fraction [14]. The solutions of the equation $P(A-P) = 1$ are $P_{\pm} = 1/2[A \pm (4-A^2)^{1/2}]$. It is easily checked that P_+ differs from $\exp[ik\tilde{h}]$ by terms at most in $\tilde{h}^6$, while the same is true for P_- when compared to $\exp[-ik\tilde{h}]$. We have intentionally used h for the propagation step because we want to consider also the possibility that the integration is along a rotated axis, in which case we have $h = h \exp(i\theta)$, with h real. The quantity A is a function of the product $\tilde{h}k = hK\exp(\theta-\beta)$.

Three types of behavior are obtained when propagating P(r) according to equation (2) :

a) $\theta > \beta$ which is the value of θ required to produce localization of the resonance wave function. P(r) converges to P_+ (irrespective of the initial value given to P provided the integration path in long enough). Since P_+ is very close to $\exp[ikh]$ which is the ratio imposed by the Siegert criterion, the outgoing-wave-only condition is automatically satisfied.

b) $\theta = \beta$. The quantity hk is real. No convergence of P(r) is obtained.

c) $\theta < \beta$, including $\theta = 0$. P(r) converges to P_- (again irrespective of the initial value). As P_- is close to $\exp[-ikh]$, we obtain the inverse of the Siegert ratio. We can therefore use the following strategy : inward propagation in the free region followed by inversion of the result to fulfill the Siegert boundary condition. Since $\theta = 0$ corresponds to integration along the real axis, we have the possibility to calculate resonance energies with the same boundary conditions as for a bound state (P(r) equal to zero at both ends of the integration range), although there is no complex rotation.

This technique will be illustrated in the case of a potential studied by Ginocchio [11] who has derived analytical expressions for its bound states and resonance energies. This potential presents the peculiarity that the usual complex rotation method is not applicable because of a singular point in the complex plane [15]. Two different numerical procedures have been previously applied to circumvent this difficulty : either a real coordinate with explicit imposition of Siegert boundary condition [15] or the use [14] of Simon's

R. LEFEBVRE AND M. GARCIA-SUCRE

exterior scaling [16] which consists in using a path which is partly along the
real axis, partly along a complex axis. We refer to [11] for a definition of this
potential which depends on two parameters λ and v. With $\lambda = 26$ and $v = 5.5$
the potential admits six bound states followed by a string of resonances.
Table I gives the energies of the resonances R_i obtained either analytically or
by the method described above. The agreement is excellent. We can also
mention that this potential can be analyzed with algebraic methods [17].

TABLE I

	(a)	(b)
R_1	$\begin{cases} 4140.5, \\ 1681.5288 \end{cases}$	$\begin{cases} 4140.4988 \\ 1681.5307 \end{cases}$
R_2	$\begin{cases} 13576.5 \\ 3502.7810 \end{cases}$	$\begin{cases} 13576.501 \\ 3502.7832 \end{cases}$
R_3	$\begin{cases} 24360.5 \\ 5314.5328 \end{cases}$	$\begin{cases} 24360.501 \\ 5314.5328 \end{cases}$
R_4	$\begin{cases} 36492.5 \\ 7268.1147 \end{cases}$	$\begin{cases} 36492.501 \\ 7268.1126 \end{cases}$
R_5	$\begin{cases} 49972.5 \\ 9399.2537 \end{cases}$	$\begin{cases} 49972.499 \\ 9399.2510 \end{cases}$

Resonances energies of the Ginocchio potential [11] with $\lambda = 26$ and $v = 5.5$.
(a) Analytical results,   (b) Calculated resonance energies with integration
along the real axis and no enforcement of boundary conditions. Resonances
energies in the form $E_v - i\Gamma_r/2$. Upper number : E_r. Lower number : Γ_r.

3. RESONANCE CALCULATIONS WITH AN OPTICAL POTENTIAL IN THE ASYMPTOTIC REGION

We illustrate now a procedure for localizing the resonance function which has been proposed by Jolicard and Austin [18] in the context of the Hazi-Taylor stabilization method [19]. It was recently shown [20] that the method can also be employed when combining propagation and matching, both in one channel and in multichannel situations. The method consists in adding in the asymptotic region an imaginary potential which tends to a constant value, say $-2iV_0$. This potential must be switched on smoothly so that the outgoing character of the Siegert wave is not affected. The energy which normally should be $E = k^2$ will be changed into $\bar{E} = \bar{k}^2 = k^2 + 2iV_0$. The resonance wave number is such that $Im(k) < 0$ and this property is directly related to the divergent character of the wave. We can choose V_0 to ensure that $Im(\bar{k})$ is positive. This ensures localization of the wave. The technique has some similarity with the complex rotation method where the wave number is transformed into $k\exp(i\theta)$, with θ such that the new wave-number has a positive imaginary part. In both case the trick is to move the wave-number from the lower to the upper half of the complex plane.

As a very simple application of this technique we consider the potential $15\, r^2\exp(-r)$, the kinetic energy operator being given the form $-d^2/dr^2$. This potential [21] serves as a bench-mark for the illustration of the various approaches to the resonance problem. One of the forms proposed by Jolicard and Austin [18] for the optical potential is $-iAr^n$ with A and n chosen to perturb the potential only in the asymptotic region. We apply the finite difference boundary value method [22, 23] which consists in replacing the Schrödinger equation by a tridiagonal matrix. With $A = 10^{-8}$ and $n = 8$ and an integration range from 0 to 20 there is obtained for the resonance energy :

$$E = 6.852763 - i\, 0.0255476$$

instead of the commonly accepted value [e.g. 24]

$$E = 6.852781 - i\, 0.0255498.$$

We have shown elsewhere [20b] on the examples of the rotational predissociation of a van der Waals complex and of Stark ionisation of the hydrogen atom that the technique of the asymptotic optical potential is easily extended to multichannel situations.

4. RESONANCE CALCULATIONS WITH COMPLEX ALTERATION OF THE POTENTIAL

We examine now another way of working with a complex potential which will produce localization of the Siegert wave. This is suggested by some recent work on the summation of the divergent series giving the energy of the hydrogen atom placed in a static electric field [25]. One of the approaches consists in summing the series for a complex value of the field intensity and to use a point Padé procedure to obtain through analytic continuation the resonance energies for real values of the electric field. We have shown elsewhere [26] that the energies for a hydrogen atom placed in a complex electric field can be obtained through a coupled channel formulation with bound state boundary conditions. This is followed by a Padé fit and analytic continuation. Good values of the resonance energies are obtained when the width is at least 10^{-4} a.u., corresponding to a field intensity of no less than 0.05 a.u.

The example given now corresponds to the radial Stark effect for the ground state hydrogen atom. In atomic units the wave equation is written :

$$[\frac{d^2}{dr^2} - \frac{l(l+1)}{r^2} + \frac{2}{r} + 2 Fr + E] = 0 \qquad (6)$$

F is the electric field intensity in a.u.. With F < 0 the equation presents only bound states and has been used to describe the so-called charmonium model [27]. For F > 0 the equation yields only resonance states. It has often been considered since Titchmarsh [28] first studied its complex eigenvalues. We consider the lowest resonance of the s channel (l=0). We have used a matching technique combined with propagation of the ratio of values of the wave function at two adjacent points of the grid (see section 2). The electric field intensity is given an imaginary value. The boundary conditions are those of a bound state since an imaginary electric field ensures that the wave function may be made to vanish asymptotically. The eigenvalues are given in Table II for a field intensity going from 0.02 i to 0.20 i by steps of 0.02 i.

TABLE II

F(a.u.)	Lowest eigenvalue (a.u.)
i 0.02	- 0.49940757 - i 0.29947450(-1)
i 0.04	- 0.49770941 - i 0.59608179(-1)
i 0.06	- 0.49508178 - i 0.88796350(-1)
i 0.08	- 0.49170963 - i 0.11742872
i 0.10	- 0.48774853 - i 0.14548494
i 0.12	- 0.48332010 - i 0.17297622
i 0.14	- 0.47851754 - i 0.19992782
i 0.16	- 0.47341240 - i 0.22637020
i 0.18	- 0.46806005 - i 0.25233474
i 0.20	- 0.46250391 - i 0.27785171

Eigenvalues obtained for the ground state hydrogen atom in an imaginary radial electric field : Parentheses represent powers of 10.

The 10 eigenvalues are used to build a [4, 5] point Padé representation [29]. Let F_p ($p = 1, 2, .., 10$) denote the field intensities chosen to calculate the eigenvalues. These eigenvalues, say $E(F_p)$ are equated to the ratio of two polynomials :

$$E(F) = \frac{P(F_p)}{Q(F_p)} \qquad p = 1, 2, \ldots, 10 \qquad (7)$$

With P being a polynomial of order 4 and Q a polynomial of order 5 we can obtain the 10 unknowns from the 10 relations (7). Only 10 coefficients are needed because the constant terms of one of the polynomials can be set equal to 1 without introducing any restriction. The function $E(F)$ which is built in this way can be used to estimate the energy for real values of F. Table III gives the energies obtained for a few real values of the electric field intensity ranging from 0.04 a.u. to 0.14 a.u.. Comparison with the

TABLE III

$F(a.u.)$	$E(F)$	Resonance energy
0.04	- 0.5631 - i 0.1592(-4)	- 0.5630 - i 0.9567(-5)
0.06	- 0.5984 - i 0.7597(-3)	- 0.5985 - i 0.7974(-3)
0.08	- 0.6366 - i 0.5167(-2)	- 0.6365 - i 0.5085(-2)
0.10	- 0.6744 - i 0.1344(-1)	- 0.6746 - i 0.1347(-1)
0.12	- 0.7113 - i 0.2459(-1)	- 0.7112 - i 0.2478(-1)
0.14	- 0.7465 - I 0.3810(-1)	- 0.7462 - i 0.3800(-1)

The column $E(F)$ gives the energies (in a.u.) obtained for various values of the (real) electric field intensity after a point Padé fit to the energies calculated for imaginary values of F. The resonance energies are calculated from a direct integration of the wave equation with complex rotation.

resonance energies obtained from a direct integration of the wave equation with complex rotation shows that the real parts of the energy are always well reproduced. The imaginary parts are reproduced satisfactorily when the field intensity exceeds 0.04 a.u.. We have not tried to improve the procedure althrough this could be done in various ways : changing the range of imaginary field intensities, improving the conditions for integration or increasing somewhat the degrees of the polynomials. It is however clear from the Table that the procedure is a sound one which could be used in other problems.

5. RESONANCE CALCULATIONS WITH A COMPLEX PERTURBATION

The last procedure to be studied consists in adding a complex perturbation to the Hamiltonian. This potential is also producing localization of the resonance wave function, as in the method described in section 3. However we do not restrict the potential to be effective only in the asymptotic region. The eigenenergies may be considerably affected. A series of calculations is made for different values A_p of a parameter A which is a factor in the potential. The eigenvalues $E(A_p)$ are used to build a point Padé representation. The final step consists in an analytic continuation to remove the perturbation. This is obtained by setting equal to zero the factor in the perturbing potential, so that the resonance energy is $E(0)$. We consider again the potential $15\ r^2\exp[-r]$. Table IV gives (column a) the stable

TABLE IV

A	(a)	(b)
0.1	6.859575 - i 0.123796	6.874781 - i 0.141837
0.2	6.867226 - i 0.221131	6.898289 - i 0.251132
0.3	6.875609 - i 0.317471	6.923136 - i 0.354878
0.4	6.884687 - i 0.412915	6.949099 - i 0.454063
0.5	6.894411 - i 0.507510	6.975967 - i 0.549385
0.6	6.904738 - i 0.601296	7.003566 - i 0.641359
0.7	6.915629 - i 0.694309	7.031752 - i 0.730378
0.8	6.927049 - i 0.786582	7.060412 - i 0.816757
0.9	6.938965 - i 0.878144	7.089455 - i 0.900748
1.0	6.951348 - i 0.969021	7.118805 - i 0.982564

The lowest resonance energy of the potential $15r^2\exp[-r]$ perturbed by a potential of the form $-iAr$ (column (a)) or of the form $-iAr^2$ (column (b)). Because of the added complex perturbation all calculations are made with bound state boundary conditions.

eigenvalue which is identified with the perturbed resonance energy for a perturbation of the form $-iAr$ with A ranging from 0.1 to 1.0 by steps of 0.1. A [4,5] point Padé procedure yields :

$$E(0) \doteq 6.85277 - i\ 0.02551$$

With a perturbation of the form $-iAr^2$ and the same values for A, there is obtained the set of eigenvalues given in column b of Table IV. The point Padé procedure gives :

$$E(0) \doteq 6.85276 - i\ 0.025539$$

It is also worthwhile noticing that two other sets of stable eigenvalues are observed in the two cases which are unambiguously associated with the two next members of the 'string of resonances' of this potential [30]. For instance with the perturbing potential of the form $-iAr^2$ these two sets of eigenvalues yield, after application of the Padé procedure

$$9.66947 - i\ 2.23581 \quad\text{and}\quad 10.55474 - i\ 6.77724$$

Instead of the reported values [30]

$$9.66962 - i\ 2.23575 \quad\text{and}\quad 10.55456 -i\ 6.77810$$

It is remarkable feature of the method that resonances with very large widths (such as the third member of the string) can be detected with this method, while the usual phase shift or stabilization procedures are completely ineffective. We have looked without success for sets of eigenvalues which could be correlated with the next resonances. Although there are other stable eigenvalues in the spectra they correspond probably to solutions of the wave equation which are not the analytic continuation of the resonance wave functions. A similar situation has previously been analyzed in detail in the case of complex rotation [31].

6. CONCLUSION

We have illustrated in various ways that there is considerable flexibility in the treatment of the wave equation to yield the resonance energies. These procedures are not limited to one-dimensional problems. It would also be worthwhile to examine further their implementation in methods using basis sets of integrable functions.

REFERENCES

[1] D. W. Noid and M. L. Koszykowski, Chem. Phys. Lett. $\underline{73}$, 114 (1980).

[2] E. Segev and M. Shapiro, J. Chem. Phys. $\underline{77}$, 5604 (1982).

[3] G. Hose, H. S. Taylor and A. Tip, J. Phys. A$\underline{17}$, 1203 (1984).

[4] I. C. Percival, J. Phys. B$\underline{6}$, 1226 (1973)

[5] D. W. Noid, M. L. Koszykowski, M. Tabor and R. A. Marcus, J. Chem. Phys. $\underline{72}$, 6169 (1980)

[6] R. M. Stratt, N. C. Handy and W. H. Miller, J. Chem. Phys. $\underline{71}$, 3311 (1979).

[7] O. Atabek and R. Lefebvre, Phys. Rev. A$\underline{22}$, 1817 (1980)

[8] P. G. Burke and C. Tate, Comput. Phys. Commun. $\underline{1}$, 97 (1969).

[9] O. Atabek, R. Lefebvre and A. Requena, Mol. Phys. $\underline{40}$, 1107 (1980).

[10] B. R. Johnson, J. Chem. Phys. $\underline{69}$, 4678 (1978).

[11] J. N. Ginocchio, Ann. of Physics, $\underline{152}$, 203 (1984).

[12] B. Numerov, Publ. Obs. Central Astrophys. Russ. V$\underline{2}$, 188 (1933).

[13] D. W. Norcross and M. J. Seaton, J. Phys. B$\underline{6}$, 614 (1973).

[14] R. Lefebvre and M. Garcia-Sucre, Ann. L. de Broglie (to be published).

[15] D. T. Colbet, R. Mayrofer and P. R. Certain, Phys. Rev. A$\underline{33}$, 3560 (1986).

[16] B. Simon, Phys. Lett. $\underline{71}$A, 211 (1979).

[17] Y. Alhassid, F. Iacchello and R. D. Levine, Phys. Rev. Lett. $\underline{54}$, 1746 (1985).

[18] G. Jolicard and E. J. Austin, Chem. Phys. Lett. $\underline{121}$, 106 (1985).

[19] A. U. Hazi and H. S. Taylor, Phys. Rev. A$\underline{1}$, 1120 (1970)

[20] M. Garcia-Sucre and R. Lefebvre, (a) J. Chem. Phys. (in press) : (b) Int. J. of Quantum Chem. (in press).

[21] R. A. Bain, J. N. Bardsley and C. V. Sukumar, J. Phys. B $\underline{7}$, 2189 (1974).

[22] D. G. Truhlar, J. Comp. Phys. $\underline{10}$, 123 (1972).

[23] O. Atabek and R. Lefebvre, Chem. Phys. Lett. $\underline{84}$, 233 (1981).

[24] H. D. Meyer and O. Walter, J. Phys. B $\underline{15}$, 3647 (1982).

[25] H. J. Silverstone, B. G. Adams, J. Cizek and P. Otto, Phys. Rev. Letts. $\underline{43}$, 1498 (1979) : W. P. Reinhardt, Int. J. of Quantum Chem. $\underline{21}$ 133 (1982).

[26] M. Garcia-Sucre and R. Lefebvre, submitted for publication.

[27] F. R. Vrscay, Phys. Rev. A$\underline{31}$, 2054 (1985).

[28] E. C. Titchmarsh, *Eigenfunction Expansions Associated with Second Order Differential Equations* (Oxford Un. Press. London, 1958).

[29] L. Schlessinger, Phys. Rev. $\underline{167}$, 1411 (1968).

[30] H. J. Korsch, H. Laurent and R. Möhlenkamp, J. Phys. B$\underline{15}$, 1 (1982).

[31] O. Atabek and R. Lefebvre, Il Nuovo Cim. $\underline{76}$B, 176 (1983).

THE SPECTROSCOPY AND PHOTOPHYSICS OF THE AMINO ACID TRYPTOPHAN IN THE GAS PHASE

Thomas R. Rizzo[*] and Donald H. Levy
James Franck Institute and Department of Chemistry
University of Chicago
Chicago, Illinois 60637, U.S.A.

ABSTRACT. The resonantly enhanced two photon ionization spectrum and the fluorescence spectrum of the amino acid tryptophan have been observed in the gas phase in a supersonic molecular beam. The gas phase spectra are well resolved as compared with solution spectra, and this resolution permits the assignment of features in the excitation spectrum to distinct conformers of the tryptophan molecule. The emission spectra of the various conformers contain both sharp and broad features, and the broad features have been assigned to emission from an intramolecular exciplex formed from an interaction between the indole chromophore and the amino acid sidechain. Excitation is initially to a non-exciplex excited electronic state that is similar in geometry to the ground electronic state. This excitation is followed by energy redistribution to the exciplex state that takes place on the timescale of the fluorescence lifetime. A prerequisite to intramolecular exciplex formation is the formation of a zwitterion in the excited electronic state. The zwitterion is formed by proton tunneling through the barrier that separates the initially excited state from the exciplex state.

I. INTRODUCTION

In solution molecular energy redistribution and relaxation are common phenomena, being induced by interactions between a solute molecule and the solvent. For example, if a solute molecule is vibrationally excited in a specific mode, interaction with the solvent can redistribute the vibrational energy into other modes of the solute and into various degrees of freedom of the solvent. Electronic energy can be shifted between various electronic states by solvent interaction or alternatively it can be converted into energy associated with solvent motion.

R. Lefebvre and S. Mukamel (eds.), Stochasticity and Intramolecular Redistribution of Energy, 133–147.
© 1987 by D. Reidel Publishing Company.

However it is often difficult to distinguish energy relaxation and redistribution properties which result from solvent interactions from those due to the intrinsic properties of the molecule. The work described in this paper is directed toward performing gas phase studies of a molecule which normally exists only in solution with the intent of distinguishing the photophysical properties which are intrinsic to the molecule from those which involve the solvent. We find that in some senses an isolated molecule *can act as its own solvent* in terms of how it affects energy redistribution. Recently we have developed experimental techniques for seeding the amino acid tryptophan into a supersonic molecular beam thus making it possible to study the spectroscopy and photophysics of the isolated molecule.[1] Interesting redistribution of electronic energy was found to occur in the excited electronic states of this molecule,[2] and this paper is a summary of the behavior of the excited electronic states of tryptophan.

II. THE EXCITATION SPECTRUM OF TRYPTOPHAN IN THE GAS PHASE

Prior to this work, the spectroscopy of tryptophan had been extensively studied, but only in condensed media.[3] In solution, the electronic spectrum of tryptophan is ~2000 cm^{-1} broad with just a hint structure superimposed on a broad envelope. The lowest excited singlet electronic state of tryptophan has been assigned as a ^1L_b state, with the transition between this state and the ground electronic state occuring at around 2900Å. A second excited electronic state (^1L_a) is believed to exist at slightly higher energy than the ^1L_b state. The energy difference between the ^1L_b and the ^1L_a states (and perhaps even their ordering) depends on the solvent, however the broadness of the solution spectra makes the identification of the second excited electronic state extremely difficult.

In contrast, the spectrum of tryptophan in a cold, supersonic molecular beam is well resolved, and vibrational detail that is hidden in solution is easily observed.[4] Using a molecular beam source describe elsewhere,[4] resonantly enhanced two-photon ionization spectra of jet-cooled tryptophan are obtained by monitoring the parent ion at the tryptophan mass as the excitation laser is scanned across the electronic transition. Mass analysis of the photoionization spectrum is important to prevent contamination of the tryptophan spectrum by the spectra of other species or decomposition products. Once the source conditions necessary for the production of the tryptophan spectrum have been established, it is possible to measure the fluorescence excitation spectrum of tryptophan as well as the

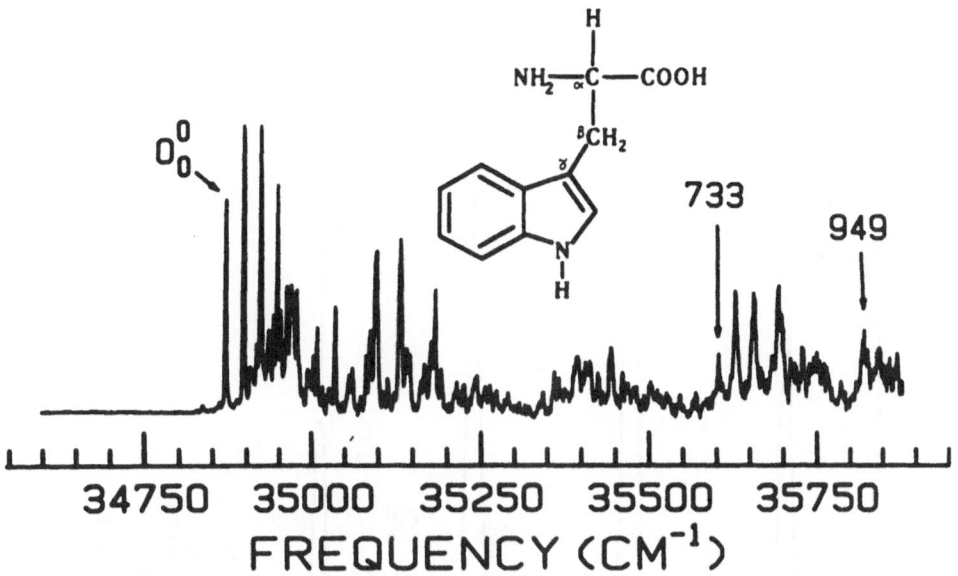

Fig. 1.--Resonantly enhanced two-photon ionization spectrum of tryptophan in a molecular beam. The source temperature is 230°C and the carrier gas is argon at a backing pressure is 1.4 atm.

photoionization spectrum. As shown in Figs. 1 and 2, the resonantly enhanced two-photon ionization spectra exhibit a wealth of structure near the origin of the transition to the first excited singlet state. The fluorescence excitation spectrum of tryptophan (not shown) is nearly identical with the ionization spectra shown of Figs. 1 and 2.

The structure in the excitation spectrum of tryptophan arises from several sources. The chromophore of tryptophan is 3-methylindole whose gas phase spectrum is well known.[5] Some of the structure shown in Fig. 1 (e. g. the bands at 733 and 949 cm^{-1}) is also observed in 3-methylindole and can be assigned to progressions in vibrational modes of the indole ring. There is a prominant 26 cm^{-1} progression indicated in Fig. 2 which does not appear in the spectrum of 3-methylindole and can be assigned to a vibrational motion of the amino acid sidechain. Other (unlabeled) low frequency progressions may also be seen in Fig. 2. In addition to these two sources of vibrational structure, there are features in the spectrum (labeled, e.g. $^B0^0_0$ in Fig. 2) which we have assigned as origins of transitions originating in different conformers of the tryptophan molecule. The indole chromophore of

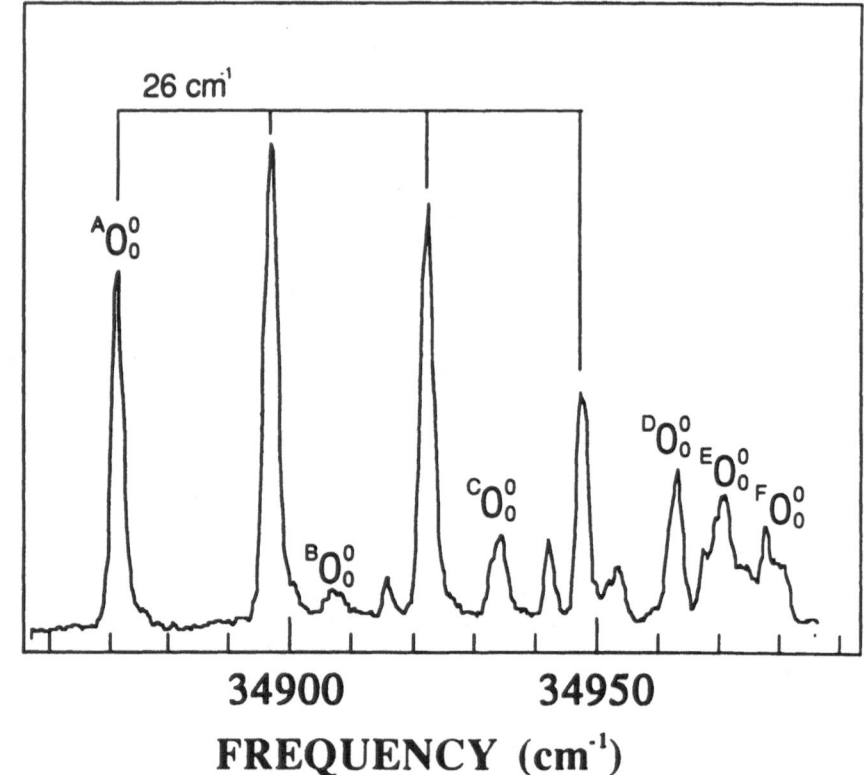

Fig. 2.--Resonantly enhanced two-photon ionization spectrum of tryptophan in a molecular beam using the pulsed source described in reference 4. Electronic origin transitions due to conformers A-F were determined by the power saturation technique described in reference 4 and are labeled $^A0^0_0$ - $^F0^0_0$. A nearly harmonic 26 cm^{-1} vibrational progression built on the origin of conformer A is indicated in the figure.

tryptophan is rigid, but the relatively floppy amino acid side-chain allows tryptophan to have many conformers which need not have identical spectra. The fact that we observe six conformer bands indicates that of all of the possible tryptophan conformations, only a small number have sufficiently different interaction between the side-chain and the chromophore to produce a distinct feature in the spectrum. The assignment of a given feature as either a vibrational progression or as a conformer origin is based on the laser power saturation behavior of the transition. The methodology which led to these assignments has been described in detail elsewhere.[4]

III.   THE EMISSION SPECTRUM OF TRYPTOPHAN IN THE GAS PHASE

The emission spectra that results from exciting various bands
in the tryptophan spectrum are shown in Fig. 3.   Of particular
interest is the top spectrum (Fig. 3a) which is produced by
exciting the feature at 34873 cm^{-1}, the lowest frequence band
in the excitation spectrum.   This emission spectrum consists
of some instrumentally sharp structure superimposed upon
broad, red-shifted background fluorescence.   The sharp
structure can easily be assigned to the origin transition and
progressions in various ring vibrational modes analogous to
the ring progressions seen in the excitation spectrum.   The
surprising feature was the broad, red-shifted emission which
has no counterpart in the excitation spectrum.

Prior to this work we had observed sharp absorption
spectra which produced broad emission spectra, and in the past
this phenomenon was attributed to internal vibrational
redistribution of energy.   For example, when jet cooled
free-base phthalocyanine is excited more than 720 cm^{-1} above
the origin, the resulting emission spectrum is broad even
though the excitation spectrum remains sharp.[6]   The model used
to explain this behavior is illustrated in Fig. 4.   This model
postulates that when broadening sets in, the density of
vibrational levels in the excited electronic state is
sufficient to allow redistribution of vibrational energy.·
Only a small fraction of the total number of vibrational
levels of the excited electronic state have significant
Franck-Condon factors with the few vibrational levels of the
ground electronic state that are populated in the cold jet.
Therefore the absorption spectrum is sharp and the initial
excitation is to those few vibrational levels with large
Franck-Condon factors.   At low excess vibrational energy where
the density of states is low, energy redistribution does not
occur, and emission occurs only from the originally populated
level, producing a sharp emission spectrum.   However with
sufficient vibrational excitation, the density of vibrational
states is high enough to allow vibrational energy
redistribution in the excited electronic state during the
fluorescence lifetime.   In this case the energy that was
originally in a single vibrational level is distributed to
many vibrational levels before emission takes place.
Therefore the total emission spectrum consists of a
superposition of spectra produced by emission from the levels
populated in the redistribution process.   If these levels have
different emission spectra, this superposition could be broad
and unresolved as was observed for phthalocyanine.

The unique feature of the tryptophan emission spectrum is
that the broadening occurs even when the molecule is excited
to what appears to be the zero-point level of the excited
electronic state (Fig. 3a).   In this case there is no excess

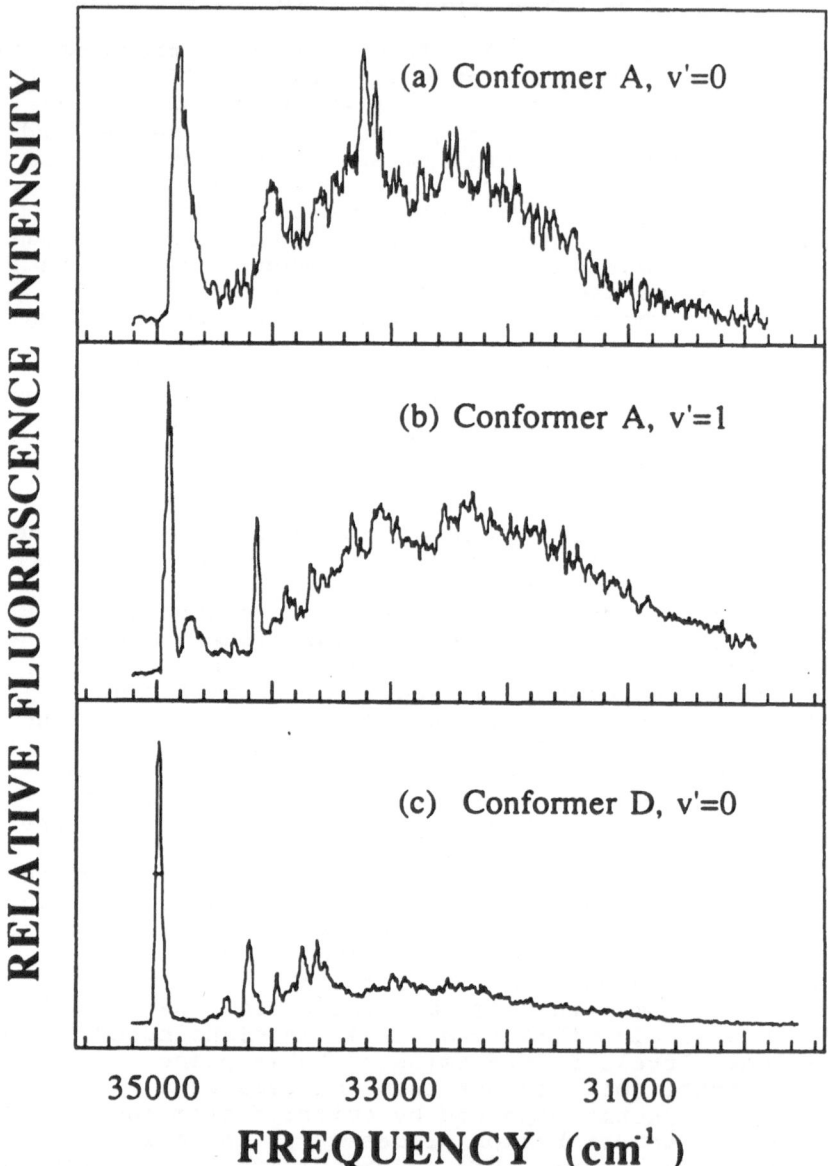

Fig. 3.--Dispersed fluorescence spectra of tryptophan subsequent to exciting (a) the electronic origin of conformer A, (b) the first member of the 26 cm^{-1} progression of conformer A, and (c) the electronic origin of conformer D. The monochromator slits are 1 mm producing a resolution of ~50 cm^{-1}.

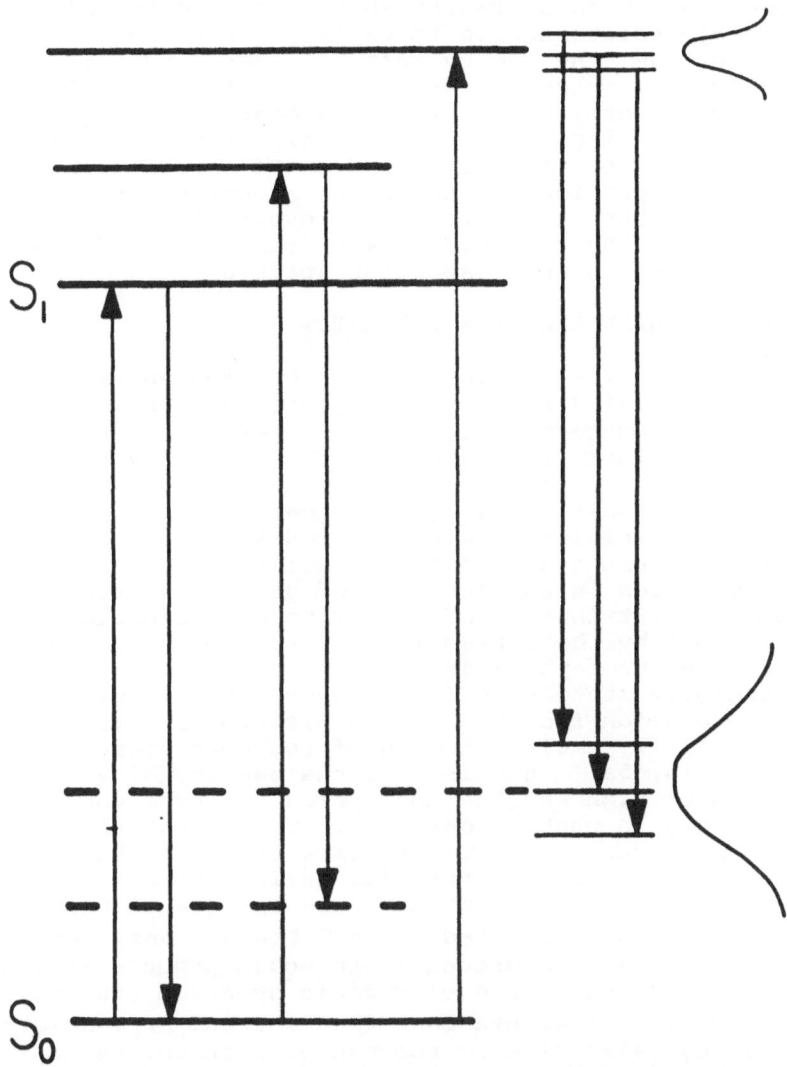

Fig. 4.--Model for producing broad emission spectrum and sharp absorption spectrum in a molecule where the potential surfaces of S_0 and S_1 are similar. Excitation is from the zero point level of S_0 (heavy solid line) to certain vibrational levels of S_1 (heavy solid lines). These levels are mixed with nearby background levels (lighter solid lines) causing redistribution of energy to background levels. Emission is from a given level of S_1 to the corresponding level of S_0.

vibrational energy to be redistributed, and to be able to
invoke energy redistribution to explain the broadening in the
emission spectrum, the energy must be associated with some
degree of freedom other than vibrations. A second electronic
state of lower energy would provide a means of energy
redistribution as indicated schematicaly in Fig. 5. If the
equilibrium position of this second state were displaced from
the equilibrium postion of the ground electronic state, direct
absorption from the low vibrational levels of the ground state
would be Franck-Condon forbidden and transitions to this state
would not be seen in the absorption spectrum.

IV.   INTRAMOLECULAR EXCIPLEX FORMATION

In solution an attractive interaction between an excited
electronic state of a solute chromophore and a solvent
molecule will produce an intermolecular exciplex state.[7]
Such a state is lower in energy than the isolated chromaphore
and its equilibrium geometry is shifted relative to that of
the ground electronic state. The signature of an
intermolecular exciplex is a broad, red-shifted emission, and
such exciplexes are quite common in solution. We suggest that
the broad emission in the spectrum of gas phase tryptophan is
produced by redistribution of energy to an *intra*molecular
exciplex formed by the attractive interaction between the.
excited indole chromaphore and the amino acid sidechain. If
only an exciplex state were formed when the molecule was
excited, absorption from the ground electronic state would be
to a vibrationally excited region of the upper state surface
as shown in Fig. 5a. In this case the density of vibrational
states would be high and the absorption spectrum would be
broad, contrary to what is observed. To account for the sharp
absorption spectrum, we must postulate that the absorption is
to a state with a similar potential surface to that of the
ground electronic state. In this case the Franck-Condon
absorption would be dominated by $\Delta v=0$ transitions, and the few
levels populated in the ground state would produce only a few
vibrational features in the electronic spectrum thus giving a
sharp spectrum. We believe that this initial $\Delta v=0$ absorption
is followed by relaxation to the exciplex state, perhaps by
tunneling through a barrier that separates the initially
populated state from the exciplex state as shown in Fig. 5b.
If the time scale for this relaxation is similar to the
fluorescence lifetime, emission can be from both the initially
populated level producing sharp features and from the many
excited levels populated in the exciplex state producing the
broad, red-shifted emission.
    We now examine in more detail the nature of the
interaction that produces the intramolecular exciplex in gas
phase tryptophan. First, consider the emission spectrum

(a)   (b)

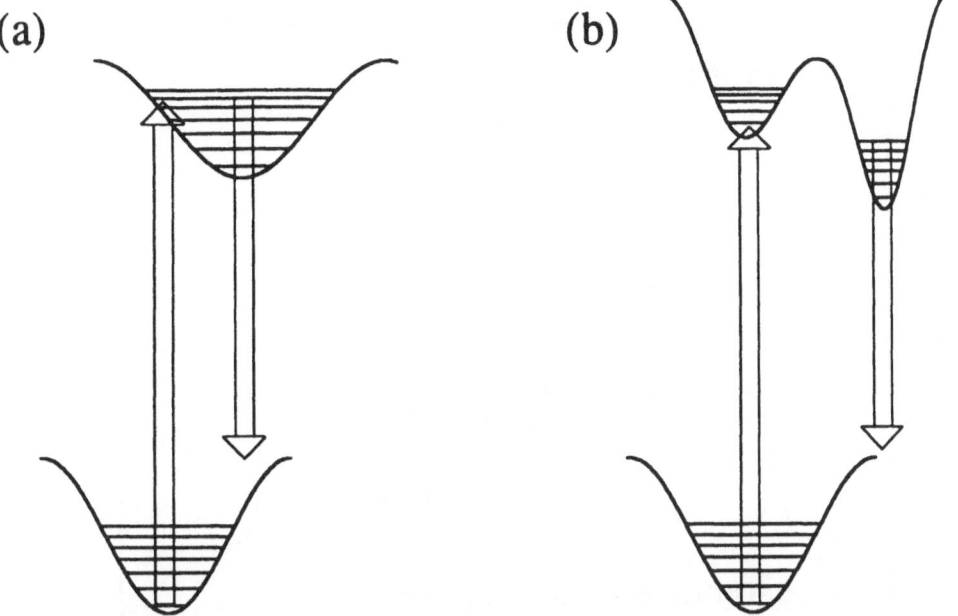

Fig. 5.--Schematic of potential curves indicating the
possibilities for excitation to an exciplex state: (a) direct
excitation of an exciplex which has a geometry shifted from
the ground state in some coordinate, and (b) excitation of
tryptophan to a portion of the potential surface which is
similar in geometry to the ground state and "protected" from
the deep exciplex well by a barrier.

produced by the other conformers of tryptophan, for example
the spectrum produced by conformer D shown in Fig. 3c.   Once
again the spectrum has sharp and broad components, but the
relative intensity of the broad component is much less than
the relative intensity of the broad component produced by
conformer A (Fig. 3a).

The fact that the relative intensity of the broad
component is different for different conformers is most easily
seen in Fig. 6.   The top spectrum (Fig. 6a) is just the
fluorescence excitation spectrum obtained by scanning the
exciting frequency and collecting emission at all frequencies;
that is, both the broad and sharp components of the emission
are collected.   As already mentioned, this conventional
fluorescence excitation spectrum is nearly identical to the
resonantly enhanced two photon ionization spectrum shown in
Fig. 1.   The lower spectrum (Fig. 6b) is an excitation
spectrum taken by scanning the exciting frequency, but only

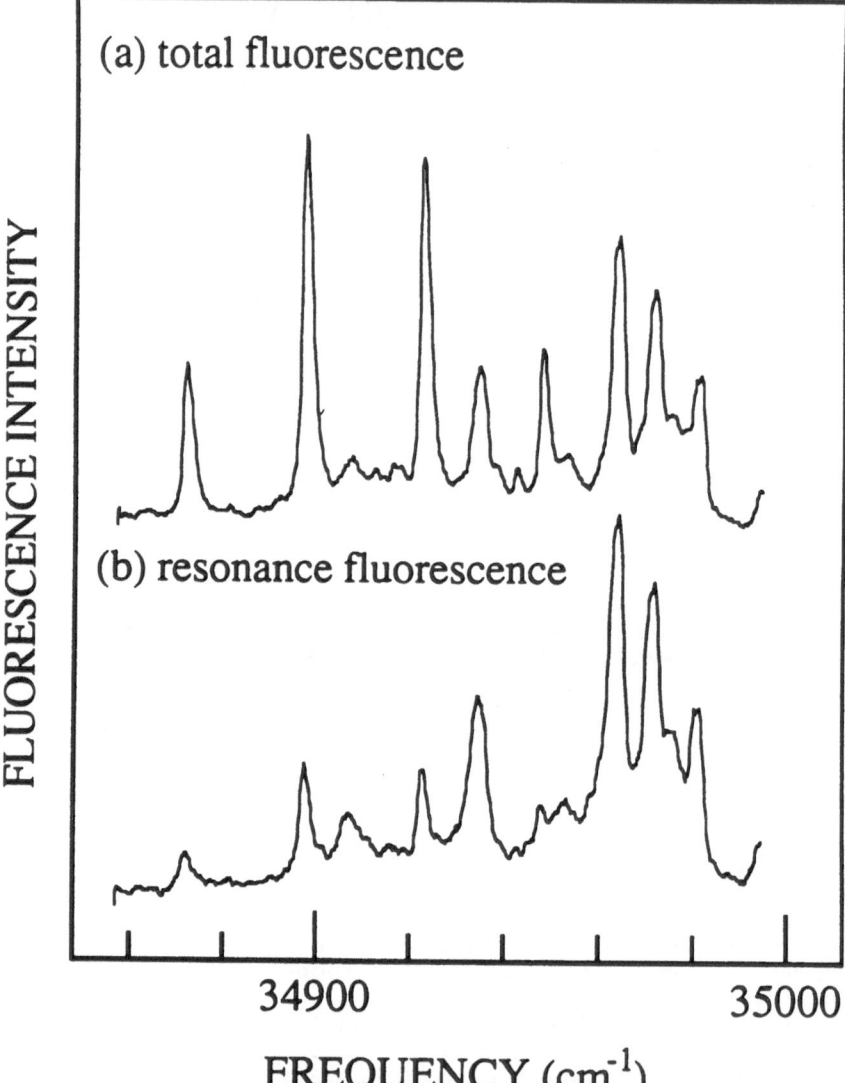

Fig. 6.--Fluorescence excitation spectrum of tryptophan monitoring (a) the total emitted fluorescence, and (b) the fluorescence emitted only in the region near the exciting laser (resonance fluorescence). The resonance fluorescence is collected by synchronously scanning the laser and monochromator, with the monochromator centered on the laser frequency. The monochromator slits are 5 mm producing a resolution of ~250 cm^{-1}.

light emitted in a band 250 cm^{-1} wide centered on the frequency of the exciting light is collected; that is, only the sharp component of the emission is collected. Therefore comparison of Fig. 6b with Fig. 6a is a comparison of the sharp to the total emission across the spectrum. The very different intensity patterns in these two spectra indicate that the relative amount of broad emission is different for different conformers. In our model, the relative amount of broad emission is a measure of the rate of relaxation to the intramolecular exciplex state. Since the exciplex interaction is between the chromaphore and the sidechain, it is not surprising that the strength of that interaction depends on the conformation of the molecule.

Further information on the nature of the exciplex interaction can be obtained from the emission spectra of various derivatives of tryptophan. In Fig. 7 we show the emission spectra obtained by exciting the lowest frequency feature in the absorption spectra of the following molecules: tryptamine (tryptophan without the carboxyl group), indole-3-propionic acid (tryptophan without the amino group), cyclo-glycyltryptophan (a cyclic dipeptide with two peptide bonds and no free carboxyl or amino groups), and N-acetyltryptophan ethyl ester (tryptophan with the amino group blocked with an acetyl group and the carboxyl blocked with and ethyl group). None of these molecules show a significant amount of broad emission, and the feature common to their structure gives an indication of what is required to form an intramolecular exciplex.

The difference between tryptophan and any of the other molecules whose spectra are shown in Fig. 7 is the ability of tryptophan to form a zwitterion. In solution at neutral pH, tryptophan exists as a zwitterion which is formed from the uncharged form shown in Fig. 1 by migration of a proton from the carboxyl group to the amino group to produce the two charged groups COO$^-$ and NH$_3^+$. All of the other molecules in Fig. 7 lack either the mobile proton or the amino acceptor site or both and exist in solution as uncharged molecules.

In its ground electronic state gas phase tryptophan is also uncharged and does not form a zwitterion. Evidence for lack of zwitterion formation in the gas phase is the fact that the tryptophan specrum occurs at almost the same frequency as the spectrum of free 3-methyl-indole, which can only exist in a molecular form. If tryptophan were a zwitterion, the existence of charged groups on the sidechain would be expected to perturb the spectrum more than is observed. The lack of ground state zwitterion formation in gas phase tryptophan is not surprising since the charge seperation necessary to form a zwitterion requires energy. In solution the solvation energy of the charged groups provides the driving force for zwitterion formation, but in the gas phase this stabilizing interaction is absent.

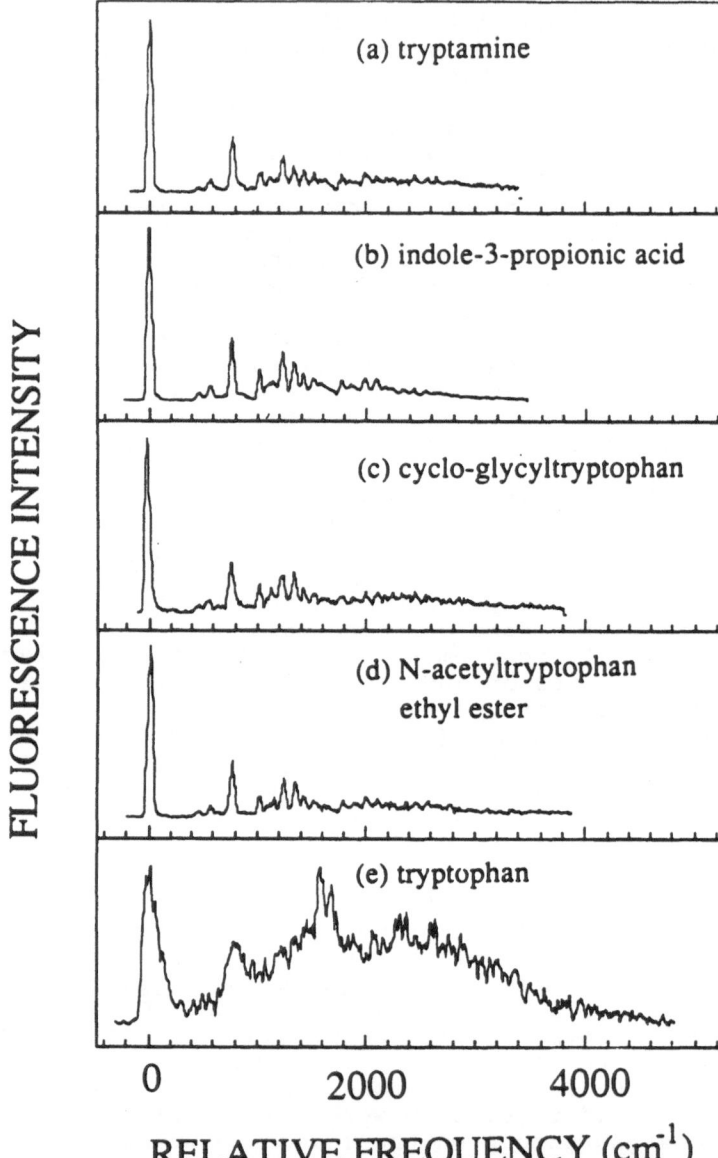

Fig. 7.--Dispersed fluorescence of (a) tryptamine,(b) indole-
3-propionic acid, (c) cyclo-glycyltryptophan, (d) N-acetyltryptophan ethyl ester, and (e) tryptophan conformer A
subsequent to exciting their respective electronic origins.
The monochromator slits are 1 mm producing a resolution of
~50cm^{-1}.

The electronic excitation spectrum of tryptophan indicates that there is little geometry change upon excitation. With the exception of one prominent 26 cm^{-1} vibrational progression, the spectrum has very little vibrational structure indicating that $\Delta v \neq 0$ transitions are unimportant, and hence the Franck-Condon region the upper state potential surface is similar to the lower state potential surface. If the lower state is in the molecular form (uncharged), the state initially populated by laser excitation must be uncharged as well.

The exciplex state, however, might well be generated by the formation of a zwitterion, and the absence of exciplex emission from molecules that cannot form a zwitterion supports this conclusion. As mentioned previously, the formation of a zwitterion requires a driving force to achieve charge separation, and in this case the driving force would be the exciplex interaction between the zwitterion and the indole ring which lowers the electronic energy of the molecule. Moreover, if the excited electronic states of the molecular and zwitterionic forms of tryptophan were separated by a barrier as required by the nature of the spectrum, the zwitterion could be formed by proton tunnelling through the barrier. The rate of such a tunnelling process would be very dependent on the proximity of the carboxyl proton to the amino group (i.e. the conformation of the molecule). This is exactly what is observed.

A final piece of information is provided by the effect of isotopic substitution on the emission spectrum. Tryptophan has four labile protons (two on the amino group, one on the carboxyl, and the proton attached to the indole nitrogen), and if one of these positions is to be deuterated, they all must be deuterated. The emission spectrum that is produced by d^4-tryptophan has both broad and sharp features, but the relative intensity of the broad feature is reduced by a factor of 1.8 relative to the spectrum of undeuterated tryptophan. If the relative intensity of the broad feature is a measure of the rate of exciplex formation, and if exciplex formation proceeds via proton tunnelling to produce a zwitterion, then it is not surprising that deuteration reduces the intensity of the broad emission.

V.   CONCLUSIONS

The electronic spectrum of tryptophan in a supersonic molecular beam is very much more highly resolved than the corresponding spectrum in solution. The increased resolution has allowed the identification of features that can be assigned to a few different conformers of tryptophan that can be formed by rotations about the bonds in the amino acid sidechain. The fact that excitation of these different conformer bands produces different emission spectra indicates

that the conformers do not interconvert on the timescale of
the fluorescence lifetime.

The emission spectra of the various conformers contains
both sharp and broad features, the ratio of the intensities of
the sharp and broad features varying from conformer to
conformer. This behavior has been attributed to the formation
of an intramolecular exciplex between the indole chromaphore
and the amino acid sidechain subsequent to electronic
excitation. The fact that the absorption spectrum is sharp
while the emission spectrum is, in part, broad indicates that
the electronic state that is initially excited is similar to
the ground electronic state and does not involve an exciplex
interaction. During the fluorescence lifetime, the
electronic energy migrates to the exciplex state producing the
broad emission. Thus, the ratio of intensity of broad to
sharp emission is a measure of the energy redistribution rate,
and the fact that this' ratio differs among the conformers
indicates that the strength of the exciplex interaction
depends on the conformation of the molecule. Experiments with
chemically and isotopically substituted tryptophans indicate
that the exciplex is formed only when a zwitterion can be
formed in the excited electronic state. The ground electronic
state and the excited electronic state initially reached upon
excitation are not ionic states, and the zwitterionic exciplex
is formed by proton tunneling between the excited state of the
neutral molecule and that of the zwitterion. These phenonena,
the formation of zwitterions and exciplexes, are well known in
solution but were previously thought to require solvent
stabilization. Certainly interactions with the solvent must
stabilize both zwitterions and exciplexes, but this work
indicates that even in the gas phase, the ability to form
zwitterions and exciplexes is an intrinsic property of the
excited electronic state of tryptophan.

This work was supported by the National Science
Foundation under Grant CHE-8311971.

References

*Present address: Department of Chemistry, University of
Rochester, River Station, Rochester, New York 14627, U.S.A.

1. T. R. Rizzo, Y. D. Park, and D. H. Levy, *J. Am. Chem. Soc.*
   107, 277 (1985); T. R. Rizzo, Y. D. Park, L. Peteanu, and
   D. H. Levy, *J. Chem. Phys.* **83**, 4819 (1985).

2. T. R. Rizzo, Y. D. Park, and D. H. Levy, *J. Chem. Phys.*
   (in press).

3. See, for example, E. H. Strickland, J. Horowitz, and C.
   Billups, *Biochem.* **9**, 4914 (1970); E. H. Strickland, C.
   Billups, and E. Kay, *Biochem.* **11**, 3657 (1972); Y. Yamamoto

and J. Tanaka, *Bull. Chem. Soc. Jpn.* **45**, 1362 (1972); G. R. Fleming, J. M. Morris, R. J. Robbins, G. J. Woolfe, P. J. Thistlethwaite, and G. W. Robinson, *Proc. Natl. Acad. Sci. U.S.A.* **75**, 4652 (1978); M. C. Chang, J. W. Petrich, D. B. McDonald, and G. R. Fleming, *J. Am. Chem. Soc.* **105**, 3819 (1983); T. C. Werner and L. S. Forster, *Photochem. and Photobiol.* **29**, 909 (1979); A. G. Szabo and D. M. Rayner, *J. Am. Chem. Soc.* **102**, 554 (1980); and E. F. Gudgin-Templeton and W. R. Ware, *J. Phys. Chem.* **88**, 4626 (1984).

4. T. R. Rizzo, Y. D. Park, L. Peteanu, and D. H. Levy, *J. Chem. Phys.* **84**, 2534 (1986).

5. J. Hager and S. C. Wallace, *J. Phys. Chem.* **87**, 2121 (1983); R. Bersohn, U. Even, and J. Jortner, *J. Chem. Phys.* **80**, 1050 (1984); and T. R. Hays, W. E. Henke, H. L. Selzle, and E. W. Schlag, *Chem. Phys. Lett.* **97**, 347 (1983).

6. P. S. H. Fitch, C. A. Haynam, and D. H. Levy, *J. Chem. Phys.* **74**, 6612 (1981).

7. J. B. Birks, *Photophysics of Aromatic Molecules* (Wiley, New York, 1970); M. Gordon and W. R. Ware, eds., *The Exciplex* (Academic Press, New York, 1975).

EXCITED VAN DER WAALS COMPLEXES AS A PROBE FOR INTERMEDIATE STATES IN COLLISIONS

W. H. Breckenridge, O. Benoist d'Azy, M. C. Duval
C. Jouvet et B. Soep.
C. N. R. S.
Bât. 213 – Laboratoire de Photophysique Moléculaire
91405 – ORSAY Cedex – France.

ABSTRACT. We report here the observation of excited states in van der Waals complexes whose dissociation lead to electronic relaxation (Hg N_2) or to chemical reaction (Hg H_2). The spectroscopic observation of those complexes gives some new information on the intermediate states responsible for relaxation or chemical reaction. This kind of experiment is a selective way of observing the collision complex in relaxation and reactive processes.

❋❋❋

Excited state van der Waals complexes allow the selective preparation of collision complexes in many relaxation processes including vibrational relaxation [1], electronic relaxation [2], chemical reactions [3]. For a collision involving excited A* with B, the excitation and dissociation of the van der Waals complex A – B is viewed as the half collision analog. The structure of the A^*, B intermediate can be studied by the optical excitation of this A – B complex and moreover the influence of the electronic configurations and the intermolecular movements upon the collision process can be deduced.

Van der Waals complexes are prepared in specific states by optical excitation in supersonic jets. The subsequent evolution of the complex is described as a photodissociation where the dynamics can be studied through the time decay of the system and the energy distribution over the fragments. It is important to study both the structure (electronic and geometry) as well as the dynamics of such systems. We shall report herein the excitation of the Hg–N_2 ($^3P_1$) complex leading the electronic relaxation in Hg ($^3P_0$) + N_2 and the chemical reaction Hg–H_2 ($^3P_1$) leading to the formation of Hg H ($^2\Sigma^+$)+H.

THE Hg – N_2 COMPLEX

At room temperature collision of Mercury in ^3P states with rare gas atoms are inefficient in quenching the ^3P states, as appears in Table 1. However collisions with diatomics (N_2) induce electronic relaxation from $^3P_1$ to $^3P_0$

149

R. Lefebvre and S. Mukamel (eds.), Stochasticity and Intramolecular Redistribution of Energy, 149–162.
© 1987 by D. Reidel Publishing Company.

with small cross sections and much greater cross section from $^3P_2$ to $^3P_1$ and $^3P_0$. It is clear that the molecular movements added by the new — degrees of freedom of the atom — diatom complex (as compared to the atom-atom case) are effective in relaxing the $^3P_1$ state. It has appeared extremely interesting to study the structure and decay of complexes formed by Hg^3P_1 and N_2, as the channel leading to N_2 (v=1) is closed energetically, in difference with collisions (10).

TABLE I

	$\sigma \overset{\circ}{A}^2$	réf.
$Hg\ (^3P_2) + N_2 \rightarrow Hg\ (^3P_{1,0})$	20	3
$Hg\ (^3P_2) + He \rightarrow Hg\ (^3P_1)$	$<10^{-2}$	4
$Hg\ (^3P_2) + Xe \rightarrow Hg\ (^3P_1)$	$<10^{-3}$	4
$Hg\ (^3P_1) + N_2 \rightarrow Hg\ (^3P_0)$	$0.72; 0.97; 1.1$	5.6.7
$Hg\ (^3P_1) + Ar \rightarrow Hg\ (^3P_0)$	$<2 \times 10^{-2}$	8
$Hg\ (^3P_1) + CO \rightarrow Hg\ (^3P_0)$	30	9
$Hg\ (^3P_0) + N_2 \rightarrow Hg\ (^1S_0)$	$<10^{-5}$	9
$Hg\ (^3P_0) + Ar \rightarrow Hg\ (^1S_0)$	$<10^{-5}$	4
$Hg\ (^3P_0) + He \rightarrow Hg\ (^1S_0)$	$<10^{-5}$	4

When excited in a state correlated to $^3P_1$ the $Hg-N_2$ complex exhibits both fluorescence to the ground state and dissociation into $^3P_0$ and N_2. The fluorescence excitation spectra and the action spectra, probing $^3P_0$ for variable excitation of the complex, reveal the efficiency of each excited level to induce the electronic relaxation to $^3P_0$. The relaxation efficiency of each vibrational level or group of rotational levels of the complex can be deduced quantitatively by measuring the time decay of those levels. The observed rate is given by $1/\tau$ = Fluorescence rate (same as mercury) + disocciation rate. On the other hand, the sole observed decay for $^3P_1$ mercury rare gas complexes (He, Ar, Ne) is fluorescence with the same lifetime as free mercury (120 ns) and no relaxation to $^3P_0$ has been detected. The specific movements characteristic of the $Hg-N_2$ system will be studied through the spectroscopy of the complex in the following.

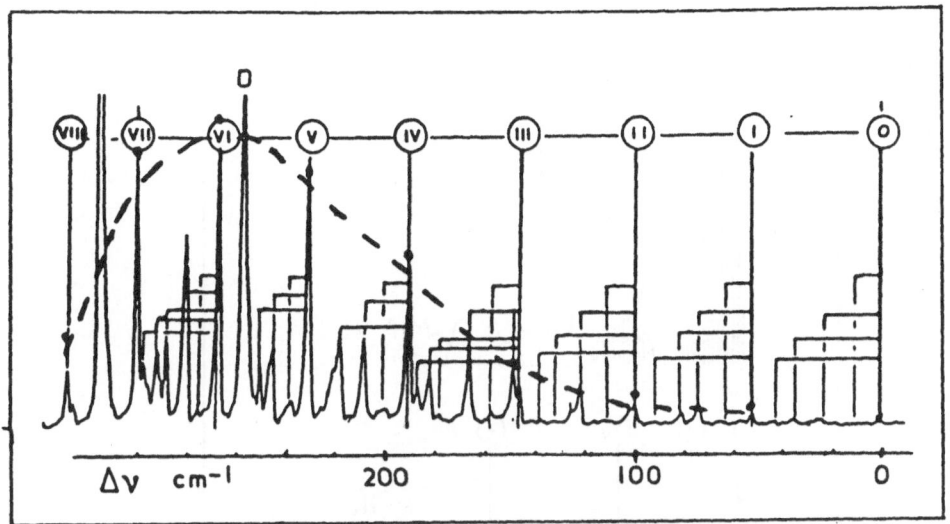

Fig. 1 - Fluorescence excitation spectrum of Hg N_2.

The fluorescence excitation spectrum is displayed in Fig. 1. There are two sets of lines as is usual in complexes involving a p atomic orbital[11] corresponding to the average orientations of this 6p orbital with respect to the Hg - N_2 axis. The blue shifted shallower state converges to a continuum due to the direct dissociation into Hg $^3P_1$ + N_2 which we have identified by double resonance of Hg ($^3P_1 \rightarrow {}^3S_1$). The red shifted state which exhibits two types of progressions. One is labeled with roman numerals in Fig. 1 and can be identified as a stretching progression of the Hg-N_2 bond. This identification comes from the similarity in frequency and appearance of the equivalent Hg-A_r progression. Therefore one is led to separate in between two types of vibrational movements in the Hg-N_2 complex, the stretching vibration and the bending vibrational movement of N_2 with respect to mercury. We shall assume, in the simplest approximation[12] that the corresponding radial and angular potentials are independant, the actual potential being the sum of a radial and angular dependence.

THE STRETCHING MOVEMENT

The intensity distribution of the pure stretching vibrations indicates a strong displacement in the radial equilibrium distances between the X ground state, and the red shifted excited state. A one-dimensional Franck Condon simulation of the fluorescence excitation spectrum yields a separation ΔR = – 1Å between the X and A equilibrium distances. Moreover we can infer that the X state equilibrium distance should be very close to the Hg Ar one and as the binding energies of both ground state complexes are close : D_0 = 125 cm^{-1} for Hg-Ar and D_0 = 100 cm^{-1} for Hg-N_2, (X) (from the separation in Fig. 1 of the Hg line and the dissociation continuum).

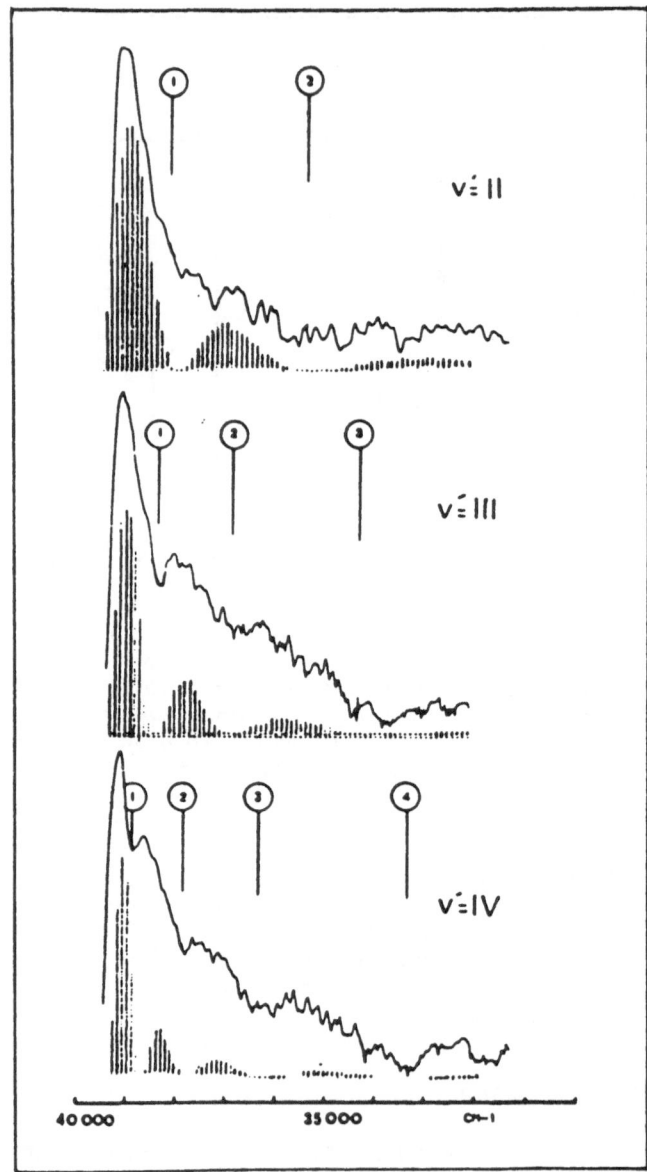

Fig. 2 – Fluorescence emission spectra originating from pure stretching quanta : from top to bottom $v' =$ II, III, IV in Log Scale. F.C. simulations are displayed under each relevant spectrum in linear scale.

The number of the stretch quanta has been assigned, as the intensity of the O-O band is very small, through the emission spectra in the bound free and bound bound regions, after excitation of the most prominent stretching peaks. The number of oscillations in these spectra will reveal the oscillations of the excited state streching wavefunction (when no torsion is excited) as was done for the Na-A$_r$ [13] or Hg A$_r$ [16] diatomics. The peaks labeled II and III assigned to 2 and 3 quanta of stretching as they display respectively 2 and 3 minima corresponding to the nodes of the excited state stretching wave function Fig. 2. A Birge-Sponer ΔG plot can be constructed with this labeling and yields for the A state the following constants listed in Table 2. The ground state constants were deduced from the onset of the blue continuum (2b). (D_o) and from the bound bound emission spectra revealing a stretching frequency of vs $\simeq$ 15 cm^{-1} (Table 2 : X STATE 90°).

Using the preceeding constants (table 2) to fit the ground state with a Morse one-dimensional potential, the resulting bound free Franck-Condon simulations were completly out of range from the experimental spectra. We used instead of a larger equilibrium distance for the potential, listed in Table 2 : X STATE 0°, to fit the experimental spectra from v'= 2, 3, 4. The fact that this potential fits the different spectra obtained for different initial levels v=2 to 4 gives reasonable confidence in it, but we have to explain why this latter potential is so different from the one deduced from the bound bound transitions. We are led to the conclusion that there is a drastic geometry change between the ground X state and the excited A state. We are now going to observe the evidence of this geometry change in the prominent torsional progressions.

TABLE 2

DIMENSIONAL Hg-N_2 MORSE CONSTANTS			
ωe cm^{-1}	ωe xe cm^{-1}	De Å	De cm^{-1}
A STATE	55	1,6	3,3	460
X STATE 90°	15	0,5	4,3	107
X STATE 0°	14	0,5	5,2	90

THE LIBRATIONAL MOVEMENT

Those modes are most interesting as they yield information on the anisotropy of the potential surface and have up to now received in optical spectra little attention in contrast to ground state spectroscopy. The fluorescence excitation spectrum in Fig. 1 and 3a displays some small lines blue shifted from the stretching peaks whose lifetime is much shorter (c. a. 15ns) as compared to the other peaks (c. a. 70 ns at the maximum). The observed efficient relaxation into $^3P_0$ of those bands is observed in the action spectrum(3b) where those band are prominent and belong to a progression starting from the stretching band onto the blue. For each stretching quantum there is a progression which seems to stop close to the next stretching quantum, we thus inferred the barrier might to free rotation to be of the order of the streching maxima separation (~50 cm^{-1}).

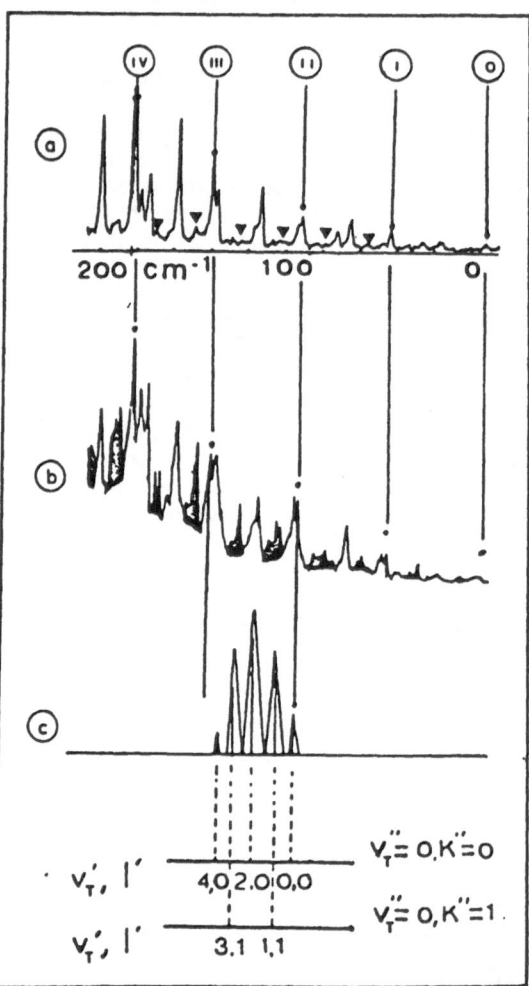

Fig. 3 – a) Fluorescence excitation spectrum 0→IV, the triangles indicate the $^3P_0$ most active modes.
          b) $^3P_0$ action spectrum, the shaded bands correspond to the most active modes.
          c) Simulation of the torsional transitions originating for a T shaped ground state (V_0=25) where the lowest populated states are K'=0, 1 (as in Nitrogen, owing to nuclear spin statistics). The excited state is linear with V_0=41 cm^{-1}.

Using a treatment developped by Ewing for A_r O_2 complexes [14] the libration is described in terms of a hindered rotor by the anisotropic potential $V_{0/2}$ (Cos 2θ), where V_0 may be parametrized as a function of the stretching quantum to account for the decrease of the rotational barrier with the stretch quanta. We produced good simulations of the band positions with $V_0 \approx 40$ cm^{-1} for the lower v' in Fig. 3c.

The intensity in the torsional modes is clearly the result of a geometry change between the A and X states as shows the former simulation where A is linear and X, T shaped, as the similar A_r N_2[14] and A_r O_2[15] complexes. The result of this geometry change is the discrepancy observed earlier in the bound free emission where in fact the potential observed is the one attained by vertical emission upon the ground state surface (Fig. 5). Owing to the T shaped equilibrium geometry in the ground state the linear geometry accessed in the emission is thus likely to be much more repulsive than the T shape in the ground state at short distances, this corresponds to the fitted potential listed in table 2 : X STATE 0°.

Bound free spectra from combination bands were also used to check the previous assignments of the torsional bands. The spectrum of the III-2 peak emission is shown in Fig. 4 together with the III peak emission and a double modulation is manifested. The observed structure in the III-2 spectrum is the result of the excitation of v Torsion = 2 (l = 0, l beeing the vibrational angular momentum) where the wave function has maxima for θ = 0˙ and α˙.

Fig. 4 — Fluorescence emission from peak III (top) compared to III-2 (bottom). The triangles indicate the double modulation observed in this combination band.

Thus schematically the emission is the sum of two contributions from the 0°
and the $\overset{\bullet}{\alpha}$ potentials as is described in Fig. 5. This assignment of the
torsional bands describes consistently the general features of the excited A
state spectrum but does enlight the fact that all the torsional bands with
efficient $^3P_0$ relaxation correspond to I ≠ 0. In these conditions for I ≠ 0,
the vibrational angular momentum is non zero corresponding to a rotation
along the Hg − N_2 axis. We also observed by time decay measurements
within the rotational contour of the peaks assigned to 0 quantum of torsion
that overall rotation was extremely efficient to induce the electronic relaxation
to $^3P_0$: the decay time between 70 ns out the peak to 15 ns at the blue
edge.

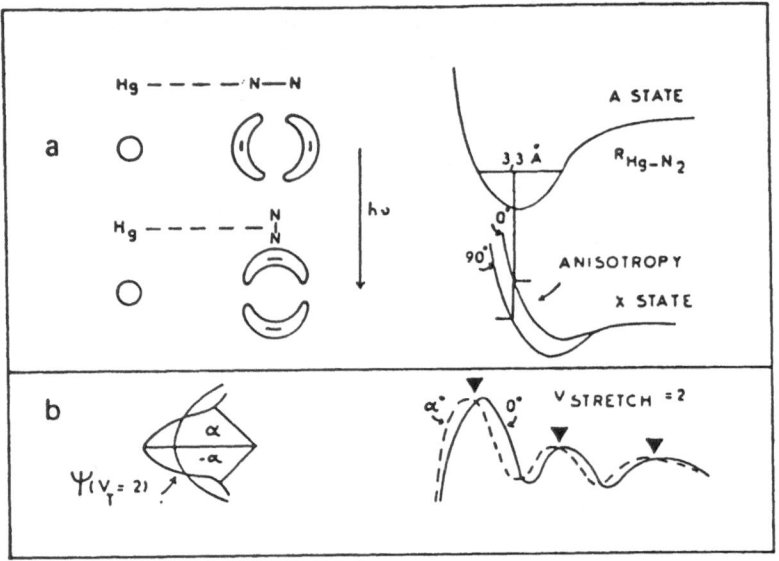

Fig. 5 − a) Schematics of the A→X bound free emission showing the influence
of the anisotropy on the emission.

   b) Bound free emission from $V_{stretch}$ = 2, V_T = 2 showing that the
actual spectrum should be a superposition of both the O° and $\overset{\bullet}{\alpha}$
contributions. The wave function ψ_T (V=2) is plotted versus the tprsion angle.

We have provided through the various spectra (fluorescence excitation,
action, emission) a wealth of consistent information on the vibrational
structure of the HgN_2 complex and its decay dynamics into $^3P_0$. The spectra
are described by stretching and bending combination bands. The bending
modes correspond to hintered rotation of N_2 in a linear equilibrium geometry
with a 40 cm^{-1} barrier in the excited state. The assignement of the bending
motions enlights the efficiency of vibrational angular momentum upon the $^3P_0$
relaxation. These findings are in agreement with a model developped by
Jouvet and Beswick [17].

CHEMICAL REACTION WITHIN THE Hg-H$_2$ VAN DER WAALS COMPLEX

The fundamental goal of chemical dynamics is a detailed understanding of the processes by which reactants are converted into primary products. Under most conditions, even for the simplest atom-molecule cases, the reactive process involves several different quantum states of both reactants and products (as well as a myriad of possible collisional trajectories), so that there has been a concerted effort in recent years to perform experiments as "state-selected" as possible. While much progress has been made, it is still extremely difficult to control some aspects of molecular collisions even in the most sophisticated of crossed-molecular-beam experiments (for example, the impact parameter and therefore the initial orbital angular momentum).

A method has been devised in this laboratory which should contribute importantly to experimental efforts to study reactive events involving state-selected reactants and products. Fundamentally, the technique consists of transforming the full-collision problem into that of half-collisions of reactants prepared at a fixed geometry. A ground state atom and a molecule are prepared as a van der Waals' complex at low temperatures in a supersonic expansion. Tunable laser radiation is then utilized to generate, selectively, an electronically excited state of the complex which can produce chemical products. This corresponds, in essence, to creating the reactive electronically excited atom in the presence of a molecule at a known distance. The collisional situation is therefore replaced by a simpler photodissociation situation, and the time evolution of the complex can be studied since the initial time is well-defined by the optical excitation. Also, the regions of the upper-state surface which are accessed can be analyzed through the characteristics of the observed spectrum (vibration, rotation, Franck-Condon factors). In a similar manner recent experiments probe the reactive surface by exciting a collision pair onto the reactive surface. This excitation process is similar as its origins from the repulsive branch of the van der Waals potential corresponding to the collision pair, instead of the attractive one as in our case, hence exploring a different portion of the reactive surface.

The reactive van der waals method appears especially attractive because the different approaches of the reactants correspond either to van der Waals' isomers or different electronic states of the complex locked by the molecular field.

It was known that in gas phase, the excitation of the resonant Hg (6 $^1S_0$-6$^3P_1$) transition in presence of H$_2$ leads to the formation of HgH($^2\Sigma^+$) through a direct mechanism (18).

The experiment can be described as following :
- formation of the Hg-H$_2$ (6 $^1S_0$) ground state complex in a supersonic expansion.
- excitation of this complex to an electronic state which correlates with Hg(6$^3P_1$).

The different products issued from the excitation were detected as follows:
1) The Hg-H$_2$ (6 $^3P_1$ - 6 $^1S_0$) fluorescence or the fluorescence of the free mercury atom issued from the dissociation of the complex when excited above the dissociation limit is collected on a photomultiplier.

2) HgH ($^2\Sigma^+$) rotational population is monitored by laser induced fluorescence exciting the $^2\Sigma^+v=0$--$^2\pi_{1/2v=0}$ transition using a second dye laser.

3) The Hg ($^3P_0$) issued from the fine structure transition induced in the complex as for Hg-N$_2$ is observed by laser induced fluorescence, the second laser being set on the 6 $^3P_0$- 7 $^3S_1$ transition.

4) The two channels Hg (6 $^1S_0$) +H +H or . Hg (6 $^1S_0$) + H$_2$ are dark in this experiment.

The intermediate state of reaction is studied through the fluorescence excitation spectrum or through the action spectrum : excitation of the complex at variable wavelength and observation of the products at a fixed wavelength.

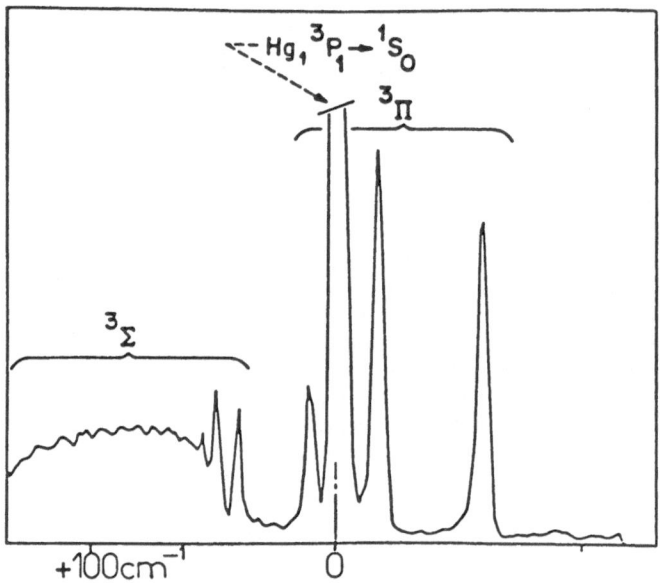

Fig. 6 – Fluorescence excitation spectrum of the Hg-Ne complex.

Let first examine the fluorescence excitation spectrum of the Hg-Ne (Fig. 6) as a guide for a non reactive van der Waals complex. Since the polarizabilities of H$_2$ and Ne are comparable the van der Waals potential curves should be very similar. The Hg-Ne spectrum contains two sets of blue and red lines shifted from the Hg (6 $^1S_0$-6 $^3P_1$) transition. These set of lines have been assigned to vibrational progressions of two electronic states, the blue shifted one being $^3\Sigma$ ($\Omega=1$ in the Hund's case C notation) and the red shifted one to the $^3\Pi$ ($\Omega=0$ in the C case). The vibrational progression in the $^3\Sigma$ state converges to the dissociation limit and the continuum to the blue corresponds to the direct excitation above the dissociation limit : in this case the complex dissociate quickly, and then the fluorescence corresponds to free Hg (6 $^3P_1$) emission.

Figure 7a shows the fluorescence excitation of Hg–H$_2$ complex. The only signal arises from the excitation of the $^3\Sigma$ state above the dissociation limit, then the observed fluorescence is the fluorescence of the free mercury atom. No emission arise from the bound states, which indicates that these states react at a rate ($>10^8$ s^{-1}) faster than the fluorescence emission (10^7 s^{-1}).

Shown in Figure 7b is the action spectrum of the HgH product. At the center of the spectrum is the Hg ($6\,^1S_0 - 6\,^3P_1$) transition mainly due to the intense Hg emission through the cut off filter. The true action spectrum of the complex exhibits only two bands : the first one shifted to higher energy by 22cm^{-1} and the second by 35 cm^{-1} from the Hg ($6\,^1S_0 - 6\,^3P_1$) transition. The second band lies just below the dissociation limit of the Hg–H$_2$ complex above which a very weak signal of HgH is observed. The same spectrum was recorded for Hg–D$_2$ revealing the same bands displaced by 35 and 36 cm^{-1} from the Hg ($6\,^1S_0 - 6\,^3P_1$) transition. With the laser resolution (0.3 cm^{-1}) some diffuse rotational structure is observed. From these two bands going to the red, a structureless continuum is observed for more than 200 cm^{-1}.

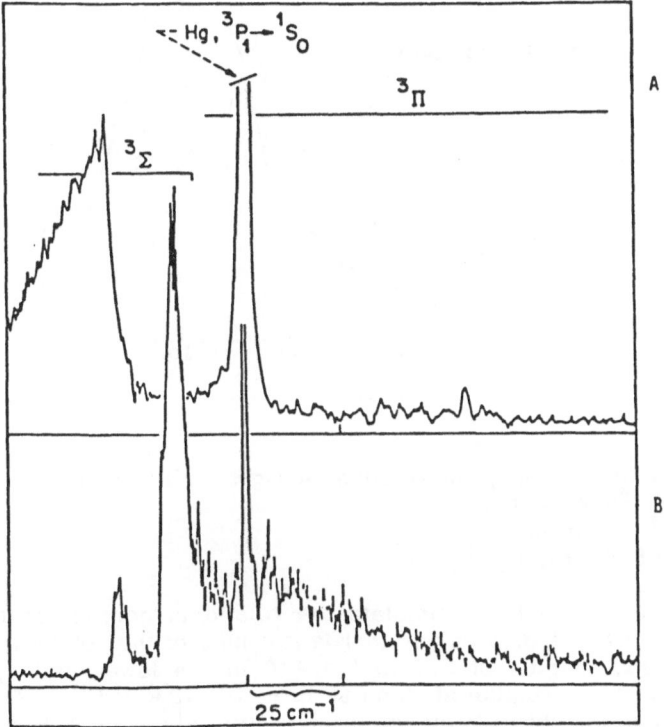

Fig. 7 – a) Fluorescence excitation spectrum of the Hg–H$_2$ complex.
         b) Action spectrum of the Hg–H$_2$ complex probing Hg–H $^2\Sigma^+ \rightarrow$ $^2\Pi_{1/2}$.

It has not been possible to obtain a Hg $(6\ ^3P_0)$ action spectrum which shows that the channel 3 is a very minor process.

What can be learnt from these spectra ?

The two bands have been assigned (by analogy with the Hg–Ne complex) to the vibrational levels of the $^3\Sigma$ state : the main band is the vibrationless level and the second one the first quantum isotope of stretching. This assignment is consistent with the deuterium effect assuming a Morse potential. The frequency of the dissociation limit corresponds to the sum of the frequency of the Hg $(6\ ^1S_0 - 6\ ^3P_1)$ transition and of the ground state binding energy which is then $D_0 = 35\ cm^{-1}$. The binding energy of the $^3\Sigma$ state can be then deduced to be $20\ cm^{-1}$. As the binding energy of these two states $^1\Sigma$ and $^3\Sigma$ is smaller than the H_2 rotational constant the diatom should rotate freely in the complex.

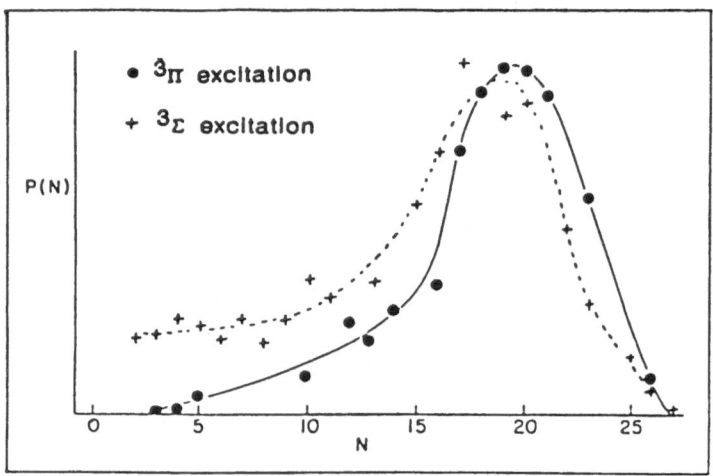

Fig. 8 – Rotational energy distribution of HgH $^2\Sigma^+$ upon :
a) o $^3\Pi$ excitation
b) + $^3\Sigma$ excitation
of the Hg–H_2 complex.

It has been possible to simulate the rotational contour of (0–0) band using reasonable and isotopically consistent values of the rotational constants (equilibrium bond lengths of 3.7 and 4.4 Å in the lower and upper states respectively) and a rotational temperature of 10°K [19]. The spectral convolution used for this simulation was in the order of 1.5 cm^{-1} : this means that initial state in the complex evolves to the product slower than 2 ps but faster than 10 ns as no fluorescence from this state is observed. Thus the intermediate state for the $^3\Sigma$ symmetry in the 4 to 5 Å region presents a small minimum at long distance and an activation barrier to the reaction greater than 25cm^{-1} as Hg($^3P_1$) rather than HgH is formed when the complex is excited above the dissociation limit. However when the bound states are excited i.e. when the elastic collision channel is closed the reaction occurs. There are two possible reaction channels from this $^3\Sigma$ well : The first is tunneling of the H atom through the barrier, the second involves a weak coupling with the $^3\Pi$ state (induced by the anisotropy of the potential) which then evolves to HgH + H. When the $^3\Pi$ (Ω=0) symmetry intermediate state is excited a structure alike the Hg-Ne one does not appear. Instead a continuum is observed this indicates that the Hg-H$_2$ complex reacts to form HgH within 0.1 ps (i.e. the life time associated with a broadening of 50cm^{-1}, the expected value of the vibrational stretching quantum in a non reactive complex). Thus the Hg($^3P_1$) + H$_2$ system in a Π entrance channel symmetry reacts with no activation barrier as a small barrier would have slown down the process. This result is consistent with earlier quantum yield measurements of Callear and al.[20] in which they determined that the quenching cross section occurs at every gas kinetic collision and with the ab-initio calculation of Millie and Bernier [21] in which no barrier is found in C_s or C_{2v} geometry for the experimental Π state (B_2 in C_{2v}).

The product energy distribution.
The rotational distribution of the HgH (v=0) is presented in figure 8a for the $^3\Pi$ complex excitation and in figure 8b for the $^3\Sigma$ excitation. Both distribution peak at fairly high rotational quantum numbers and since there is nearly no rotational energy when the complex is excited all the rotational energy comes from the dynamics on the reactive surface, then the energy release comes from a non linear triatomic species. This is also consistent with the ab-initio calculations [21] and with other similar systems such as Mg-H$_2$ [22] or Zn-H$_2$. As it can be seen from the Fig. 8, the HgH rotational distribution for the $^3\Sigma$ excitation is quite similar to the Π one, except that there is a minor component at low quantum numbers. This strongly suggests that there are two competitive channels for the reaction begining in the Σ symmetry : the main one due to a coupling to the Π state and the second one Hg-H-H linear configuration, thus producing no or very little rotation in the exit channel.

This example shows that the van der Waals Hg-H$_2$ excitation provides essential and direct information upon the reactive surface such as barrier height, wells and the effect of the symmetry of the entrance on the reactivity.

REFERENCES

[1] D.H. Levy, *Photoselective Chemistry Advances in Chemical physics XLVII* 1981, page 323.
[2] a) C. Jouvet, B. Soep, *J. Chem. Phys.* **75** 166 (1981)
    b) C. Jouvet, B. Soep, *J. Chem. Phys.* **80** 2229 (1984)
    c) A. Gato, M. Eujli, N. Mikami, M. Ito
       J. Phys. Chem. **90** 2370 (1986).
[3] a) C. Jouvet, B. Soep, *Chem. Phys. Lett.* **96** 426 (1983)
    b) W.H. Breckenridge, C. Jouvet, B. Soep
       J. Chem. Phys. **84** 1443 (1986).

[4] J.A. Kolts and D. W. Setser, page 202, *Reactive Intermediates in gas phase*. Ed. D. W. Setser - Academic press 1979.

[5] J. Pitre, K. Hammond, L. Krause
Can. J. Phys. 6 2101 (1972).

[6] E. Samson, *Phys. Rev.* 40 940 (1932).

[7] H. Horiguchi, S. Tsuchiya
Bull. Chem. Soc. Japan 44 1213 (1971).

[8] A. B. Callear, P.M. Wood
Trans. Faraday Soc. 67 272 et 598 (1971).

[9] H. Horigushi, S. Tsuchiya
Bull. Chem. Soc. Japan 50 1657 et 1661 (1977).

[10] a) E.E. Nikitin, *Theorie of elementary atom and moleculars process in gases*. Clarendon press. Oxford 1974.
b) H. Horiguchi and S. Tsuchiya
Jour. of Chem. Soc. Faraday Trans. I, 71 1164 (1975).

[11] a) K. Fuke, T. Saito, K. Kaya
J. Chem. Phys. 81 2591 (1984).
b) M.C. Duval, C. Jouvet et B. Soep
Chem. Phys. Lett. 119 317 (1985).

[12] S. Holmgrem, M. Waldman, W. Klemperer
J. Chem. Phys. 67 4414 (1977) and ref. therein.

[13] a) R.E. Smalley, P.A. Auerback, P.S.H. Fitch, D.H. Levy and L. Wharton
J. Chem. Phys. 66 3778 (1977).

[14] G. Henderson, G. Ewing, Ar-O$_2$
J. Chem. Phys. 59 2280 (1973).

[15] G. Henderson, G. Ewing, Ar-N$_2$
Mol. Phys. 27 903 (1973).

[16] M.C. Duval, O. Benoist d'Azy, W. H. Breckenridge,
C. Jouvet, B. Soep *J. Chem. Phys.* In press.

[17] C. Jouvet, A. Beswick
To be published.

[18] A.C. Vikis, D.J. Le Roy
Can. J. Phys. 51 1207 (1973) and Ref. therein.

[19] W.H. Breckenridge, M.C. Duval, C. Jouvet, B. Soep
Journal de Chimie Physique. In press.

[20] A.B. Callear, J.C. Mc Gurk
J. of Chem. Society Faraday Trans II, 62 289 (1972).

[21] A. Bernier, P. Milllie
To be published.

[22] W.H. Breckenridge, *In reaction in small transient species*
Ed. by A. Fontigne, M.A.H. Glyne (Academic London 1983)
and ref. therein.

VAN DER WAALS COUPLING BETWEEN INTERNAL ROTATION AND MOLECULAR VIBRATIONS. THE METHYL ROTOR EFFECT ON IVR.

George E. Ewing, Robert J. Longfellow, David B. Moss, and
Charles S. Parmenter
Department of Chemistry
Indiana University
Bloomington, Indiana 47405

ABSTRACT. New intramolecular vibrational redistribution (IVR) lifetimes
for S_1 p-fluorotoluene (pFT) confirm the accelerating effect of the
methyl rotor on IVR. A theory is outlined, based on van der Waals
collisional interactions between the methyl hydrogens (internal
rotation) and ring atoms. These interactions are modulated by ring
vibrations leading to internal rotation-vibration coupling. The
calculated coupling matrix elements and selection rules are sufficient
to account for the special state mixing that is a prerequisite to the
accelerated IVR dynamics.

1. INTRODUCTION

A study of intramolecular vibrational redistribution (IVR) within
the S_1 state of p-difluorobenzene (pDFB) has shown that the IVR lifetimes are markedly sensitive to the vibrational identity of the
initially pumped S_1 level[1]. Further explorations have allowed us to
look also at the sensitivity of IVR dynamics to a modest perturbation of
chemical structure, namely the replacement of a fluorine atom by a
methyl group to make p-fluorotoluene (pFT)[2]. The effect on IVR was
dramatic. IVR lifetimes, which in pDFB ranged between 300 and 100 psec,
were reduced to ten psec or less for equivalent initial levels in pFT.
Furthermore, the onset of the extensive vibrational level mixing
associated with IVR as excitation climbs the S_1 vibrational ladder in
these 300K systems changes accordingly. Whereas one needs to reach
levels near 1600 cm^{-1} on pDFB before seeing such congestion in
collision-free fluorescence spectra, in pFT it is already present in
abundance at levels near 400 cm^{-1}.

Such pronounced effects of this simple chemical change were not
expected. The group of Smalley, for example, could see little effect of
methyl substitution in benzene during their extensive studies of fluorescence from alkyl benzenes in seeded supersonic jets[3]. Similarly
Stewart and McDonald[4] were unable to see special consequences of methyl
substitution in their jet-cooled IR fluorescence study of S_0 state
mixing in a large number of small organic molecules. We now know the

163

R. Lefebvre and S. Mukamel (eds.), Stochasticity and Intramolecular Redistribution of Energy, 163–169.
© 1987 by D. Reidel Publishing Company.

reason. It appears experimentally that thermal population of methyl
rotor levels, such as occurs in 300K systems, is an essential precursor
to its special behavior[2].

The methyl substitution creating pFT from pDFB was expected to
produce only modest changes in IVR characteristics and state mixing for
several specific reasons. First, the rotor is, by the usual standards
of internal rotation, almost free with barriers of only 5 cm^{-1} and 34
cm^{-1} in the S_O and S_1 states, respectively[5]. Consequently, the substitution produces minimal changes in the 30 "ring" fundamentals so that
the level structure due to these "ring" modes changes only modestly in
quantitative detail and not all in qualitive character. No profound
effects on IVR would be expected to ensue from this cause. Second, the
eight methyl internal vibrational modes are of high frequency[6], the
lowest five being between 1000 and 1500 cm^{-1} and the remaining three
above 2900 cm^{-1}. Thus, these modes cannot contribute appreciably to the
S_1 level density below 2000 cm^{-1}, the region where most of the IVR
changes have been characterized[2].

By default, the qualitative changes in pFT level mixing and IVR
dynamics appear connected with the remaining degree of freedom to be
considered, namely the rotational motion of the methyl group. In this
discussion, we present briefly the results of some new measurements of
IVR dynamics in pFT that confirm the acceleration of IVR suggested by
the preliminary measurements[2]. We will then outline a probable cause of
the methyl effect. The effect turns out to be the consequence of
internal energy transfer (or level-mixing) between rotor-vibrational
levels. That transfer is caused by simple collisional van der Waals
interactions of the rotor hydrogens and the adjacent ring carbons and
hydrogens.

2. RESULTS

Chemical timing of dispersed fluorescence spectra after pumping
each of three initial S_1 pFT levels has lead to new measurements of IVR
dynamics. The experimental details are given elsewhere[1,7-9]. The data
analysis is similar to that used in the most recent pDFB measurements[1]
except that some of the corrections for minor artifacts have not yet
been incorporated. The ensuing pFT IVR lifetimes are somewhat shorter
than our preliminary values[2], and we do not expect them to differ
greatly from the final values that will be calculated after the data
necessary for full correction becomes available.

None of the new pFT IVR lifetimes exceed ten psec. Specifically,
for the initial level 3^1 (ϵ_{vib} = 1195 cm^{-1}), we observe τ = 9 psec;
for 5^2 (ϵ_{vib} = 1596 cm^{-1}), τ = 2 psec; for $3^1 5^1$ (ϵ_{vib} = 1988 cm^{-1}), τ = 2 psec. We can compare these values with those observed after
pumping the same vibrational energy in S_1 pDFB. For the pDFB level 3^1,
evidence of spectral congestion associated with IVR is so limited that
its IVR lifetime may be taken as essentially infinite. The values for
the pDFB levels 5^2 and $3^1 5^1$ are τ = 290 and 97 psec, respectively[1].
Thus the new pFT data confirm the comparison made with the preliminary
data[2]: IVR appears at lower energies in pFT, and where IVR timescales

can be compared for identical levels, those in pFT are shorter by more
than an order of magnitude.

3. DISCUSSION

We seek to describe how the presence of the nearly free methyl
rotor in pFT can enhance vibrational state mixing relative to pDFB and
to concommitantly increase the IVR rate. The problem is most easily
addressed with reference to a pFT level much below those where the IVR
dynamics have been studied. The published comparison[2] of the collision-
free fluorescence spectra from the pFT level 6^1 (ϵ_{vib} = 410 cm^{-1}) with
that of the corresponding pDFB level (ϵ_{vib} = 410 cm^{-1}) makes the
experimental issue clear. Abundant congestion indicative of vibrational
state mixing occurs in the pFT spectrum where none is present with pDFB.
Furthermore, the pFT congestion disappears when the fluorescence
spectrum is obtained in a cold supersonic jet expansion, so that thermal
population of the rotor levels in the 300K experiment appears to be
essential for state mixing.

One can see from a vibrational level diagram why vibrational state
mixing at this low energy (410 cm^{-1}) is absent in pDFB and also absent
in pFT with the rotor in its zeroth level. The nearest vibrational
level interactions would involve energy separations of 10 cm^{-1} or more
so that any significant level mixing requires big matrix elements
(comparable in magnitude to the level spacing). It is inconceivable
that more than a few levels could participate in such mixing.

Rotor excitation in pFT completely changes this situation. While
the experimental details of rotor levels in both S_0 and S_1 pFT have
recently become available from the work of Okuyama, Mikami and Ito[5], for
our purposes it is good enough to use the approximate term value of a
free rotor $E_m \approx$ (5 cm^{-1})(m^2) with the quantum number m = 0, ±1,
±2... If we excite $\nu_i = 6^1$ pFT with m_i = 6, for example, the total
S_1 rotor-vibrational energy is about 410 + 180 = 690 cm^{-1}. If this
rotor-vibrational level $|6^1$, m_i = 6> can couple with m_f = 0 rotor
vibrational levels, it is coupling with the S_1 vibrational manifold in
the region of ϵ_{vib} = 690 cm^{-1}. Here the vibrational level density is
beginning to pick up, and opportunities for much closer level interactions are present. Consider excitation of $\nu_i = 6^1$ pFT molecules
with a distribution of rotor levels extending from m_i = 0 to somewhat
beyond kT in rotor energy up to say, m_i = ±10. Opportunities occur for
numerous close level matches between $|6^1$, m_i> and $|\nu_f$, m_f> with m_f =
0, ±1, ±2... and ν_f some low lying fundamental or combination
frequencies. Such mixings can in principle produce the vibrational
congestion present in the 300K 6^1 pFT fluorescence spectrum. The job
now is to establish a viable rotor-vibration interaction mechanism and
to learn if selection rules and the size of the matrix elements are
sufficient to allow such mixing.

Interaction between states with high and low rotor energy, such as
$|6^1$, m_i = 9> and $|\nu_f$, m_f = 3>, is effectively a transfer of rotor
energy to vibrational energy that must necessarily involve modulation
of the vibrational motion of the ring by the methyl rotor motion. One

can rule out contributions from bond dipole-dipole forces since the C_{3v} methyl top will have its net dipole along the top axis, invariant to the rotor motion. Hyperconjugation, partial double bond character of the rotor-ring C-C bond, may be responsible for the small rotor barrier, but these electronic effects are commonly associated with much larger barriers. A third source of interaction, which we favor for reasons to be given below, is the van der Waals collision interaction between the rotor hydrogens and the ring carbon atoms with their hydrogens alpha to the rotor ring attachment.

We have explored the van der Waals mechanism in detail[10] and find that it is of sufficient magnitude to account for the observed congestion (and associated state mixing) in the 300K 6^1 fluorescence spectrum of pFT. Whether it is the only significant interaction mechanism remains hidden in the ambiguities of the calculations, but it would be surprising if it were not the dominant contributor. Full arguments will be presented elsewhere[10]. Here we outline the approach and some results.

The interatomic distances can now be estimated fairly reliably in both the S_1 and S_0 electronic states from an ensemble of spectroscopic studies[5,11,12]. In both electronic states, the overlap of rotor-ring atomic radii (covalent and van der Waals radii) is substantial. At the position of maximum overlap, for example, the rotor-ring hydrogen distance is about 2.1 Å whereas the van der Waals contact distance would be about 3 Å. The rotor hydrogen-ring carbon distances is about 2.4 Å compared to a contact distance of about 3.4 Å. As the methyl rotates, its hydrogens are thus interacting with ring atoms on the repulsive wall of an interatomic potential. The interaction is accordingly extremely sensitive to the vibrational modulation of the ring geometry. By this means it is easy to see how rotor motion and ring vibrations are coupled.

With a molecular mechanics program[13], it is possible to calculate the van der Waals interaction energies between the rotor hydrogens and the ring atoms in the S_0 and S_1 states for the maximum and minimum rotor configuration. The difference between these values is the barrier height from this specific contribution.

The calculations show substantial net interaction energies (repulsive) in each state for both configurations. The small barriers are the consequence of a small change in the interactions with rotor motion. In the S_0 state, the interaction energies are 123 cm^{-1} (max) and 115 cm^{-1} (min) to give a barrier of 8 cm^{-1}. In the S_1 state, we find 711 cm^{-1} and 672 cm^{-1} for a barrier of 39 cm^{-1}. These calculated barriers compare favorably with the spectroscopically measured[5] barriers of 5 and 34 cm^{-1}, respectively. The exercise suggests that van der Waals forces are serious contenders for the source of the dominant terms in the torsional potential of pFT.

If we assume that rotor-ring interactions are due to van der Waals forces, a potential suitable for calculating coupling matrix elements for rotor-vibrational levels can be given as the product of two terms:

$$V = V_o e^{-ar}(1-\cos 6\phi).$$

In this potential, r is an effective distance between the methyl hydrogens and ring atoms and Φ is the torsion angle. The six-fold torsional barrier is scaled by V_O. The van der Waals interaction is represented only by the repulsive wall whose steepness is scaled by the range parameter a. If r is expressed in terms of an equilibrium geometry r_0 and a vibrational displacement Δr, the potential becomes

$$V = V_0 e^{-ar_0} e^{-a\Delta r} (1-\cos 6\Phi)$$

where $V_0 e^{-ar_0}$ is one-half of the observed torsional barrier of 34 cm^{-1}.

The matrix element for coupling initial, $|v_i m_i\rangle$, and final, $|v_f m_f\rangle$, rotor-vibrational states now separates into vibrational and torsional terms

$$V_0 e^{-ar_0} \langle v_f|e^{-a\Delta r}|v_i\rangle\langle m_f|(1-\cos 6\Phi)|m_i\rangle$$

Selection rules emerge immediately from this expression, with that for the change in rotor quantum number being $\Delta m = 0, \pm 6$ for ring vibrations that are symmetric with respect to a C_2 rotation about the top axis. For antisymmetric ring vibrations the reduced symmetry generates a 3-fold barrier with $\Delta m = 0, \pm 3$ allowed. If $e^{-a\Delta r}$ is expanded in a power series, the matrix element decreases monotonically as the change in ring vibrational quantum number, Δv, increases. The vibrational term is familiar as the standard collisional integral of SSH theory[14]. Moreover, our treatment here of vibrator-rotor mixing via van der Waals coupling bears a close analogy to the recent Fermi resonance calculation of Ewing[15] where chemical bond vibrational motions and van der Waals bond vibrations are mixed.

The vibrational integral has been evaluated numerically by expressing Δr not in terms of the normal coordinates but in a less tedious approach that uses Cartesian coordinates

$$a\Delta r = a_x\Delta x + a_y\Delta y + a_z\Delta z.$$

The range parameters a_x, a_y and a_z are determined by calculating van der Waals energies when the whole CH_3 group is displaced by given Δx, Δy or Δz values. The vibrational integral can now be expressed as a product of terms in each normal mode, with each term being a standard harmonic oscillator integral. For example, a low frequency mode v_{16a} = 188 cm^{-1} is an out-of-plane motion for which $\Delta y = \Delta z = 0$ giving for a $\Delta v = 1$ change, the intergral

$$\langle v_{16a}|a_x\Delta x|(v-1)_{16a}\rangle$$

which is a vibrational transition moment matrix element.

Numerical evaluation of similar integrals for various modes and various Δv changes gives the magnitudes of the rotor-vibration matrix elements. The average values are instructive. For $\Delta v = 1$ we find 4 cm^{-1}, for $\Delta v = 2$ we find 1.3 cm^{-1}, for $\Delta v = 3$ we find 0.44 cm^{-1} and so forth.

By direct counting of rotor-vibrational states in energy regions

appropriate for 6^1 excitation in pFT (with a distribution over initial
states with $m_i = 0$ to ± 10), we can now identify a typical selection of
interacting levels. The selection is of course restricted since we must
pay attention to the rotor and vibrational selection rules and, of equal
importance, to the requirement that energy spacings with respect to the
magnitude of the coupling matrix elements are commensurate with
substantial mixing. The count of levels is about fifty, with the
average coupling matrix element from this ensemble being about 0.6 cm^{-1}.

The spectroscopic implications of these calculations are consistent
with the observations. Most of the $|6^1, m_i = 0 >$ to $|6^1, m_i = \pm 10>$
rotor-vibration states excited by the laser pump in the 300K experiment
are coupled to other rotor states having a different vibrational
description, i.e. $|v_f, m_f>$. In emission, each state will produce
vibrational structure of the common initially pumped level 6^1 so that
the structure characteristic of 6^1 emission will dominate the spectrum.
Additionally, each excitation will contribute to a lesser degree the
vibrational structure from its own small set of vibrational identities
present in its coupled levels. Since this set differs for each m_i state
pumped in excitation, the ensemble of coupled vibrational identities is
broad, and it will produce a richly structured background perceived as
congestion.

The comparisons of observed and calculated 6^1 spectroscopies have
been made on a semi-quantitative basis[10]. The agreements reenforce the
proposition that van der Waals rotor-vibration energy transfer is an
important contributor to state mixing in this molecule and plays a
significant role in the accelerated IVR dynamics. There is nothing
specific to pFT in the origins of these interactions. Thus we would
expect that these consequences of methyl stubstitution would have rather
wide generality among 300K polyatomics.

ACKNOWLEDGEMENT

Financial support from the National Science Foundation and from the
Donors of the Petroleum Research Fund, administered by the American
Chemical Society is greatly appreciated. C.S. Parmenter is also
grateful for a Senior U.S. Scientist Award from the Alexander von
Humboldt-Stiftung, held with Prof. E.W. Schlag at the Technische
Universitat Munchen.

REFERENCES

1. K.W. Holtzclaw and C.S. Parmenter, J. Chem. Phys., 84, 1099 (1986).
2. B.M. Stone and C.S. Parmenter, J. Chem. Phys., 84, 4710 (1986).
3. J.B. Hopkins, D.E. Powers, and R.E. Smalley, J. Chem. Phys., 71,
3886 (1979); J.B. Hopkins, D.E. Powers, and R.E. Smalley, J. Chem.
Phys., 72, 5049 (1980); J.B. Hopkins, D.E. Powers, and R.E. Smalley,
J. Chem. Phys., 73, 683 (1980); 74, 745 (1981).
4. G.M. Steward and J.D. McDonald, J. Chem. Phys., 78, 3907 (1983).
5. K. Okuyama, N. Mikami, and M. Ito, J. Phys. Chem., 89, 5617 (1985).

6.   J.H.S. Green, Spectrochim. Acta 26A, 1503 (1970); J.K. Wilmshurst
and A.J. Bernstein, Can. J. Chem., 35, 911 (1957).
7.   R.A. Coveleskie, D.A. Dolson, and C.S. Parmenter, J. Phys. Chem.,
89, 645 (1985).
8.   R.A. Coveleskie, D.A. Dolson, and C.S. Parmenter, J. Phys. Chem.,
89, 655 (1985).
9.   R.A. Coveleskie, D.A. Dolson, D.B. Moss, S.C. Munchak, and C.S.
Parmenter, Chem. Phys., 96, 191 (1985).
10. D.B. Moss, C.S. Parmenter and G.E. Ewing, (submitted)
11. T. Cvitas, J.M. Hollas, and G. Kirby, Mol. Phys., 19, 305 (1970).
12. T. Cvitas and J.M. Hollas, Mol. Phys., 20, 645 (1971).
13. N.L. Allinger and Y.H. Yuh, QCPE Program No. 395, Indiana University
(1980).
14. R.N. Schwartz, Z.I. Slawsky and K.F. Herzfeld, J. Chem. Phys., 20,
1591 (1952); F.I. Tanczos, J. Chem. Phys., 25, 439 (1956).
15. G.E. Ewing, J. Phys. Chem., 90, 1790 (1986).

ROTATION VIBRATION AND ELECTRONIC RELAXATION

A. Amirav and J. Jortner
School of Chemistry
Tel-Aviv University
Tel Aviv 69978,
Israel

ABSTRACT. In this paper we present experimental results demonstrating the manifestation of Intramolecular Vibrational energy Redistribution (IVR) on the vibrational energy dependence of the emission quantum yield in anthracene and 9-cyanoanthracene. Strong rotational effects on the fluorescence quantum yields from vibrational states in the S_1 manifold of 9-cyanoanthracene were observed, which serve as the fingerprints of Coriolis coupling, serving as the dominant vibronic coupling mechanism leading to IVR. We discuss the possible manifestation of intratriplet rotational induced IVR in pyrazine as the dominant channel that controls its controversial S_1 excited state dynamics, invoking Vibrational Crossing as a new intra and interstate mixed intramolecular radiationless process. We conclude that rotational effects play a central role in IVR, which can strongly enhance interstate electronic relaxation.

I. INTRODUCTION

Radiationless transitions in electronically-vibrationally and rotationally excited states of large isolated molecules are traditionally separated into two categories; interstate Electronic Relaxation (ER) between different electronic states, and Intrastate Vibrational energy Redistribution (IVR). The rotational degrees of freedom were considered to be of little or no influence on the excited-state dynamics [1]. Recently there has been a considerable experimental [2-4] and theoretical [5,6] work on the effect of rotation on electronic intrastate as well as interstate radiationless processes. This recent work focused mostly on intermediate sized molecules, like benzene [2] or pyrazine [3,4], and has demonstrated that the rotational degree of freedom can play a major role, dominating the dynamics of the electronically excited state.

In this paper we shall describe some recent experimental results shedding new light on the manifestation of Intramolecular Vibrational energy Redistribution (IVR) on interstate electronic relaxation in large poliatomic molecules. We shall describe the strong rotational effect

R. Lefebvre and S. Mukamel (eds.), Stochasticity and Intramolecular Redistribution of Energy, 171–184.

found on this IVR leading into rotational effect on interstate electronic relaxation. We shall also describe a simple general model that will
unify our results on the large 9-cyanoanthracene molecule with the
various conflicting experimental results existing for the intermediate
sized pyrazine molecule.

2. EXPERIMENTAL

Our experimental techniques [7] for the measurement of absolute fluorescence quantum yields were extended to allow for a better spectral resolution. Absorption spectra and fluorescence excitation spectra of 9
cyanoanthracene and anthracene cooled in planar supersonic expansions
were simultaneously determined using a pulsed xenon lamp and a monochromator. Pulsed planar jets were generated by expansion of the molecule
seeded in Ar through a nozzle slit. Two nozzle slits were used in these
experiments having the dimensions of 0.22 x 33 and 0.27 x 90 mm. The
repetition rates of both nozzles were 6 Hz and the width of gas pulses
was 300 μs. Anthracene and 9-cyanoanthracene were heated in the nozzle
chamber to 130°C or 150°C respectively and mixed with Ar at the stagna-
tion pressure of p = 80-120 Torr. Light from a pulsed simmered Xe
flashbulb (pulse duration 24 μs) was passed through a 0.75 m Spex monochromator equipped with a 2400 lines/mm grating. The spectral resolution was 0.11 Å for 15 μ slits. The light emerging from the monochromator focused onto the jet parallel to the slit at a distance of x = 10 mm
from it. The light beam was split by a sapphire window and monitored by
two vacuum photodiodes. The attenuation ΔI of the light beam due to
absorption was determined from the difference in the light intensity
before and after crossing the planar jet. The lamp-induced fluorescence
(LMIF) intensity I_F was monitored by a photomultiplier. The absorption
signal $\Delta I / I_o$ and the LMIF signal I_F / I_o were normalized to the
incident light intensity I_o. The relative quantum yield, q, is given by
$q = I_F / \Delta I$. Absolute quantum yields Y are measured in comparison with
molecule with known y in a binary mixture [7]. The q values across the
rotational contour were obtained from the simultaneous scans of fluorescence and absorption and normalized to that of the electronic origin
which was measured as being close to 1.00 [7]. q values from different
vibronic levels were measured at the center of the rotational contour.

3. EXPERIMENTAL RESULTS

A considerable insight into the dynamics of IVR can be gained from the
vibrational energy dependence of both the emission lifetime and the
quantum yield. The Ev dependence of the interstate Electronic Relaxation (ER) rate or the emission quantum yield is typically shown for
anthracene in Fig. 1.

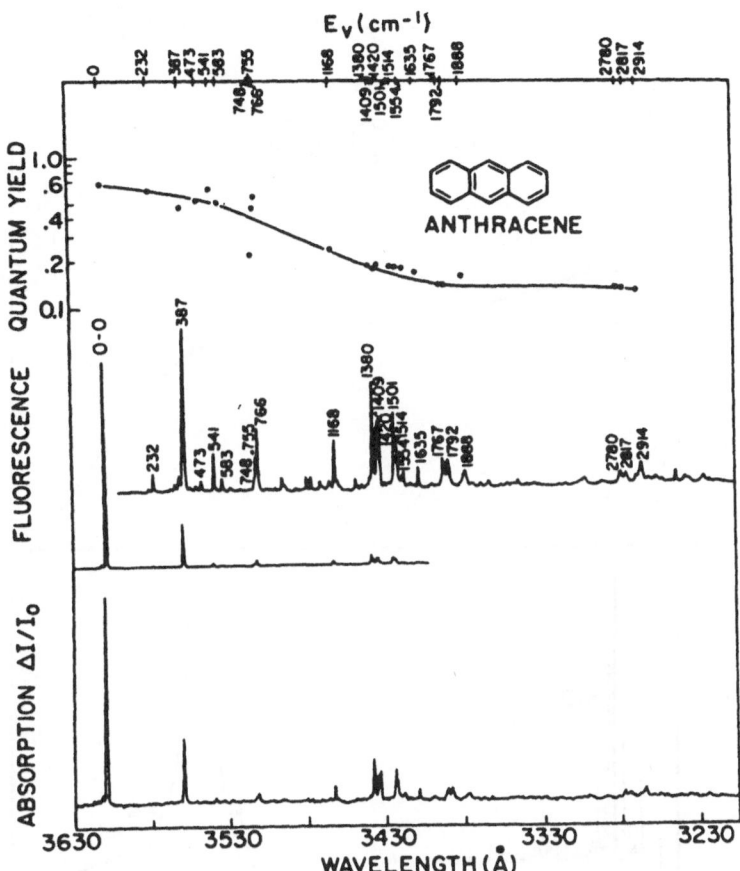

Figure 1.   Absorption spectrum, fluorescence excitation spectrum and absolute fluorescence quantum yields of anthracene.   Pulsed nozzle temperature is 140°C.   The numbers represent the vibrational excitation above the S_1 electronic origin (in cm^{-1}).

It reveals three energy domains:
A)   The low energy region $0 < E_v < 800 cm^{-1}$, where the emission quantum yield Y is relatively high and some mode specificity of the ER rate is exhibited.
B)   The intermediate energy region $800 < E_v < 2000 cm^{-1}$, where both the lifetime and the emission quantum yield Y decrease monotonically, while the ER rate increases with increasing E_v.
C)   The high energy domain $E_v > 2000 cm^{-1}$ where Y and the ER rate exhibit a weak dependence on the excess vibrational energy.

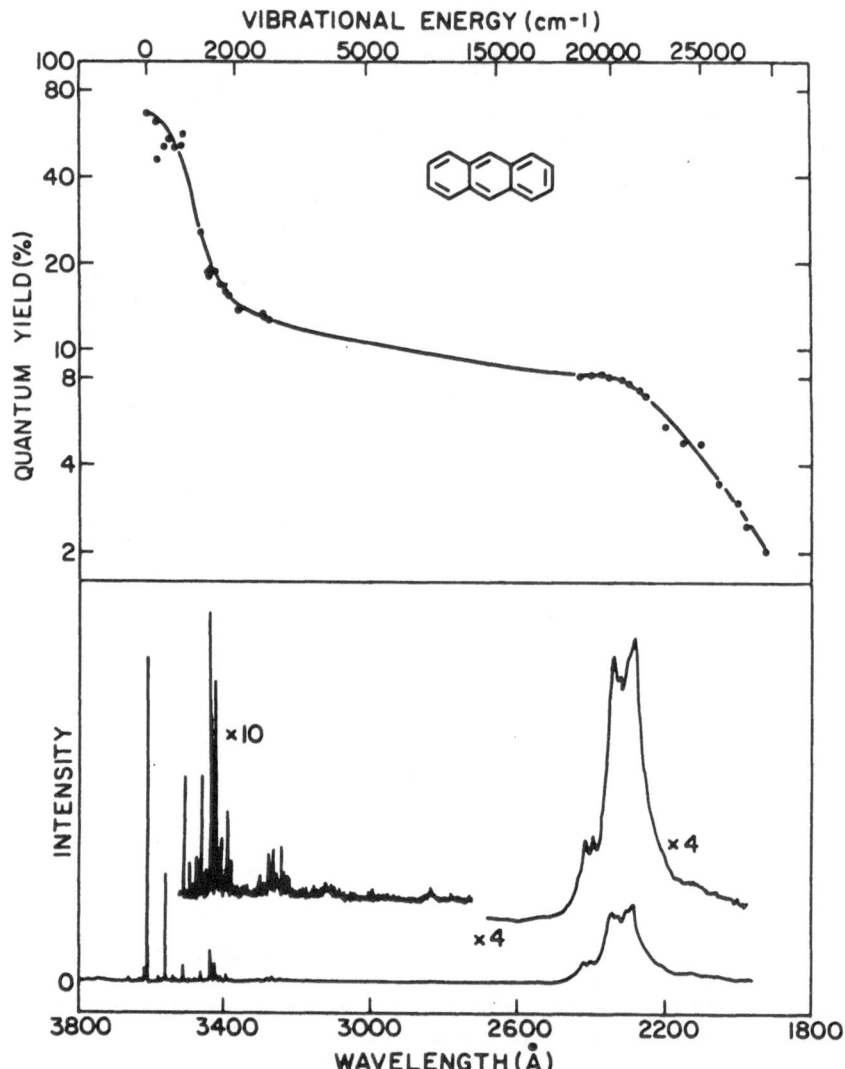

Figure 2.   Fluorescence excitation spectrum and absolute fluorescence quantum yields of anthracene over an extended spectral range including higher electronic states.   Monochromator resolution is reduced to ~ 1Å to increase the relative amplitude of the higher electronic states. Quantum yield measurements are taken point by point at a low nozzle temperature i.e. 120°C to eliminate dimer effects.

In Fig. 2 we show the vibrational energy dependence of the emission quantum yield Y in an increased energy range of 25000cm⁻¹ including higher electronic states. In comparison with lifetime measurements [10] it is clear that the pure radiative lifetime is approximately electronic and vibrational energy independent [8]. From this experimental result and from energy-resolved emission data [9] it is clear that the emitting state throughout the excitation range shown in Fig. 2 is the S_1 state, which is populated by internal conversion from higher excited states. In Fig. 2 we show that the ER rate is vibrational energy independent over a broad spectral range of up to $16000 cm^{-1}$ excitation vibrational energy and only then internal conversion (S_1-S_0) to the ground state starts to dominate [10]. In the first $16000 cm^{-1}$ vibrational energy the ER is dominated by intersystem crossing to T_1 which is mediated by higher triplet states lying below the origin of S_1 [8].

The specific characteristics of region (A), the energy limits of range (A) (B) and (C) and the difference between range (A) and (C) depend on the nature of the specific molecule. The existence of energy domains (B) and (C) seem to be universal and was observed in the S_1 manifold of many large aromatic molecules [8]. The vibrational energy dependence in range (B) is considered to be an important manifestation of the initiation of IVR [11] and a simple model explaining it is shown in Fig. 3. According to this model [12] we invoke intrastate vibronic interaction (anharmonic or Coriolis) that couples the S_1 doorway state, which carries oscillator strength from the electronic ground S_0 state, to background S_1 state which do not carry oscillator strength from S_0 (dark vibrational states). This scrambled S_1 manifold exhibits ER to the statistical triplet manifold. The major cause for the decrease of the emission quantum yield with increasing Ev in range (B) is attributed to the difference in the nonradiative decay rates of the doorway state and of the background states. Different symmetries of these levels will affect the Franck-Condon factors for these modes. The high energy range Ev > $2000 cm^{-1}$ (range (C)) corresponds to the statistical limit of IVR where the S_1 doorway states, which carry oscillator strength from $S_0(0)$ decay nonradiatively to the dense manifold of background S_1 states, which in turn decay to the triplet manifold. The fluorescence decay rate and emission quantum yield is then determined by the total (radiative and nonradiative) decay rate of the S_1 background manifold. Thus IVR may serve as the rate determining step for interstate electronic relaxation. This model, which rests on the coupling between IVR within S_1 and $S_1 \longrightarrow T$ ER provides a semiquantitative description of several experimental observations, such as nonexponential decay with the absence of quantum beats in range (B) [12], which converge to a fast single exponential decay in range (C) [12]. The relevant level scheme for IVR in conjunction with ER is shown in Fig. 3. It contains a single doorway state ¦ s >, the background S_1 manifold (¦1>) and the (¦T>) "statistical" quasi-contininum. The total decay width γ_s and γ_1 of the discrete states consist of radiative (denoted by r) and nonradiative ER (denoted by nr) contributions, i.e

$$\gamma_s = \Gamma_s^r + \Gamma_s^{nr} \; ; \; \gamma_\ell = \Gamma_\ell^r + \Gamma_\ell^{nr}$$

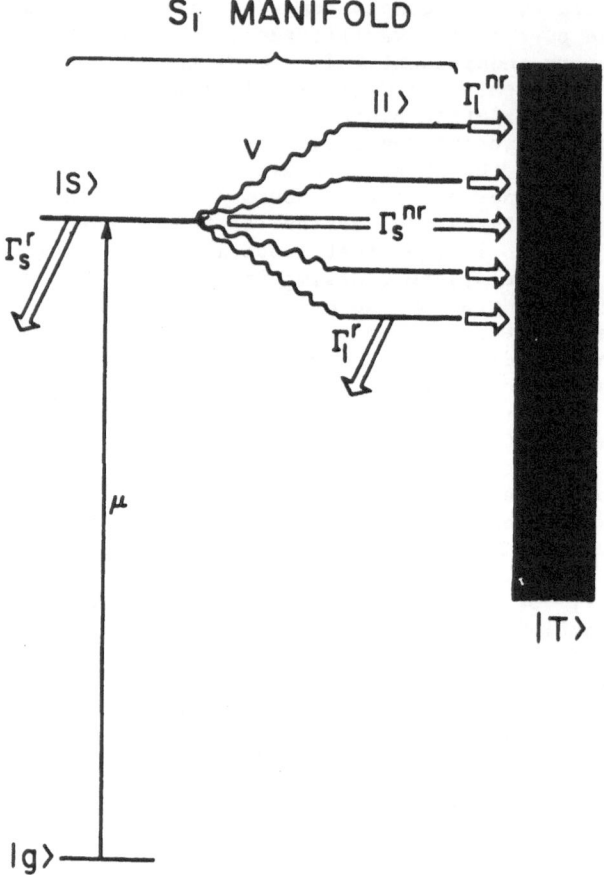

Figure 3. A. level scheme for IVR modulation of S_1-(T) inter-system crossing in a large molecule.

Radiative decay occurs from both |s> and (|l>) states to final states in the vibrational manifold of the ground electronic state. The radiative decay rates of all states in the S_1 manifold are equal, i.e., $\Gamma_s^r = \Gamma_l^r$ for all l. On the other hand, as the |s> and (|l>) states are of distinct vibronic symmetry, we expect that $\Gamma_s^{nr} \neq \Gamma_l^{nr}$ and on the basis of the experimental data which reveal the decrease of Y with increasing Ev when the relative contribution of the (|l>) state increases we assert that $\Gamma_s^{nr} < \Gamma_l^{nr}$. It has been shown that for the intermediate level structure when |Em-Em'|>Ym, Ym', the total photon counting rate I(t) from the coherently excited state is given by [13]

$$I(t) = (\Gamma_s^r - \Gamma_\ell^r)\left|\sum_m a_s^m \mu_{gm} \exp(- iE_m t - \gamma_m t/2)\right|^2 + \Gamma_\ell^r \sum_\ell |\mu_{gm}|^2 \exp(- \gamma_m t) \tag{1}$$

where μ_{gm} is the transition moment from $S_o(0)$ to $|m\rangle$.

The first term in Eq.(1) involves interference and dephasing contributions. In view of the equality of the radiative width $\Gamma_s^r = \Gamma_\ell^r$ this contribution vanishes and the total fluorescence photon counting rate is given by:

$$I(t) = \Gamma_\ell^r \sum_m |\mu_{gm}|^2 \exp(- \gamma_m t) \tag{2}$$

The two major predictions from this model are:
a)   The total time resolved emission in range (B) can be bi- or multiple-exponential due to IVR.
b)   Quantum beats will not be exhibited in the total temporal fluorescence decay.

As the multiple exponential decay result from differences in Γ_s^{nr} and Γ_ℓ^{nr} the proper molecular choice do demonstrate that this effect should have a large reduction of the emission quantum yield in range (B) and preferably constant high Y in range A. 9-cyanoanthracene fulfills these requirements. In Fig.4 we show its absorption spectrum and its lamp induced fluorescence spectrum together with the absolute fluorescence quantum yields.

The general trends are as in anthracene (Fig. 1) but the ratio of the quantum yield in range (A) to that of range (C) is increased to 10 and Y is constantly high in range (A).   We have collaborated on this problem with E.C. Lim and S. Okajima of Wayne University who have measured the time-resolved emission from several vibrational states of 9-cyanoanthracene demonstrating [12] the mentioned bi- or multiple-exponential decay in range (B) converging to a single fast exponential decay in range C.

The 1160cm^{-1} vibration is very interesting as from density of states estimates (based on that calculated for anthracene [14]) and from its location in the middle of range (B) as shown in Fig.4, (its emission quantum yield is the average of range (A) and (C)) one can assert that it represents the simple case of one doorway state coupled to one or very few background dark vibrational states. We consider a two-level system (Nl=1) and ignore a possible small modification of the energy gap ΔE between the $|s\rangle$ and the $|l\rangle$ states (this assumption is based on the weak rotational dephasing of the quantum beats in the energy and time resolved emission of anthracene shown by Felker and Zewail [14]).   The time resolved emission should be bi-exponential as is experimentally observed [12]. The short component originating from near-resonant interaction should be $\tau_+ = [(Y_\bullet + Y_1)/2]^{-1}$.   In case of off-resonance behaviour the short component should be shorter than $[(Y_\bullet + Y_1)/2]^{-1}$ and its amplitude should be reduced. The long component in case of near-resonant interaction should be identical with the short component but in

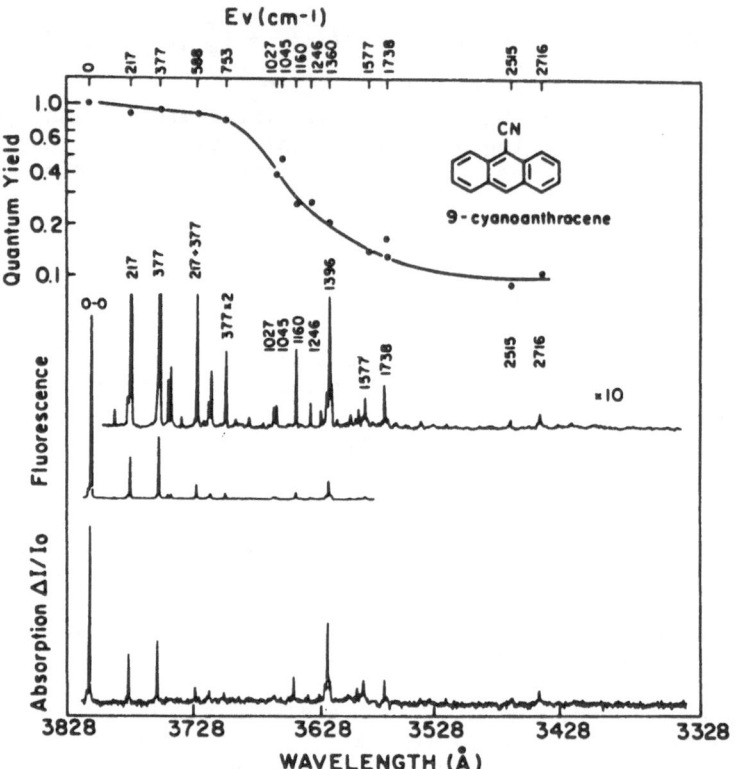

Figure 4. Absorption spectrum, fluorescence excitation spectrum and absolute fluorescence quantum yields of 9-cyanoanthracene (9 CNA) over the range 3828-3328 A. 9-CNA is seeded in Ar at P=120torr and at a nozzle temperature of 150°C. The numbers represent the vibrational excitation above the S_1 electronic origin (in cm^{-1}).

case of off resonance behaviour the long component value should be close to that of range (A), $\gamma_e^{-1} \gg \tau_- \gg [(\gamma_e + \gamma_1)/2]^{-1}$. From this naive two level-system analysis we expect to observe in the experiment a biexponential decay curve with A^+/A^-, the amplitude ratio between the short- and the long-time decay being smaller than 1. Any additional background state can further reduce the values of τ_+ and τ_- and increases the A^+/A^- ratio. The experimental results have revealed a long decay with a lifetime value close to that of range (A), and a short decay with a lifetime close to and only slightly shorter than $[(\gamma_e + \gamma_1)/2]^{-1}$, together with $A^+/A^- = 1.5$. These results cannot be accounted for in terms of our simple picture that predicts on the basis of the lifetimes a value of $A^+/A^- \ll 1$. This apparent contradiction can however be rationalized if this behaviour is rotationally dependent, i.e. the excitation is inhomogeneous and rotationally unresolved. In Fig.5 we show the excitation spectrum measured simultaneously with the direct

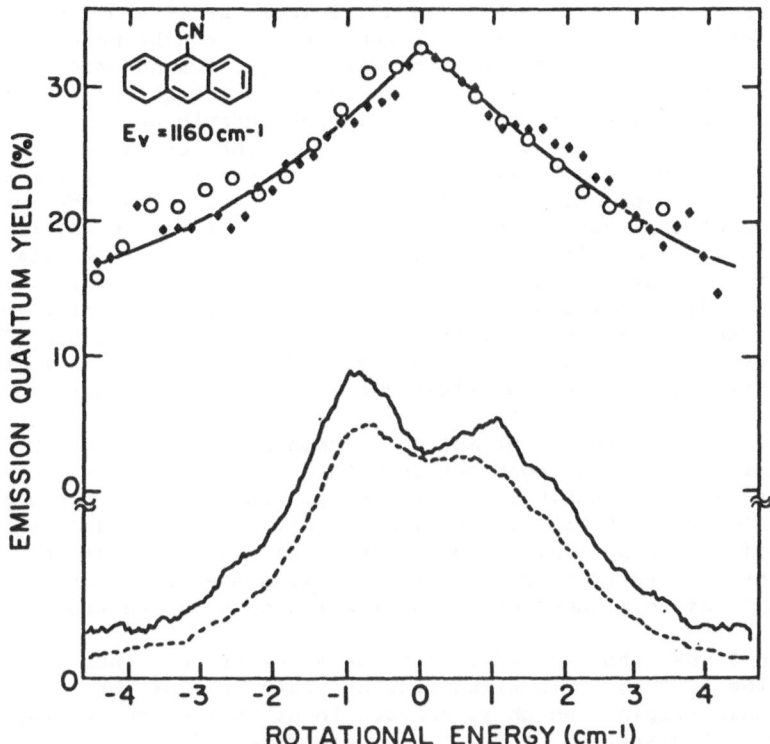

Figure 5. Absorption spectrum fluorescence excitation spectrum and absolute fluorescence quantum yield of the rotational contour of the 1160cm^{-1} vibration of 9-cyanoanthracene. Spectral resolution is ~0.11 A (0.8cm^{-1}). Nozzle temperature is 140°C and the argon backing pressure is 84 Torr resulting in an estimated rotational temperature of ~30°K. the points on the absolute quantum yield are +++ for the spectrum shown (Tr~30K) and 000 for a lower rotational temperature using an argon backing pressure of 115 Torr. The good agreement between the two sets of data indicates that the effects of dimer background or of sequence congestion are negligible.

absorption of the 1160 cm^{-1} vibration. One can clearly observe that the ratio of fluorescence signal to that of the absorption maximizes at the center of the spectrum for low or zero rotational energy and gradually decrease as the rotational energy increases. The ratio of the fluorescence to absorption signal calibrated to the absolute emission quantum yield is shown above these spectra. While the absolute emission quantum yield Y of the electronic origin is close to 1.0 and that in range (C) is about 10% (Fig.4), Y of the low rotational states of the 1160 cm^{-1} vibrational state is 0.33 and gradually dropping down to ~0.17 at 4 cm^{-1}

in the R or P branch. This effect is quite large as it implies that the rotational energy of a few or a few tens of cm^{-1} can bring us to the statistical limit of IVR in a way that requires a few hundreds cm^{-1} of vibrational energy. In other words, we have found a strong rotational effect on the radiationless transitions in 9-cyanoanthracene. The magnitude of this effect monotonically increase with the rotational energy in the same way in the R and P branches. As it is unlikely that the energy gap between the optical active doorway state ($|s\rangle$) and the optical inactive vibrational background states ($|l\rangle$) is altered in any significant way [14], we are left with Coriolis coupling as the best rationalization for this effect. We note that there is a strong similarity to IVR in the intermediate sized benzene, as was beautifully demonstrated by Riedle, Neusser and Schlag [2]. As Z-type Coriolis coupling has $V^{\sim}K$ [2] and the precessional K state can span the values $K = -J...+J$, it can be asserted that for higher J values one can have a larger coupling strength.

Thus the observed bi- or multiple-exponential decay in the time resolved emission is inhomogeneous in nature. The molecules in K=0 or low K states may show mostly long time decay, while those molecules in high J,K states may exhibit mostly single short-time decay due to an increased coupling width. This argument rationalizes the relatively large amplitude of the short component together with the value of $\tau-$ being close to γ_{ℓ}^{-1} as an inhomogeneous combination of emission from several J,K states.

In Fig.6 we show the absorption measurement of the rotational contour of the 1160 cm^{-1} vibration together with that of the vibrationless electronic origin, which is redrawn to match the same energy scale of the 1160 cm^{-1} vibration. A few trends are observed, most prominent is the fact that the "shoulder" of the P and R branches observed in the electronic origin is almost missing in the 1160 cm^{-1} vibration rotational contour. In addition, the contour of the 1160 cm^{-1} vibration is narrower than that of the electronic origin and only in the far "edge" of the contour does it become stronger than that of the electronic origin. The erosion of the dip between the P and R branches can easily be rationalized assuming IVR with coupling width (unharmonic or Coriolis) of 0.5-1 cm^{-1}. This assumption, however, fails to explain the narrowing of the contour as it results in a slightly broader rotational contour of the 1160 cm^{-1} vibration. A crude simulation shown as full circles in Fig.6 shows that both effects as well as the increased height at the far "edge" of the 1160 vibration can be obtained assuming a monotonically increasing coupling strength. The full circles shown in Fig.6 represent this crude simulation where the absorption contour of the 1160 cm^{-1} is constructed from that of the electronic origin, assuming that the coupling width increase monotonically with J. In this case we have taken our experimental resolution as 1.0 cm^{-1} and assumed that the vibronic coupling resulted in a spectrally unresolved splitting of the two Molecular Eigenstates (MEs) resulting in reduced resolution and increased broadening. Introducing the splitting W as $W = 1.0+0.14($ $\Delta E)^{3/2}$ where ΔE is the spectral energy difference from the center of the contour, we obtained the full circles shown in Fig.6 which are in qualitative and reasonable quantitative agreement with the experimental

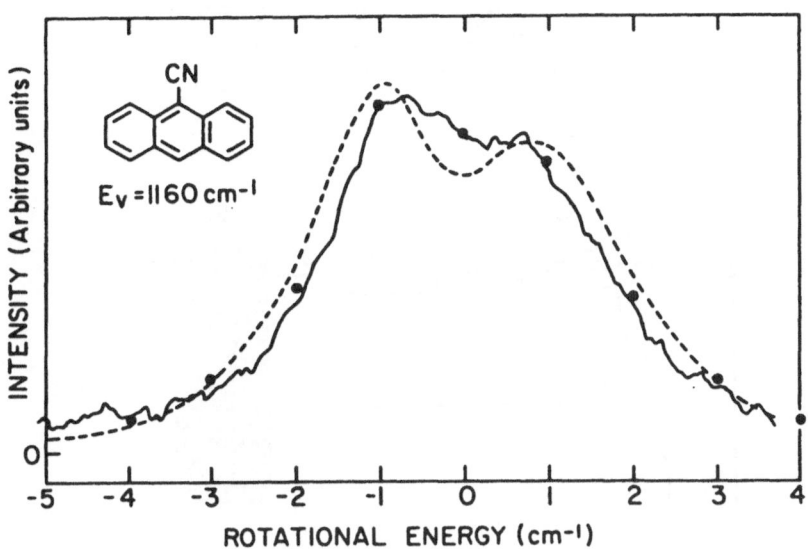

Figure 6. Absorption spectrum of the rotational contours of the electronic origin (dashed line) redrawn to match identical energy scale, together with that of the rotational contour of the $1160 cm^{-1}$ vibration of 9-cyanoanthracene. Zero rotational energy is arbitrarily assigned at mid point of the P and R branches. The solid circles are the results of a model simulation (see text).

results. Again this piece of experimental information suggests that Coriolis coupling is the dominant mechanism leading into IVR in large polyatomic molecules like 9 cyanoanthracene.

Finally, we would like to comment that by using similar arguments we can address ourselves to the complicated problem of pyrazine dynamics. A full account on this explanation is given elsewhere [6]. Consider the level scheme shown in Fig.7. The electronic origin doorway state is strongly (spin-orbit) coupled to the sparse triplet T manifold resulting in a bunch of several Molecular Eigenstates (MEs). From here on the picture is similar to that of Fig.3. The MEs can be vibronically coupled to other triplet states, which are not coupled to the singlet doorway state through spin orbit coupling. In other words, the MEs can undergo intrastate IVR within the triplet manifold due to the dominant triplet character of each ME.

As Riedle and Neusser have demonstrated that Coriolis coupling induces IVR in benzene at $3000 cm^{-1}$ vibrational energy, and we have demonstrated the same in 9 cyanoanthracene at $1160 cm^{-1}$ vibrational energy, it was proposed [6] that in pyrazine at $4000-4500 cm^{-1}$ triplet

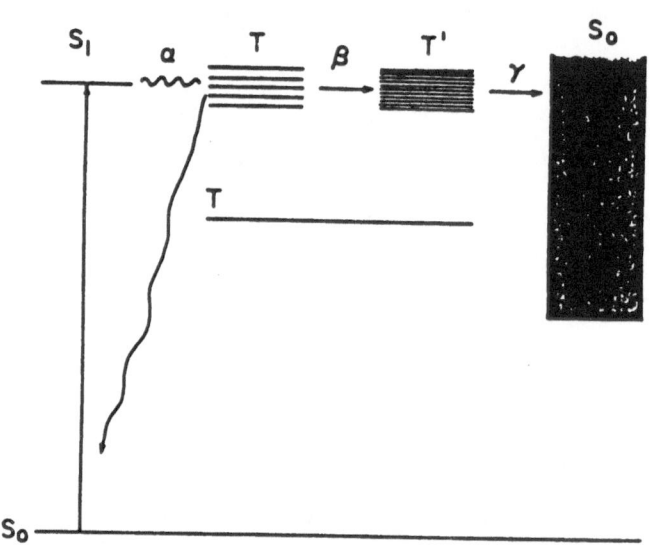

Figure 7.   Energy level diagrams in pyrazine for   intratriplet   (T-T')
IVR, which result in interstate S_1-T' electronic relaxation. S_0 is
radiatively coupled to S_1 with $\Delta J=0\pm 1$, $\Delta K = 0$ selection rules. S_1 is
strongly coupled to the sparse triplet T manifold, via spin-orbit
coupling. The spin-orbit selection rules are $\Delta N=0$ and no ΔK selection
rules. T is nonradiatively coupled due to K" selective Z-type Coriolis
coupling to T'.   T' contain energy levels of T that due to selection
rules are not allowed in spin-orbit coupling.   T' decay nonradiatively
to S_0 in $\tau \geqslant 350$ Nsec.

vibrational energy IVR can occur, and this IVR is also dominated by
Coriolis coupling leading into a strong distinction between   MEs
(J",K"=0)   and MEs (J",K"$\neq$0),as Z type Coriolis coupling vanishes for
MEs with K"=0 (K" is the precessional quantum number of the triplet
component of the MEs). In this case intratriplet state IVR result in
singlet-triplet ER and thus IVR can be considered as ER in pyrazine.
This explanation results in a semiquantitative estimate of several
experimental results and in qualitative agreement with all the various
many "conflicting" experimental results concerning pyrazine $S_1(^1B_{3u})$
dynamics.[3,4,6] According to this model [6]:

1. The ultra-high resolution spectra shown by Kommandeur and colleagues
[4] is due to MEs (J"K"=0) alone, while MEs (J"K"≠0) will result in low
amplitude (10^{-3}-10^{-4}) background "grass" underneath the sharp spectral
feature observed.
2. The short and long decay time observed [4] is inhomogeneous [6], thus
accounting for the "problem of the missing states" [3]. The amplitude
ratio A^+/A^- of the short-to long-time decay is quantitatively given by
the statistical ratio of the number of MEs (J",K≠0) to MEs (J"K"≠0), as
the former result in short time decay alone and the latter which is only
weakly Coriolis coupled result in long time decay alone.
3. Any incoherent or coherent light source used so far should result in
the same A^+/A^- after correction to the detection width, thus eliminating
the problem of the nature of the short component under Nsec incoherent
excitation.
4. The J dependence of the emission quantum yield and its absolute
value [3] can be quantitatively constructed [6] from the ultra-high
resolution spectra and the statistics of MEs(J"K"=0) and MEs(J"K"=0).

CONCLUSIONS

We have shown that IVR constitutes an important radiationless transition
which is intimately related to interstate ER. In several cases it can be
considered by itself as ER. In the case of pyrazine intertriplet states
vibrational energy redistribution increased the singlet dilution and
hence resulted in interstate electronic relaxation. This mechanism is a
combination of both internal conversion and intersystem crossing and can
be called Vibrational Crossing (VC).

    In addition we have demonstrated that Coriolis coupling and not
anharmonic coupling is the dominant vibronic coupling for IVR in the S_1
state of in 9-cyanoanthracene and possibly in the S_1-(T) coupled
manifold of pyrazine. The fingerprints of Coriolis coupling are the
strong rotational dependence. Thus we conclude that the low energy
rotational excitation can induce IVR in the medium energy vibrational
states, which can strongly enhance the high energy interstate ER.

ACKNOWLEDGEMENT

    We thank Prof. E.C. Lim for many stimulating discussions and a most
rewarding collaboration on this problem. This research was supported
(AA) by the Fund for Basic Research of the Israel Academy of Sciences.

REFERENCES

1.   F.A. Novak and S.A. Rice, J. Chem. Phys. 71, 4680 (1979) and ibid
      73, 858 (1980)
2.   a) E. Riedle and H.J. Neusser, J. Chem. Phys. 80, 4686 (1984).
      b) E. Riedle, H.J. Neusser and E.W. Schlag, J. Phys. Chem. 86,
          4847 (1982).

3.   A. Amirav and J. Jortner, J. Chem. Phys. 84, 1500 (1986).
4.   a) K.E.   Drabe and J.   Kommandeur.   To be published in "Excited
          State" Editor E.C. Lim.
      b) J. Kommandeur, B.J. van Der Meer, H. Th. Jonkman in
          "Intramolecular   Dynamics"   eds. J. Jortner and B. Pullman, D.
          Reidel Publ. Dordrecht(Holland) 1982, p. 259.
5.   E.   Riedle, H.J. Neusser, E.W. Schlag and S.H. Lin, J. Phys. Chem.
      88, 198 (1984).
6.   A. Amirav, Chem. Phys. (in press).
7.   M.   Sonnenschein, A. Amirav and J. Jortner, J. Phys. Chem. 80, 1050
      (1984)
8.   A. Amirav, C. Horovitz and J. Jortner, submitted to J. Chem. Phys.
9.   W.R. Lambert, P.M. Felker and A.H. Zewail, J. Chem. Phys. 81, 2209
      (1984).
10.   C.S.   Huang,   J.C.   Hsieh and E.C.   Lim, Chem. Phys. Lett. 28, 130
      (1974).
11.   A. Amirav, U. Even and J. Jortner, 75, 3370 (1981).
12.   A.   Amirav,   J. Jortner, S.Okajima and E.C. Lim, Chem. Phys. Lett.
      126, 487 (1986).
13.   D. Scharf, M.Sc. Thesis, Tel Aviv University (1983).
14.   P.M. Felker and A.H. Zewail, J. Chem. Phys. 82, 2994 (1985).

A QUANTITATIVE DETERMINATION OF THE
QUANTUM YIELD OF THE FLUORESCENCE
OF THE $|B_{34}(0-0)$ STATE OF PYRAZINE

Jan Kommandeur

Laboratory for Physical Chemistry
The University of Groningen
Nijenborgh 16, 9747 AG GRONINGEN
The Netherlands

Abstract

Knowledge of the Molecular Eigenstate spectrum of pyrazine and of
the lifetimes of the molecular eigenstates permit a quantitative
determination of the quantum yield of the P(0) member of the $^1B_{34}(0-0)$
transition of this compound. The results are generalized to higher J'
states of pyrazine.

1. Introduction

Quantum yields of the fluorescence can be simply calculated if of
any one state the radiative and the non-radiative lifetimes are known.
Comparison with experiment then requires that only that particular state
is excited by the light source used. In an intermediate level structure
molecule like pyrazine this condition is, in general, not fulfilled,
because the Molecular Eigenstate spectrum is so dense that most lasers or
other light sources will (coherently) excite a number of states.
For pyrazine, the ME spectrum belonging to the P(1) transition is
known through the use of an extremely narrow laser and an only $10MH_z$
Doppler broadened doubly skimmed molecular beam (1). Recently of a number
of ME's the lifetimes have been measured (2). The results are generalized
to higher J. states. First Coriolis coupling in S_1-S_0 decay is important,
then (J'> 4) the number of triplets coupled starts to increase.

2. The quantum yield of P(1)

A recent high sensitivity excitation spectrum of the ME's belonging
to the J' = 0, K' = 0 state of pyrazine has been obtained (2). Over a
spread of 7,6 GH_z 36 ME's were found. The density of triplet vibronic
states determined in this way is still of the same magnitude as reported
before (2), only then only 12 states were found, but they were spread
over a much narrower energy width.

185

R. Lefebvre and S. Mukamel (eds.), Stochasticity and Intramolecular Redistribution of Energy, 185–191.
© *1987 by D. Reidel Publishing Company.*

Eight of the 36 states found in the excitation spectrum were sufficiently intense to allow a determination of their lifetime. The results are given in Table I.

Table 1 Properties of ME's of Pyrazine

Energy ME (MH$_z$)	Excitation intensity	Lifetime (nsec)
-1456	1666	200
- 535	1278	512
- 353	3891	443
- 221	8168	342
- 44	1305	437
62	4031	560
765	10000	280
867	1503	529

Using the algorithm of Lawrance and Knight (3) the ME-spectrum could be deconvoluted into the zero-order states, their widths and their coupling constants. These results are given in Table II.

Table II Zero order states, their couplings, their widths

Type	Energy (MH$_z$)	coupling element (MH$_z$)	width (MH$_z$)
T	-1286	462	5
T	- 502	119	1.6
T	- 308	105	1.6
T	- 98	150	2.7
S	- 43	-	5
T	13	117	0.6
T	463	457	3
T	848	67	1

It is clear that both zero-order triplet and singlet decay occur. Therefore the quantum yields of the various ME's are not the same. They can be determined by using the expression $|C_S^2| \gamma_r^s T_{ME}$, where $|C_S^2|$ is the magnitude of the singlet amplitude in the ME, γ_r^s the singlet radiative width (3,3 MH$_z$) and T_{ME} the lifetime of the ME. Table III gives an oversight of the various quantities involved.

<u>Table III</u> Molecular Eigenstates, singlet amplitude quantum yield

Energy (MH$_z$)	$|c_s^i|^2$	Q_i
-1456	0.118	0.08
- 535	0.065	0.11
- 353	0.122	0.18
- 221	0.200	0.23
- 44	0.071	0.10
62	0.110	0.21
765	0.245	0.23
867	0.069	0.12

It is clear that the quantum yield varies from one ME to another by almost a factor of three!

If one wants to speak of the quantum yield of the P(1) transition, one most use a laser, which spans all the ME's of P(1) and no others. Assuming that a laser is "white" over the relevant energy region, we can write for the singlet amplitude in ω-space:

$$C_s(\omega) = \frac{1}{\omega - \sum\limits_T \dfrac{V_{ST}^2}{\omega - \omega_T + i\gamma_T} + i(\gamma_{nr}^s + \gamma_r^s)} \qquad (1)$$

where the radiationless width of the triplets (γ_T) and of the singlet γ_{nr}^s have been accounted for.

For the eight ME's of which the lifetimes were measured, we know all the relevant parameters and $C_s(\omega)$ can be numerically fourier transformed into $\hat{C}_s(t)$. The quantity $\gamma_r^s |\hat{C}_s(t)|^2$ can be integrated over all time to give the total light emitted. Dividing it by γ_r^s, which is the absorption of the state S_1 gives the quantum yield, which for the J' = 0, K' = 0 state of pyrazine turns out to be 0.15, which in view of the inaccuracies involved in both the quantum yield and the lifetime measurements is in good agreement with the experimental value of 0.22 reported by Amirav and Jortuer (4). This number gives us a favorable starting point for the discussion of quantum yields at higher J'.

3. The quantum yield as a function of J'

The lifetime of pyrazine appears to be constant with J', as reported by Lim at al (5) and Pratt et al (6). As was pointed out by Jortner and Amirav (4) this must mean that at higher J' the lifetime is dominated by triplet decay, since they find in their quantum yield experiment that at high J' it goes as $(2J'+1)^{-1}$. Apparently what happens is, that at high J' K-mixing occurs in the triplet manifold, and if it is sufficiently complete, the number of triplet states coupled to one J', K'-state goes up as the number of K-states, therefore as (2J'+1). This means that the magnitude of the singlet amplitude in an ME goes down as $(2J'+1)^{-1}$ and if the triplet state lifetimes are not J', K' dependent, the quantum yield of the ME's goes down as $(2J'+1)^{-1}$.

This appears to be a correct interpretation for high J'-values. For J' ≤ 4, however, the ME-experiments (7) permit counting of the number of ME's and it was convincingly shown that in this range it is constant per J', K' state.

This conclusion also follows from symmetry arguments. At low J' most of the K' states are of different nuclear symmetry and therefore cannot mix. At higher J' this limitation does not apply any more.

If K-mixing is not important for the low J' states, and if the quantum yield is still dependent on J' there must be some other J'-dependent radiationless decay channel. Since by the argument of the J' independent lifetime this cannot be the triplet decay, it must be the singlet decay.

The most obvious interaction is a S_1 - S_0 Coriolis coupling, which we explore in the next section.

4. S_1 - S_0 Coriolis coupling

Figure Ia shows the experimentally determined rotational spectrum of the $^1B_{3n}$ transition. Comparison with fig. Ib shows that assuming constant triplet and singlet decay, i.e. calculating the spectra with a Boltzmann distribution and the Höhnl-London factors gives very bad agreement.

We therefore assumed Coriolis coupling in the singlet state according to:

$$\Gamma_z = A_z K^2$$
$$\Gamma_+ = B_+ (J - K)(J + K + 1)$$
$$\Gamma_- = B_- (J - K)(J + K + 1)$$

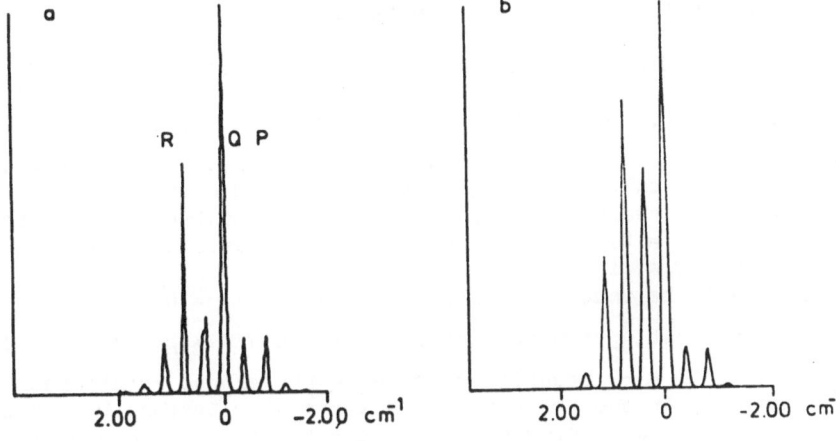

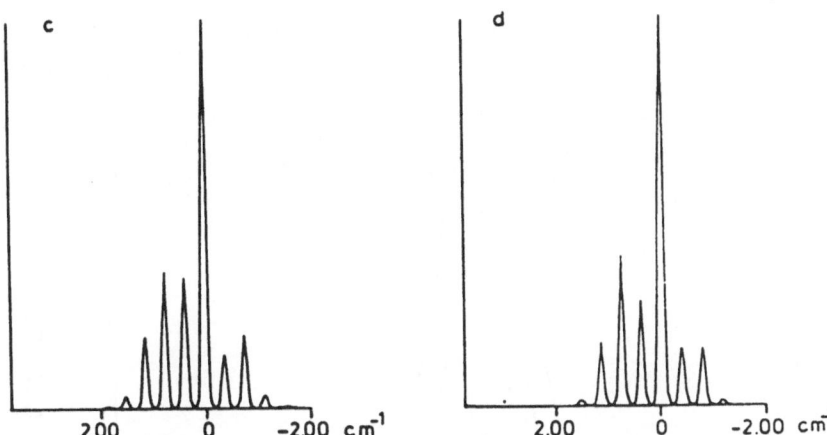

<u>Fig. 1</u>
a) The rotational spectrum as determined experimentally
b) The rotational spectrum calculated without any J', K' dependent
quantum yield

c) The rotational spectrum calculated with singlet Coriolis coupling
d) The rotational spectrum calculated with Coriolis coupling and J'' = 0,
   K'' = 0 frozen in at 20%

where the Γ represent the radiationless rates given by the diagonal and off-diagonal Coriolis Coupling (8).

The best fit is displayed in fig. Ic for the values Γ_z = 1,5 MH$_2$, Γ^+ = 1,7 MH$_2$ and Γ^- = 0,3 MH$_2$. Although a considerable improvement has occurred, the agreement is not all together satisfactory. In particular, the R(0) transition is calculated to be relatively too high. This discrepancy is largely removed in fig. Id, where it has been assumed that the (ground) state J'' = 0, K'' = 0, from which R(0) arises has not reached equilibrium, i.e. is underpopulated.

This can easily be understood, since J'' = 0, K'' = 0 has A$_g$ symmetry and the closest A$_g$-symmetry state to it is J'' = 2, K'' = 0. It therefore takes a ΔJ'' = -2 transition to reach the ground rotational state, and apparently if one tries to reach a very low temperature in the beam, these transitions are impeded. In agreement with this, measurements at higher beam temperatures did not show the anomalous behavior of R(0).

It can be concluded that Coriolis coupling gives a satisfactory explanation for the radiationless decay of pyrazine at low J'.

Fig. 2 shows a calculation of the quantum yield of pyrazine as a function of J' up to high values of that quantum number.

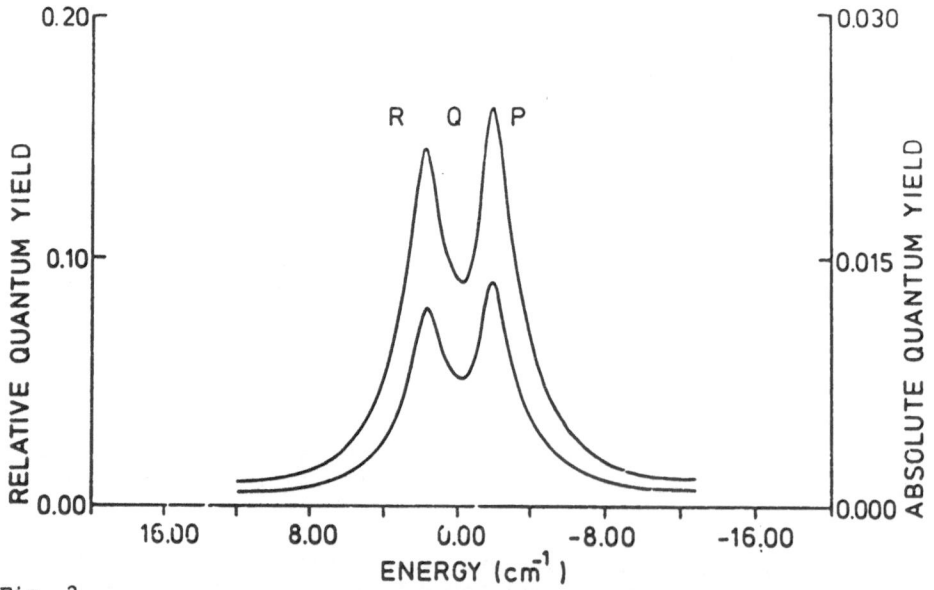

Fig. 2
The quantum yield of pyrazine at a temperature of 20 K as calculated with Coriolis coupling only for the highest and the lowest estimates of the Coriolis coupling constants.

It is quite similar to the experimental dependence published by Jortner and Amirav (4), but close inspection shows that at high J' it falls off more rapidly as J^{-2}, while they find $(2J + 1)^{-1}$. It is clear then, that between J' = 4 and say, J' = 10 the regime changes from Coriolis coupling to triplet dominated decay.

5. Summary

The quantum yield of the rotationless and vibrationless first excited electronic state of pyrazine can be quantitatively determined from the ME-spectrum and the lifetimes of these ME's.
At low J' the singlet decay is modified by Coriolis coupling, but at higher J' this effect becomes invisible, because more and more triplet states are coupled through K-mixing and the decay becomes triplet dominated.

References

1. B.J. van der Meer, H.Th. Jonkman, J. Kommandeur, W.L. Meerts & W.A. Majewski, Chem. Phys. Lett. 92, 565, (1982)

2. W.M. van Herpen, W.L. Meerts, K.E. Drabe & J. Kommandeur (submitted to J. Chem. Phys.).

3. W.D. Lawrance & A.E.W. Knight, J. Phys. Chem. 89, 917 (1985).

4. A. Amirav & J. Jortner, J. Chem. Phys. 84, 1500 (1986).

5. H. Saigusa & E.C. Lim, J. Chem. Phys. 78, 91 (1983).

6. Y. Matsumoto, L.H. Spangler & D.W. Pratt, Chem. Phys. Lett. 98, 333 (1983).

7. B.J. van der Meer, H.Th. Jonkman & J. Kommandeur, Laser Chem. 2, 77 (1983)

8. E. Riedle, H.J. Neusser & E.W. Schlag, Faraday Disc. Chem. Soc. 75, 387 (1983).

ROTATION SELECTIVE INTRAMOLECULAR VIBRATIONAL RELAXATION IN THE S_1-STATE OF BENZENE

S. F. Fischer and W. Dietz
Technische Universität München,
Institut für theoretische Physik T38,
D-8046 Garching, Fed. Rep. Germany

ABSTRACT: An oscillator carrying (light) rovibronic state is coupled to symmetry like (dark) states via anharmonic interactions and to (dark) states with different symmetry via combined anharmonic and Coriolis interactions. For the model of equally spaced but randomly distributed dark states analytic expressions for the width of the light state are derived. As an example the linewidth of the $14^1 1^2$-state of benzene is discussed as function of the rotational quantum numbers J and K.

1. INTRODUCTION

One of the central problems in molecular gas phase spectroscopy of larger molecules concerns the onset of intramolecular vibrational relaxation processes. In particular the role of Coriolis coupling is not well unterstood. Recent sub-Doppler experiments by E. Riedle and H.J. Neusser /1/ and H. Stepp et al. /2/ on benzene show that the width of selected rovibronic states of the first excited singlet depends on the rotational quantum numbers. For the $14^1 1^2$ S_1-S_0 rotational band it is found that not all states are resolved. The onset of missing lines goes along with a drop of the fluorescence quantum yield and has been refered to as a decay channel III /3/, to be distinguished from the radiative decay channel to the ground state and nonradiative transitions to the triplet or ground state singlet.

There are regimes in the spectrum of C_6H_6 which point towards strong Coriolis coupling with rotation around the symmetry (z)-axis, since only K = 0 states are resolved for J values up to about 15. For larger J and K-values around 30 those states with J $\approx$ K are best resolved and there exist a certain probability that other states with $| 1-J/K| \ll 1$ can be resolved.

Here we want to present a model which allows us to make predictions about the presence or depression of such lines. The theory is based on a specific coupling model and incorporates some statistical elements as well. The statistical nature is brought into the theory as a result of a random distribution between states with different symmetry. First we present the model then we discuss analytic solutions

R. Lefebvre and S. Mukamel (eds.), Stochasticity and Intramolecular Redistribution of Energy, 193–201.

for limiting cases and finally we apply it to spectra of benzene.

2. THE MODEL

We consider one light state denoted as s-state which carries oscillator strength and couple this state to two sets of dark states, which carry no oscillator strength. The dark states are equally spaced and they have a width γ_0 much larger than the width γ_{so} of the light state. The two sets of dark states differ in the coupling to the light state. We denote the two sets as l-states and their coupling as v_a and v_c, respectively. Thus we arrive at the following hamiltonian

$$H = E_s| s><s| + \epsilon_0 \sum_l l| l><l| +$$
$$+ v_c \sum_l (| s><l| + | l><s|) +$$
$$+ (v_a - v_c) \sum_l m_l (| s><l| +| l><s|)$$

(1)

The summation goes from $-(N-1)/2$ to $+(N-1)/2$ and m_l is zero or one, such that the constraint

$$\sum_l m_l = N_a$$

(2)

holds. Any set of m_l-values corresponds to a distribution of the two sets of differently coupling states. These states have the width γ_0 and contribute for small γ_{so} to the width of the light eigen-state γ_s according to their energy location and coupling to give

$$\gamma_s = \gamma_{so} + \gamma_c + \sum_l m_l \frac{v_a^2 - v_c^2}{l^2 \epsilon_0^2 + \gamma_0^2} \gamma_0$$

(3)

with

$$\gamma_c = \sum_l \frac{v_c^2}{l^2 \epsilon_0^2 + \gamma_0^2} \gamma_0 = \pi \frac{v_c^2}{\epsilon_0} \coth(\pi \frac{\gamma_0}{\epsilon_0})$$

(4)

The problem now is to evaluate this sum for all sets (m_l) which fulfill the constraint (2). We will get this way a distribution of widths and we will evaluate this distribution for the statistical limit ($N \to \infty$, N_a/N=const.) with regard to its average value and its spread around this mean value.

3. ANALYTICAL RESULTS

Since m_l takes only the values 1 or 0 we have a nice analogy to a Fermi system. Let us consider N_a particles with total energy E which occupy one-particle states $| l>$ with energy ϵ_l. Then the density of states $N(E)$ can be calculated from

$$N(E) = \sum_{(m_l)} \delta(E - \sum_l m_l \epsilon_l)$$

(5)

where the sum runs over all sets (m_1) that fulfill the constraint (2). In the statistical limit ($N \to \infty$, N_a/N = const.) the density can be expressed in terms of the entropy S and the fluctuation ΔE of the internal energy as /4/

$$N(E) = \exp(S/k_B)/\Delta E \tag{6}$$

k_B is the Boltzmann constant. The expressions for S and ΔE are well known for a Fermi system /4/. The analogy is obvious. Instead of (5) we have to evaluate a distribution function $w(\gamma)$ for the widths for $v_a^2 > v_c^2$

$$w(\gamma) = \binom{N}{N_a}^{-1} \sum_{m_1} \delta(\gamma - \gamma_{s0} - \gamma_c - \sum_1 m_1 \gamma_1) \tag{7}$$

with

$$\gamma_1 = \gamma_0 \frac{v_a^2 - v_c'^2}{\epsilon_0^2 1^2 + \gamma_0^2} \tag{8}$$

instead of S from Eq.(6). We evaluate our generalized entropy Φ for the width distribution by switching to the grand partition function and get

$$\Phi = \beta(\gamma - \gamma_{s0} - \gamma_c) + \sum_{1=-\infty}^{+\infty} \ln[\frac{N_c}{N} + \frac{N_a}{N} \exp(-\beta\gamma_1)] \tag{9}$$

with $N_c = N - N_a$, and the distribution parameter β is determined via a saddle point equation $\partial\Phi/\partial\beta = 0$.

$$\gamma - \gamma_{s0} - \gamma_c - \sum_{1=-\infty}^{\infty} \gamma_1[1 + \frac{N_c}{N_a} \exp(\beta\gamma_1)]^{-1} = 0 \tag{10}$$

and the square of the fluctuation term gives instead of ΔE^2

$$(\Delta\gamma)^2 = 2\pi \, \partial^2\Phi/\partial\beta^2 = 2\pi \frac{N_c}{N_a} \sum_1 \gamma_1^2 \frac{\exp(\beta\gamma_1)}{[1 + \frac{N_c}{N_a} \exp(\beta\gamma_1)]^2} \tag{11}$$

If the width γ_0 is larger than the spacing ϵ_0 we can replace the sums in (9), (10) and (11) by integrals. Switching to the variable $x = (\beta\gamma_{1=0})^{1/2}$ we get

$$\Phi = \beta(\gamma - \gamma_{s0} - \gamma_c) - 2\alpha\beta^{\frac{1}{2}} \int_0^{(\beta\gamma_{1=0})^{\frac{1}{2}}} dx[1 - x^2/(\beta\gamma_{1=0})]^{\frac{1}{2}} /(1 + \frac{N_c}{N_a} e^{x^2}) \tag{12}$$

with

$$\alpha = 2\gamma_0 \epsilon_0^{-1} \gamma_{1=0}^{\frac{1}{2}} \tag{13}$$

For the limit $\beta\gamma_{1=0} \gg 1$ we can extend the upper integration limit to infinity and get

$$\Phi = \beta(\gamma - \gamma_{s0} - \gamma_c) - 2\alpha\beta^{\frac{1}{2}} f_0 \tag{14}$$

with

$$f_0 = \int_0^\infty dx \frac{1}{1 + \dfrac{N_c}{N_a} \exp(x^2)} \tag{15}$$

Substituting these results in the analogous expression to (6) for the width distribution (7) we get

$$w(\gamma) = \pi^{-\frac{1}{2}} \alpha f_0 (\gamma - \gamma_{s0} - \gamma_c)^{-\frac{3}{2}} \exp[-\alpha^2 f_0^2 / (\gamma - \gamma_{s0} - \gamma_c)]$$

$$\tag{16}$$

The saddle point equation $\partial\Phi/\partial\beta = 0$ gives for β

$$\beta = \alpha^2 f_0^2 (\gamma - \gamma_{s0} - \gamma_c)^{-2} \tag{17}$$

and after some further approximations which hold for $f_0\gamma_0/\epsilon_0 \ll 1$ and $N_a \ll N_c$ one gets the following analytic result (Fig.1):

$$w(\gamma) = \begin{cases} (\frac{\delta}{\pi})^{\frac{1}{2}} (\gamma - \gamma_{s0} - \gamma_c)^{-\frac{3}{2}} \exp[-\delta/(\gamma - \gamma_{s0} - \gamma_c)] & \text{if } \gamma > \gamma_{s0} + \gamma_c \\ 0 & \text{else} \end{cases}$$

$$\tag{18}$$

where

$$\delta = \pi \frac{v_a^2 - v_c^2}{\gamma_0} \left(\frac{N_a \gamma_0}{N_c \epsilon_0}\right)^2 \tag{19}$$

The most probable value of γ is

$$\gamma_{mp} = \gamma_{s0} + \gamma_c + 2/3\delta \tag{20}$$

and the width

$$\Delta\gamma = (2e/3)^{\frac{3}{2}} \pi^{\frac{1}{2}} \delta \approx 4.32\delta \tag{21}$$

As we extract from the experiment the values for γ_{mp} and $\Delta\gamma$ we can extract information about the microscopic paramters v_a^2 , v_c^2 and γ_0 .

A second limiting case is defined for small β , $_{1/2}$ $\beta\gamma_{1=0} \ll 1$. We introduce in (12) the integration variable $q=x/(\beta\gamma_{1=0})^{1/2}$, expand the integrand of (12) in $\beta\gamma_{1=0}$ and get in lowest order

$$\bar{\Phi} = \beta(\gamma-\gamma_{mp}) + \Delta\gamma^2\beta^2/(4\pi) \tag{22}$$

with

$$\gamma_{mp} = \gamma_{s0}+\gamma_c+ \pi\frac{N_a}{N_c} \frac{v_a^2 - v_c^2}{\varepsilon_0} \tag{23}$$

and

$$\Delta\gamma^2 = \pi^2(v_a^2-v_c^2)^2(\frac{N_a \ N_c}{\varepsilon_0\gamma_0 N^2}) \tag{24}$$

substituting   this   expression   in (22)   we   obtain   for   the   width-distribution function (Fig.2)

$$w(\gamma) = \begin{cases} \Delta\gamma^{-1}\exp(-\pi(\gamma-\gamma_{mp})^2/\Delta\gamma^2) & \text{if} \quad \gamma>\gamma_{s0}+\gamma_c \\ 0 & \text{else} \end{cases} \tag{25}$$

The distribution parameter β is now related to γ by the simple expression

$$\beta = -2\pi(\gamma-\gamma_{mp})/\Delta\gamma^{-2} \tag{26}$$

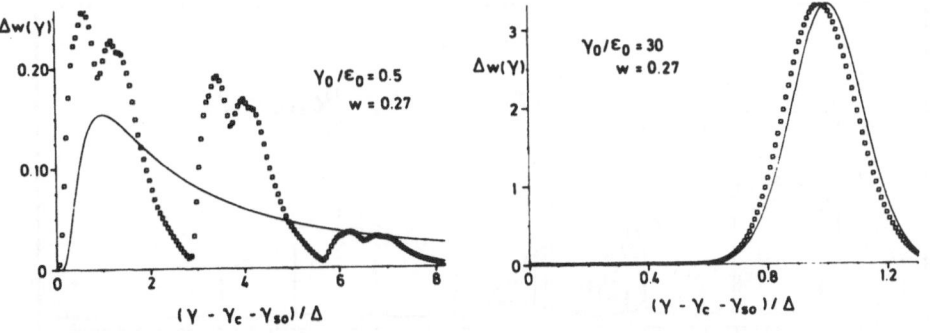

Figures 1-2.   Linewidth distributions $w(\gamma)$ in the case of small γ_0 in Fig.1 (left) and large γ_0 in Fig.2 (right).   Smooth versions (18) and (25) in Fig.1 and Fig.2 respectively are drawn as solid lines. The quasiperiodic structure from the exact count (7) (squares) is due to the equal spacing of the l-states. $w:=N_a/N$, $\Delta:=\gamma_{mp}-\gamma_c-\gamma_{s0}$

4. RESOLVABLE LINES - DISCUSSION AND INTERPRETATION OF EXPERIMENTS

To make contact of our model with the level structure of benzene we identify the s-states with rovibronically excited rigid rotor harmonic oscillator (RRHO) states $|14^1 1^2, J, K\rangle$ (vibrational symmetry B_{2u}). The couplings v_a and v_c are anharmonic and combined anharmonic with first order Coriolis /5/ respectively. The l-states are of vibrational symmetry B_{2u} and E_{2u} respectively. The dependence of v_c on the rotational quantum numbers is given by the first order Coriolis term /5/,

$$v_c^2 = v_{c0}^2 \ F(J,K) \tag{27}$$

with

$$F(J,K) = [(J \mp K)(J \pm K+1)]^{\frac{1}{2}} \approx J^2 - K^2 \qquad \text{if } K' = K \pm 1 \tag{28}$$

v_{c0} is independent from J' and K.

The $14_0^1 1_0^2$ band of $^{13}C_6H_6$ benzene of Ref./2/ shows about 40 resolved rotational lines in the spectral range Δv from 76 GHz to 150 GHz which are nearly all assigned to rotational quantum numbers J,K (Fig.3)

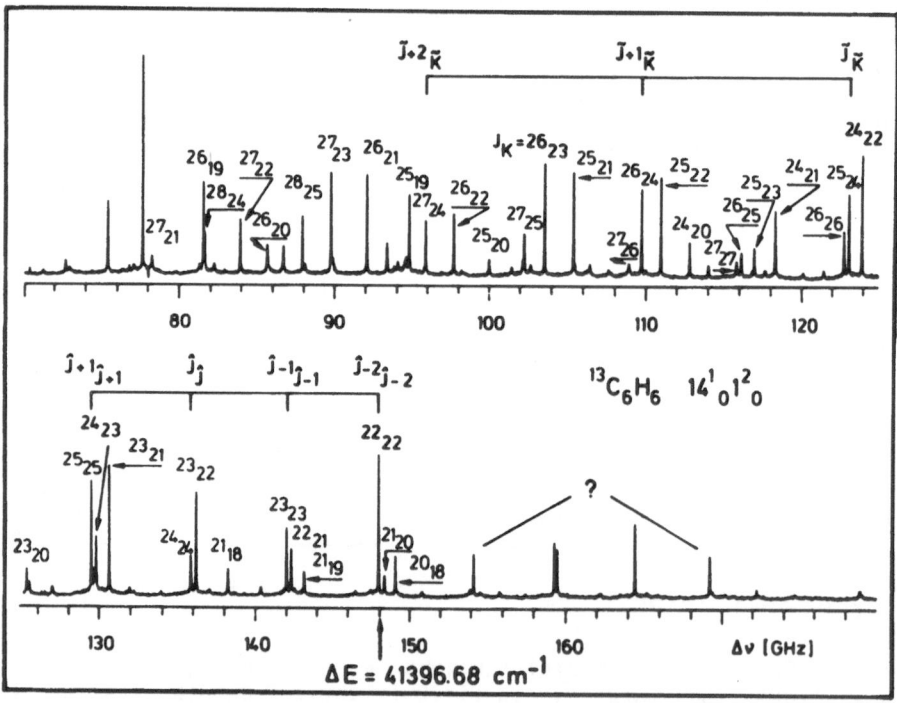

Figure 3. Experimental spectrum from Ref./2/ showing the resolved lines in the $14_0^1 1_0^2$ band of $^{13}C_6H_6$.

We counted the number N_{res} of lines which are experimentally resolved for regimes of J^2-K^2 -values between 0-50, 50-100, 100-150, etc. . Further we estimated the number N_{tot} of lines which should occur within these F intervals if the lines were completely resolvable. The fraction of experimentally resolved lines N_{res}/N_{tot} is shown in the histogram of of Fig.4-7.

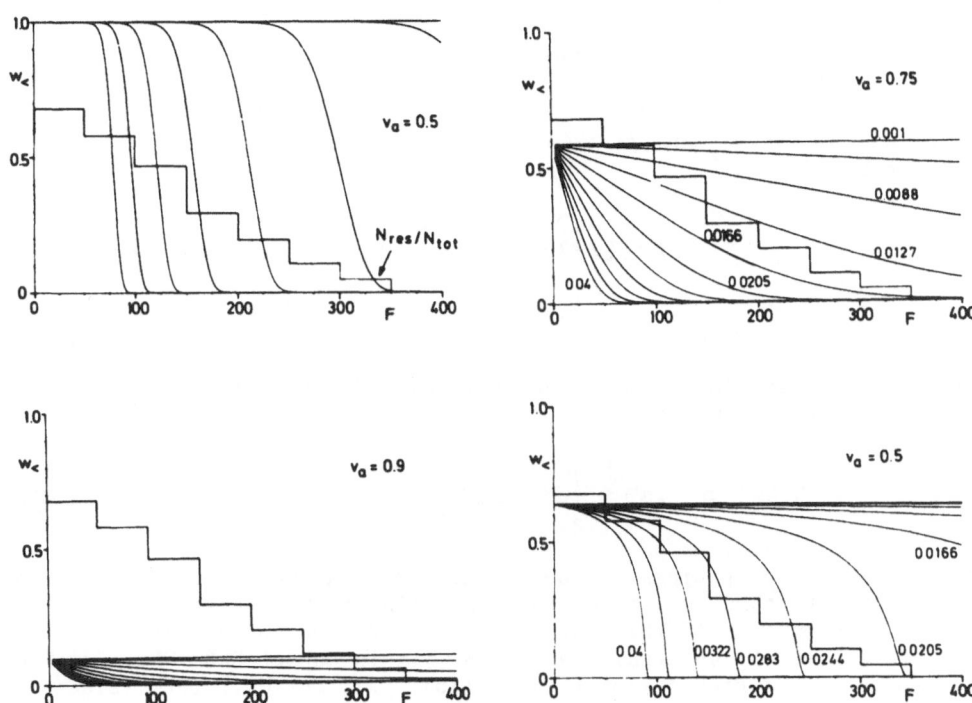

Figures 4-7. The smooth curves give the probability $w_<$ that the width of a rovibronic level is smaller than a threshold γ_{th} as function of F ($=J^2-K^2$) calculated from (30) in Figs.4-6 and from (31) in Fig.7 . $\gamma_0=30$, $\gamma_{th}=.5$ in Fig.4 (top left), Fig.5 (top right) and Fig.6 (bottom left). $\gamma_0=.5$, $\gamma_{th}=.5$ in Fig.7 (bottom right). v_{c0} is 0.04 for the steepest slope and decreases in equidistant steps with decreasing steepness of the slope to 0.001 for the flattest curve. Some v_{c0} values are drawn for the curves of Fig.5 and Fig.7 .
The histogram is the same in Figures 4-7. It shows for $F\in[0,50]$, $[50,100],...$ the number of experimentally /2/ resolved lines in the spectrum of Fig.3 in the frequency range from $\Delta v=76GHz$ to $\Delta v= 150GHz$ normalized to the total number of lines which should occur at complete resolution.

Inserting (25) or (18) in (29) one obtains the probability $w_<$,

$$w_< = \int_0^{\gamma_{th}} w(\gamma)d\gamma \tag{29}$$

that the width of the light eigen-state γ_s does not exceed a threshold γ_{th}. The integration gives for small β

$$w_< = \begin{cases} \frac{1}{2} \, \text{erfc}[(\gamma_{mp}-\gamma_{th})/\Delta\gamma] & \text{if} \quad \gamma > \gamma_{so}+\gamma_c \\ 0 & \text{else} \end{cases} \tag{30}$$

with (23,24) for γ_{mp} and $\Delta\gamma$, and for large β

$$w_< = \begin{cases} \text{erfc}([\delta/(\gamma_{th}-\gamma_c-\gamma_{so})]^{\frac{1}{2}}) & \text{if} \quad \gamma_{th} > \gamma_{so}+\gamma_c \\ 0 & \text{else} \end{cases} \tag{31}$$

with (4) and (19) for γ_c and δ respectively. "erfc" refers to the complementary error-function,

$$\text{erfc}(z) = (2\pi)^{-\frac{1}{2}} \int_z^\infty dt \, \exp(-t^2) \tag{32}$$

We find from the spacings of the l-states /2/ the parameters $\varepsilon_0 \approx 1 Gz$ and $N_a/N_c \approx .38$. Further we take for the linewidth of these vibrational states values from $^{12}C_6H_6$ benzene /6/, which are in the GHz - region. The threshold of spectral resolution γ_{th} is about 500 MHz /7/. The smooth curves of Figs.4-7 show $w_<$ of (30), (31) for different values of γ_0, γ_{th}, v_a and v_{co} for $\varepsilon_0 = 1 GHz$ and $N_a/N_c = .38$. At fixed γ_0 and γ_{th} $w_<$ decreases at F=0 with increasing anharmonic interaction v_a as shown in Figs.4-6. Further the slope of the $w_<$ - curve increases with increasing Coriolis interaction constant v_{co} (Fig.5). If both v_a and v_{co} are weak $w_<$ takes the value 1 (Fig.4). That means, the spectrum is completely resolved. With increasing v_a and small v_{co} the spectrum gets less resolved but the fraction of resolved lines shows no dependence on F (Figs.4-6, v_{co}= 0.001 - curve). At higher v_{co}, however, $w_<$ decreases sharply with increasing F, Fig.5 . In this case only lines with F$\approx$0 (J$\approx$K) appear in the spectrum. This situation is also observed experimentally for the $14_0^1 1_0^2$ rotational band of $^{12}C_6H_6$ benzene /1/.

The experimental results for N_{res}/N_{tot} are best fitted with the parameters $v_a \approx .5-1$ GHz, $v_{co} \approx 10-40$ MHz. γ_0 can be variied between 0.5 GHz and 100 GHz. If γ_{th} is varied within one order of magnitude from .5 to 5 GHz, the upper limits for v_a and v_{co} change from 1 GHz to 2.5 GHz and from 40 MHz to 120 MHz respectively. The curvature of the experimental curve is best reproduced at high γ_0-values (Fig.5).

The results are supported by estimates for v_a and v_c based on the calculation of anharmonic and Coriolis constants from semi-empirical force fields /8,9/. We find for v_a few GHz and for v_{co} about 1MHz /5/.

5. SUMMARY

The IVR treated in this work is induced by two different interactions, anharmonic and Coriolis. Both interactions are treated as perturbations in a basis set of RRHO-states endowed with finite lifetimes. For a simplified model one obtains in the limiting cases of large and small zero order linewidth simple analytical functions for the linewidth-distributions. This way it is possible to understand the selective suppression of rovibronic lines. The model is applied to benzene. The analysis gives values for effective intramolecular interactions, which are consistent with independent estimations.

6. ACKNOWLEDGEMENT

We wish to thank Prof. Neusser, Dr. Riedle and Prof. Schlag for many stimulating discussions and for providing us with experimental data prior to publication.

REFERENCES

1. E. Riedle and H.J. Neusser, J.Chem Phys. 80, 4686 (1984).
    E. Riedle, H.J. Neusser, and E.W.Schlag, Faraday discussions Chem.Soc.75, 387 (1983).

2. H. Stepp, E. Riedle, H.J. Neusser private communication.

3. U. Schubert, E. Riedle, and H.J. Neusser, J.Chem.Phys. 84, 5326 (1986).

4. L.D. Landau, E.M. Lifschitz, Statistische Physik (Akademie-Verlag, Berlin, 1975).

5. W. Dietz and S.F. Fischer, submitted to J.Chem.Phys. .

6. M. Sumitani, D.V. O'Connor, Y. Takagi, N. Nakashima K. Kamogawa, Y. Udagawa and K. Yoshihara, Chem.Phys.93, 359 (1985).

7. H.J. Neusser, private communication.

8. M.J. Robey and E.W. Schlag, J.Chem.Phys.67,2775 (1977).

9. P. Pulay, G. Fogaresi, J.E. Boggs, J.Chem.Phys.74, 3999 (1981).

DOPPLER FREE EXCITATION OF LARGE MOLECULES AND NONRADIATIVE DECAY OF INDIVIDUAL LEVELS

E. Riedle and H. J. Neusser
Institut für Physikalische und Theoretische Chemie
Technische Universität München
Lichtenbergstr. 4
D-8046 Garching, West Germany

ABSTRACT.   Doppler-free two-photon spectroscopy with cw and pulsed Fourier transform limited light sources allows the resolution of the rotational structure of the electronic spectrum of large molecules. Besides a very detailed analysis of the spectrum and a precise determination of spectroscopic constants, the observation of the decay behavior of individual levels is made possible in this way. Examples are presented for the prototype molecule benzene, C_6H_6. While regular behavior prevails in the 14^1_0-band at low vibrational excess energy, it is shown that strong coupling to background states occurs in the $14^1_0 11^1_0$-band at intermediate excess energy. A detailed model for this coupling is presented.

1.   INTRODUCTION

Ever since the first observations of nonradiative decay in large molecules, the question of the dependence of the decay rate on the excess energy in the electronic state and the nature of the states excited has been the topic of numerous investigations /1/. Generally, an increase in the nonradiative decay rate with excess energy is observed which may be quite dramatic as in the case of the onset of the so called "channel three" in benzene /2,3/.

The excitation of single vibronic states is found to be nontrivial in room temperature gas phase studies of large molecules that can display nonradiative decay, due to the rotational inhomogeneous broadening and the presence of sequence bands in the spectrum. Experiments in supersonic jets with their extremely low rotational temperature allowed great progress in this respect in the last few years, however, at the expense of loss of knowledge about the influence of the molecular rotation on the decay behavior.

R. Lefebvre and S. Mukamel (eds.), Stochasticity and Intramolecular Redistribution of Energy, 203–216.
© 1987 by D. Reidel Publishing Company.

All decay time measurements made so far /4/ have been selecting bunches of rovibronic states with a common vibronic
identity rather than single rovibronic states. In recent
years, however, there have been some indications from mode-
rately resolved measurements, that the decay behavior may
vary over the rotational contour of a vibronic transition
/5/. This makes it desirable to observe the decay of single
rovibronic states. For the large molecules that show nonradiative decay this is not easily done. Doppler-broadening
in the gas phase makes the resolution of single lines in
the spectrum impossible, since many rotational lines are
located within the Doppler width. With the advent of tunable, narrow band, high power laser light sources a number
of Doppler-free techniques /6/ like spectroscopy in a col-
limated molecular beam, saturation and polarization spectroscopy and Doppler-free two-photon spectroscopy became
available. These methods were first successfully applied in
atomic physics but now they are also used in molecular
spectroscopy /7/. In this contribution we will show that
Doppler-free two-photon spectroscopy allows the complete
resolution of the rotational line structure of vibronic
bands of the large prototype molecule benzene and the investigation of the nonradiative decay of individual rovibronic
levels.

2.   EXPERIMENTAL TECHNIQUES

The experimental set up used for the recording of Doppler-
free two-photon spectra and the investigation of the decay
of individual levels has been described in detail previously /8,9/. Only a brief review will be given here.

Doppler-free excitation takes place if a molecule simultaneously absorbs two photons of equal energy and opposite
direction of propagation from a standing wave light field
/10/. Such a light field can be generated by reflecting a
laser beam back into itself or it exists inside a cavity,
for example a Fabry-Perot resonator. The laser light used
has to be extremely narrow band since the experimental resolution that can be obtained is limited by this frequency
width. In our experiments the light of a cw frequency stabilized single mode ring dye laser (CR 699/21) pumped by a Kr^+
laser is used. At a wavelength around 5000 A typically 250
mW of tunable light with a frequency width (FWHM) of 1 MHz
are available. For extremely high spectral resolution and
very high sensitivity the Doppler-free experiment is performed within a concentric external cavity whose length is
locked to the laser frequency /8/. The UV-fluorescence
following the excitation of the molecules is monitored by
single photon counting. The use of the external cavity

increases the sensitivity by two orders of magnitude as
compared to the simpler arrangement of the backreflected
laser beam. A resolution of better than 10 MHz can be obtained with this cw set up /8/.

For the investigation of the decay of individual levels
pulsed excitation of the molecules has to be used. The
experimental set up is shown in Fig. 1 /9/. Extremely narrow
band pulsed laser light is produced by pulsed amplification
of the cw light. With three stages of excimer laser pumped
amplifiers we can generate light pulses of 500 KW peak
power and nearly Fourier transform limited bandwidth. With
an additional parasitic cavity around the second amplifier
the pulse length can be varied between 2.5 ns and 10 ns
and the frequency width accordingly between 50 MHz and 180
MHz. After passing through the sample cell the laser beam
is reflected back into itself to allow the Doppler-free
absorption. The two beams are counterclockwise circularly
polarized to suppress the Doppler-broadened background
/11/. The resulting UV-fluorescence signal is either integrated for the recording of spectra or its time behavior
is recorded with a transient digitizer. It is worth mentio-
ning that in Doppler-free two-photon absorption all molecules regardless of their velocity contribute to the observed signal. Molecules are only excited to one single
level if the laser frequency is set to a resolved rotational
line in the spectrum regardless of the number of lines
within the Doppler width. This allows the observation of
the decay of an individual level /9/.

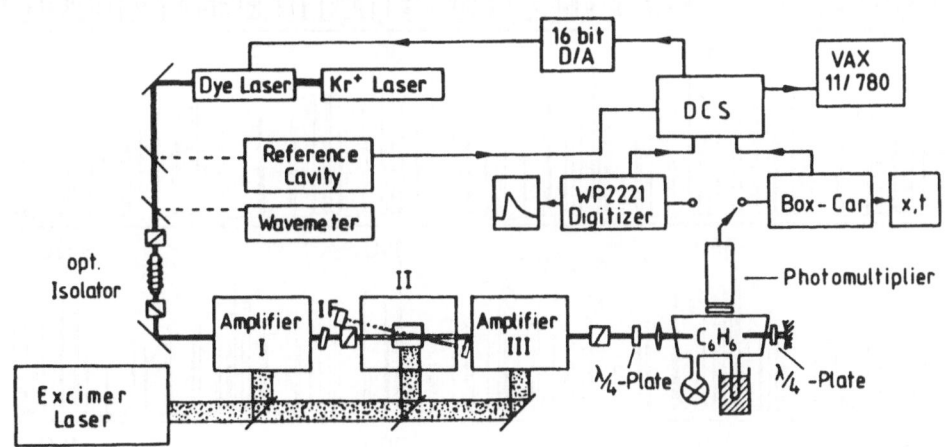

Figure 1.   Experimental set up for decay time measurements
of individual rotational levels of S_1 benzene. Fluorescence
decay is observed after pulsed Doppler-free two-photon excitation with the amplified light of the cw laser.

3. EXPERIMENTAL RESULTS AND DISCUSSION

3.1. 14^1_0-Band of C_6H_6, typical for low excess energy

3.1.1. <u>Spectroscopic results</u>
The first 6 cm^{-1} of the Doppler-free spectrum of the Q-branch of the 14^1_0-band of benzene, C_6H_6, at 39656.90 cm^{-1} are shown in Fig. 2. The spectrum has been recorded with a sample pressure of 0.7 Torr in our cw set up. The lines observed in the spectrum correspond to well resolved single rotational lines. The Doppler-width of 1.7 GHz would not allow the resolution of the individual rovibronic transitions since typically 10 lines are located within the Doppler-width. All lines in the spectrum in Fig. 2 can be assigned within the model of a semirigid symmetric top. A fit to the observed line posi-

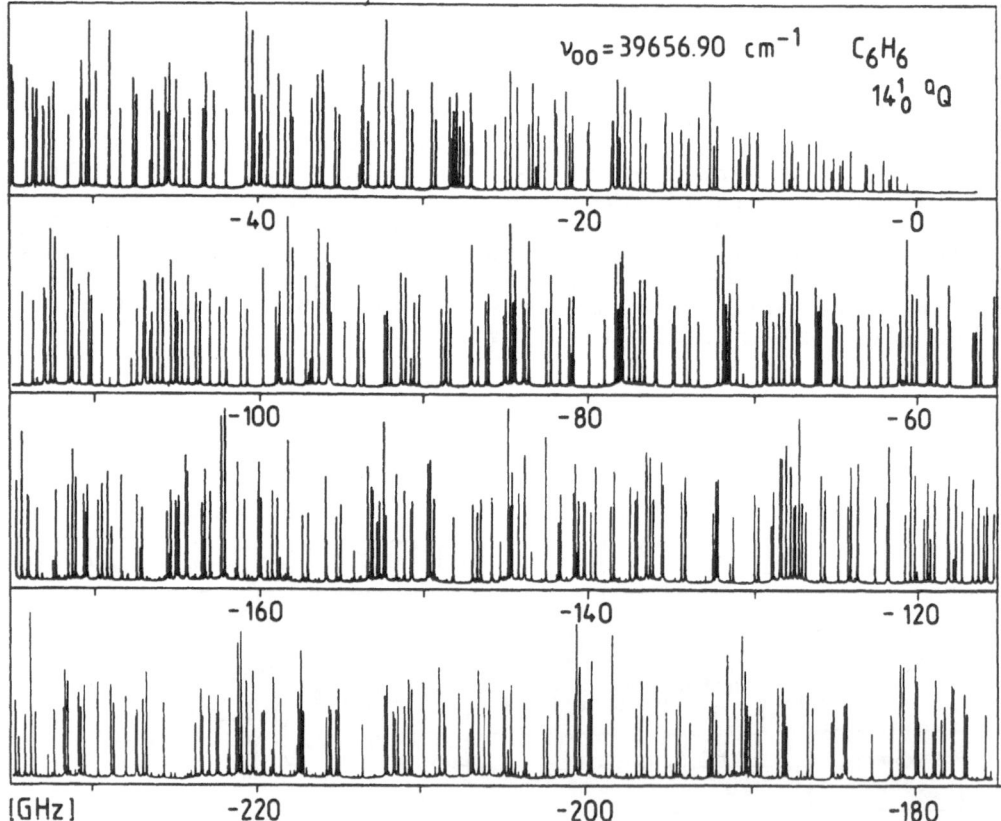

Figure 2.   Part of the Doppler-free room temperature spectrum of the Q-branch of the 14^1_0-band of benzene, C_6H_6. Every line corresponds to an individual rovibronic transition. All of the lines have been assigned.

tions renders the rotational constants of the excited state
with an accuracy of 10^{-7} cm^{-1} and the quartic centrifugal
distortion contants with an accuracy of 10^{-10} cm^{-1} /13/.
This accuracy is higher by two orders of magnitude than
previously obtained with Doppler-limited spectroscopy. The
remaining deviations (residuals) between the calculated
line positions and the observed ones are about 10 MHz.
These results clearly show that Doppler-free two-photon
spectroscopy allows the complete resolution of the electronic spectrum of benzene and the spectrum is extremely well
reproduced with the simple model of a semirigid symmetric
top.

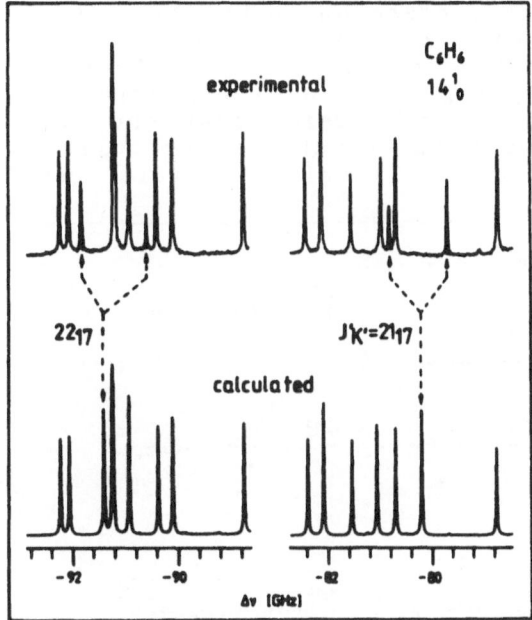

Figure 3.   Two small parts of the Q-branch of the $14^1{}_0$-band
under Doppler-free resolution. Upper traces: measured spectra; lower traces: calculated spectra. Instead of the predicted peaks $J'_{K'}=21_{17}$ and $J'_{K'}=22_{17}$ two peaks appear shifted from the calculated position in each case. (taken from
ref. /14/.)

In some parts of the spectrum, at higher rotational energy
differences between the experimental spectrum and the calculated one are observed. This is shown for two examples
in Fig. 3. While most of the observed lines (top part of
Fig. 2) are well reproduced by the calculation (bottom
part of Fig. 3) single lines of the calculated spectrum
are not found in the experimental one /14/. Instead two
smaller lines are observed in each case. If these pairs of

lines are labeled with the rotational quantum numbers of the
missing lines the dependence of the deviations on the quantum number J of total angular momentum can be plotted for
each value of K, where K is the quantum number of the projection of $\vec{J}$ on the figure axis of the molecule. The result
is shown in Fig. 4. The typical J-dependence of the deviations found is that of an avoided crossing. A careful analysis shows /14/ that the observed perturbations are caused
by the coupling of light rotational states of the 14^1 vibronic state to dark rovibronic states in the electronic S_1
state. As a result two eigenstates with mixed vibronic character result and both can be seen in the spectrum. These
are the two lines observed in the experimental spectrum as
indicated in Fig. 3. From the positions of the two lines
the coupling matrix element can be calculated for each pair
/13/. The coupling shows a strong dependence on J and K.
This leads to the conclusion that the observed coupling
must be caused by perpendicular Coriolis coupling rather
than by anharmonic or parallel Coriolis coupling /13/.

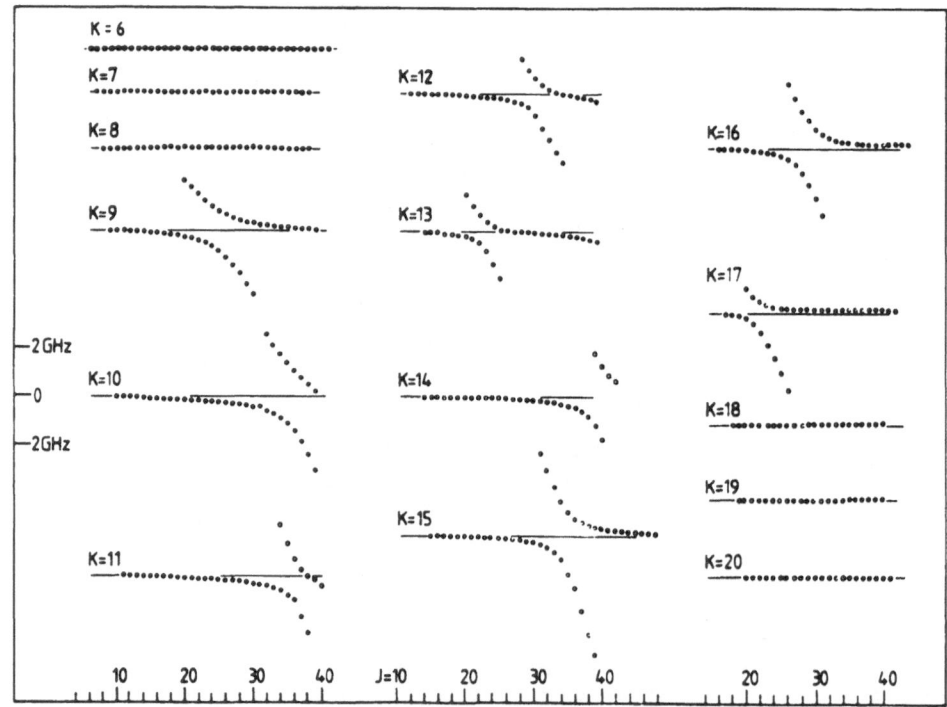

Figure 4.   Residuals (calculated - observed) of the frequencies of rotational lines in the spectrum of the 14^1_0-
band as a function of the final state quantum number J'
for several values of K'.

3.1.2. Decay behavior of individual levels

With the pulsed set up individual rotational levels in the 14^1 vibronic state can be populated and their decay behavior observed under collisionless conditions /9/. For all states that were found unperturbed in the spectroscopic analysis a single exponential decay is found and the decay time of $\tau = 135$ ns is independent of the rotational quantum numbers J and K. This is in good agreement with the assumption that the decay of the levels is determined by the radiative decay and more strongly by the coupling to the T_1 triplet state in the statistical limit /15/. On the contrary, for states found to be perturbed a significantly shorter decay time is found /9/. For example, the $J_K = (24_9)_a$ state resulting from the coupling of the $J_K = 24_9$ rotational state of the 14^1 vibronic state to a dark rovibronic state has a decay time of only 87 ns. The decay is again single exponential. It can be concluded that the dark zero order state decays much faster and due to the mixed vibronic nature of the observed eigenstate its shorter decay time results. It is seen that there exist vibrational states in S_1 at the vibrational excess energy of the light state that possess much faster decay rates. These states can not be directly excited and their observation is only possible through the analysis of perturbations. It will turn out that the existence of shortlived background states is important for the interpretation of observations at higher excess energy that are discussed below.

3.2. $14^1_0 1^2_0$-Band of C_6H_6, typical for intermediate excess energy

The nonradiative decay rate of S_1 states generally increases with vibrational excess energy. This increase is found to be fairly slow in benzene for energies below 3000 cm^{-1} but at about this energy a sudden strong decrease of the fluorescence quantum yield was found from low resolution experiments /15/. It is accompanied by the indication of severe broadening in the absorption spectrum /2/. Since none of the known radiative and nonradiative decay channels of benzene was believed to explain this behavior, it was attributed to an unknown "channel three" /2/. None of the Doppler limited measurements was able to resolve the rotational structure of the vibronic bands and therefore no information on the rotational dependence of "channel three" was available from the above mentioned experiments.

Doppler-free two-photon spectra in the vicinity of the onset of "channel three" can be recorded for progression and sequence bands of the 14^1_0-band discussed above. For a progression band the rotational structure should be iden-

tical to that of the fundamental band except for small
changes in the exact positions of the rotational lines due
to small changes in the rotational constants. This is for
example the case for the 14^1_0-band (at 1571 cm^{-1} excess
energy) and the $14^1_0 1^1_0$-band (2492 cm^{-1}). The transitions
of both bands lead to states below the onset of "channel
three". On the contrary, the upper state of the $14^1_0 1^2_0$-band
is the $14^1 1^2$ state at 3412 cm^{-1} just above the onset of
the postulated "channel three". A precise analysis of this
band should therefore render important information about
the origin of the fast nonradiative decay.

3.2.1. <u>Disappearance of rotational lines</u>   The Doppler-
free spectrum of the blue edge of the Q-branch of the
$14^1_0 1^2_0$-band is shown in Fig. 5 /8/. This spectrum was measured with the cw set up at a resolution of about 15 MHz.
The spectral range shown and the frequency scale is identical with the part of the 14^1_0-band shown at the top of
Fig. 2. A comparison of the two spectra immediately shows
that most of the rotational lines are missing in the
$14^1_0 1^2_0$-band /8,16/. To understand this surprising result

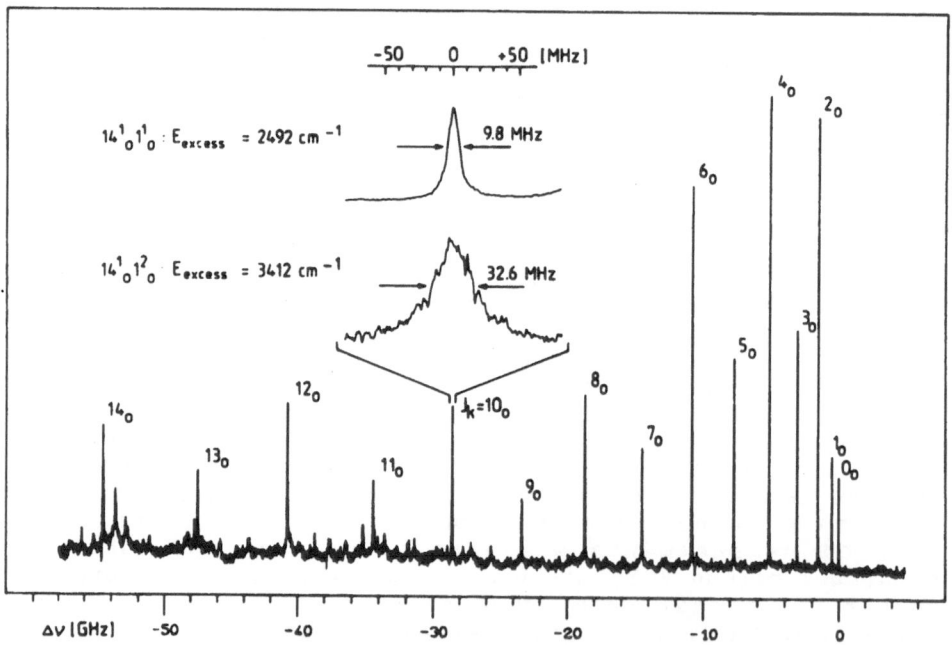

Figure 5.   Part of the Doppler-free two-photon spectrum
of the $14^1_0 1^2_0$-band of $C_6 H_6$ (taken from ref. /8/). The
lineshapes of the $J_K = 10_0$ line of the $14^1_0 1^1_0$-band and the
$14^1_0 1^2_0$-band are shown on an expanded scale.

it has to be kept in mind that the Doppler-free spectra reported are not absorption spectra but fluorescence excitation spectra. The two kinds of spectra would be identical if the fluorescence quantum yield were identical for all states excited. For this reason the disappearance of lines in the $14^1_0 12_0$-band is interpreted as a lack of fluorescence from the states populated, i.e. a fast rotationally dependent nonradiative decay.

From the position of the lines and the alternating intensity the remaining lines in the spectrum of the $14^1_0 12_0$-band were identified as K=0 lines /16/. From the disappearance of the K≠0 states we concluded that they are coupled to dark background states that rapidly decay. The coupling was identified as parallel Coriolis coupling with strength proportional to K /16/.

3.2.2. Homogeneous linewidths and decay time measurements of K=0 states

The extremely high resolution of the cw set up allows the measurement of homogeneous linewidths at very low sample pressure /8/. If the homogeneous linewidth of identical rotational lines from the $14^1_0 11_0$-band and the $14^1_0 12_0$-band are compared, a significant increase in linewidth is found for the $14^1_0 12_0$-band. This is shown for the $J_{K=10_0}$ line in the inserts of Fig. 5. While the linewidth is still experimentally limited for the $14^1_0 11_0$-band, homogeneous linewidths were measured for K=0 lines in the $14^1_0 12_0$-band for different J values /8/. These are found to increase from 2 MHz for J=0 to 46 MHz for J=14. All the observed lines have a Lorentzian lineshape within experimental accuracy.

The pulsed set up allows the measurement of decay curves for the same states (K=0) in the $14^1 12$ vibrational state /12/. For the states with low J value pulses whose width is close to 10 ns were used to ensure the highest possible resolution while for higher J values pulses of 2.5 ns duration allowed the resolution of fast decays. Typical results are shown in Fig. 6. All decay curves are single exponential and the decay times range from $\tau \gtrsim 55$ ns for J=0 to $\tau = 7.1$ ns for J=8. The observed decay rates closely agree with the relaxation rates obtained from the linewidth measurements discussed above.

The strong dependence of the decay rate on J can not be due to a rotational dependence of the pure electronic radiative or nonradiative decay rate of the $14^1 12$ state since such a dependence was not found for the 14^1 state and for this reason is not expected for the $14^1 12$ state either. Instead it has to be interpreted as a rotationally dependent IVR process which mixes the optically light state with

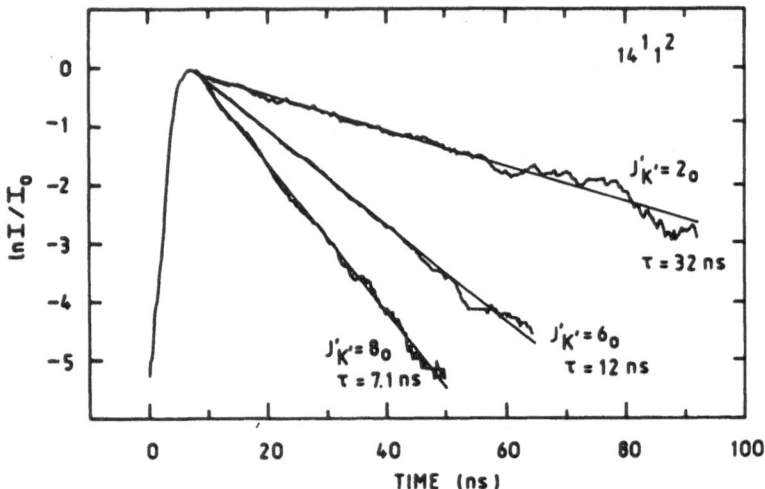

Figure 6.   Decay curves for K=0 rotational states (differing in J) of the $14^1 1^2$ vibronic state of C_6H_6 (taken from ref. /12/).

dark states in S_1 that rapidly decay nonradiatively. From the exact J dependence seen the underlying responsible coupling process is found to be perpendicular Coriolis coupling /8,12/.

3.2.3.   Model for the observed coupling in the $14^1 1^2$-state

From the experimental results reported above a model for the coupling of the $14^1 1^2$-state and its relaxation behavior evolves that is shown systematically in Fig. 7. The rotational states ($0 \leq J \leq 14$) of the $14^1 1^2$ vibronic state (shown in the middle of Fig. 7) are coupled to the rotational states of two different dark vibrational states in S_1 by different coupling mechanisms. Parallel Coriolis interaction leads to a coupling with a vibrational state of b_{1u} symmetry (shown in the right part of Fig. 7) and perpendicular Coriolis interaction leads to a coupling with a vibrational state of e_{2u} symmetry (shown in the left part of Fig. 7). The background states themselves are strongly coupled to the quasicontinuum of vibrational states in the S_0 and/or T_1 electronic state and therefore decay very fast and show strong broadening /17/. The probable reason for this strong coupling is that the dark background states are combination states containing quanta of out of plane modes /8,12,16/. These modes are the lowest frequency modes in benzene and are contained with a high probability

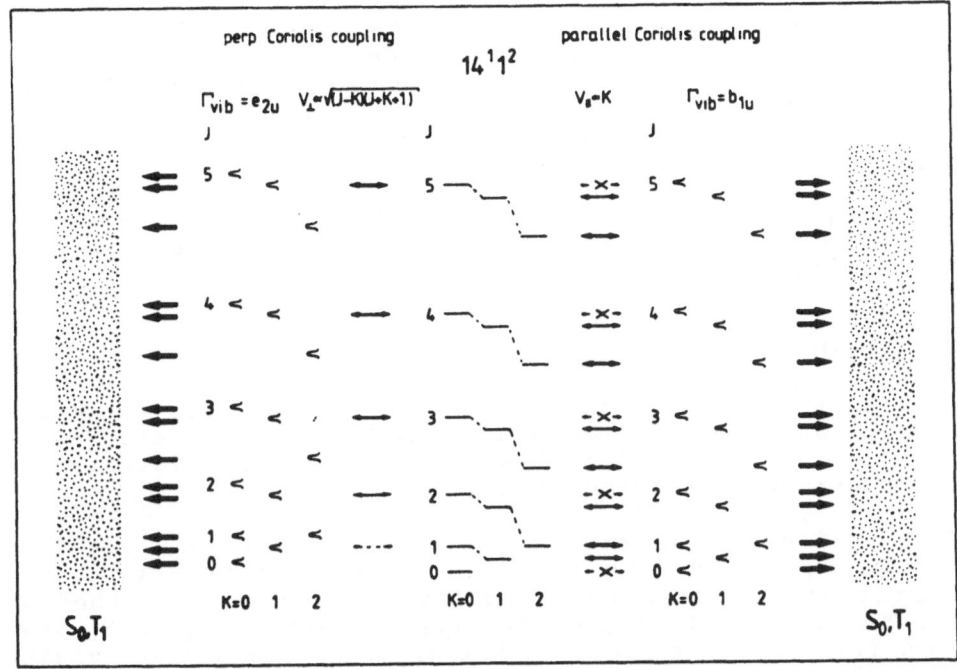

Figure 7. Schematic representation of the coupling of the low J,K rotational states of the $14^1 1^2$ vibronic state of C_6H_6 (middle part) to dark broadened states in S_1 (outer parts). The coupling is due to parallel Coriolis coupling (right part) and perpendicular Coriolis coupling (left part).

in the background states at the excess energy of 3412 cm^{-1}. They are known to strongly increase the nonradiative decay of a state as they are good accepting modes [18]. The eigenstates that result from the mixing with the rapidly decaying dark background states therefore decay themselves very rapidly. Their fluorescence quantum yield is very low and the $K \neq 0$ lines are not seen in the spectrum. $K=0$ lines are not affected by this coupling since the strength of parallel Coriolis coupling is proportional to K. Due to the $\Delta J, \Delta K = 0$ selection rule of parallel Coriolis coupling [19] the resonance condition for the coupling is automatically fullfilled for all states with different J and K if there exists a single vibrational background state at resonance with the $14^1 1^2$ zero order state. Only for very high J,K states the resonance condition is lost due to the slightly differing rotational constants of both coupled states. This is indeed observed for $J > 14$ and a different strong coupling process (perpendicular Coriolis coupling) is active in this range of the spectrum [8].

The second coupling mechanism in the low J range is weaker
and only seen for the remaining K=0 states. These states
are supposed to be in additional resonance with the rotational states of a vibrational state of e_{2u} symmetry (shown
in the left part of Fig. 7) which are again strongly broadened due to the reasons discussed above. The pairs of
states are coupled by perpendicular Coriolis coupling which
for K=0 is proportional to $\sqrt{J(J+1)}$ /19/. This causes a J
dependent mixing of the states and explains the observed
J(J+1) dependence of the decay rate of the K=0 states (see
Fig. 6) /8,12/.

4.   SUMMARY AND CONCLUSION

Doppler-free two-photon spectroscopy is shown to allow the
resolution of individual rotational lines in the electronic
spectrum of large molecules. This result is demonstrated
for the prototype molecule benzene.

In the low excess energy regime the rotational spectrum
can be explained with a simple semirigid symmetric top
Hamiltonian. Isolated perturbations can be explained by
the selective coupling of the light states to dark S_1 states. This coupling causes a mixing of the zero order states
and makes both resulting eigenstates observable. This permits the characterization of the dark state and its decay
behavior.

Above the onset of the postulated "channel three" (in the
$14^1_0 1^2_0$-band) most rotational lines are missing. From the
analysis of the spectrum and from linewidth and lifetime
measurements a detailed model about the relaxation of the
different states is found. It is concluded that rotational
dependent intramolecular vibrational redistribution (IVR)
induced by Coriolis coupling is the primary process for
the nonradiative decay of the $14^1 1^2$-state. Other vibrational states in the same excess energy range will   be
affected in a similar way, however, the exact appearance
of the vibronic bands will depend on the accidental positions of states, the strength of the coupling matrix elements and the selection rules of the particular coupling
mechanism active. The analysis of many bands and different
molecules with the Doppler-free high resolution technique
presented in this work /20/ should allow the complete understanding of the nonradiative behavior of large molecules and
in particular of the "channel three" phenomenon of benzene.

The results presented show the importance of rotation for
the nonradiative decay of large molecules. As the rotational structure of the electronic spectrum can only be

resolved by Doppler-free techniques, Doppler-free spectroscopy is seen to be of great importance for an exact understanding of nonradiative decay.

5.   ACKNOWLEGEMENT

We wish to thank Prof. Dr. E. W. Schlag for his contributions to the results presented in this paper. Valuable experimental support by H. Stepp, U. Schubert and H. Sieber is greatfully acknowledged.

REFERENCES

1.   For a review see, K. F. Freed in: Topics in Applied
      Physics, ed. F. K. Fong, Vol. 15, p. 23 ff., Springer,
      Berlin, 1976

2.   J. H. Callomon, J. E. Parkin, and R. Lopez-Delgado,
      Chem. Pys. Lett. 13, 125 (1972)

3.   L. Wunsch, H. J. Neusser, and E. W. Schlag, Z.
      Naturforsch. Teil A 36, 1340 (1981)

4.   Compare for example: K. G. Spears and S. A. Rice, J.
      Chem. Phys. 55, 5561 (1971); L. Wunsch, H. J. Neusser,
      and E. W. Schlag, Chem. Phys. Lett. 32, 210 (1975);
      M. Sumitani, D. V. O'Connor, Y. Takagi, N. Nakashima,
      A. Kamagawa, Y. Udagawa, and K. Yoshihara, Chem.
      Phys. 93, 359 (1985)

5.   B. E. Forch, K. T. Chen, H. Saigusa, and E. C. Lim,
      J. Phys. Chem. 87, 2280 (1983); P. M. Felker and A.
      H. Zewail, J. Chem. Phys. 82, 2994 (1985); A. Amirav
      and J. Jortner, J. Chem. Phys. 84, 1500 (1986)

6.   For a review see, W. Demtröder, "Laser Spectroscopy",
      Springer, Berlin, 1981

7.   E. Riedle, H. J. Neusser, and E. W. Schlag, J. Chem.
      Phys. 75, 4231 (1981); K. H. Fung and D. A. Ramsay,
      J. Phys. Chem. 88, 395 (1984); A. Kiermeier, K.
      Dietrich, E. Riedle, and H. J. Neusser, J. Chem.
      Phys., in press

8.   E. Riedle and H. J. Neusser, J. Chem. Phys. 80, 4686
      (1984)

9.   U. Schubert, E. Riedle, and H. J. Neusser, J. Chem.
      Phys. $\underline{84}$, 5326 (1986)

10.   L. S. Vasilenko, V. P. Chebotayev, and A. V. Shishaev,
      JETP Letters $\underline{12}$, 113 (1970)

11.   E. Riedle, R. Moder, and H. J. Neusser, Opt. Commun.
      $\underline{43}$, 388 (1982)

12.   U. Schubert, E. Riedle, H. J. Neusser, and E. W.
      Schlag, J. Chem. Phys. $\underline{84}$, 6182 (1986)

13.   E. Riedle and H. J. Neusser, in preparation

14.   E. Riedle, H. Stepp, and H. J. Neusser, Chem. Phys.
      Lett. $\underline{110}$, 452 (1984)

15.   For a review see, C. S. Parmenter, Adv. Chem. Phys.
      $\underline{22}$, 365 (1972)

16.   E. Riedle, H. J. Neusser, and E. W. Schlag, J. Phys.
      Chem. $\underline{86}$, 4847 (1982)

17.   A. Nitzan, J. Jortner, and P. M. Rentzepis, Proc. R.
      Soc. London Ser. A $\underline{327}$, 367 (1972); F. Lahmani, A.
      Tramer, and C. Tric, J. Chem. Phys. $\underline{60}$, 4431 (1974)

18.   H. Hornburger and J. Brand, Chem. Phys. Lett. $\underline{88}$,
      153 (1982)

19.   I. M. Mills, Pure Appl. Chem. $\underline{11}$, 325 (1965)

20.   Compare for example recent measurements by our group,
      for which a theoretical interpretation is presented
      by S. F. Fischer and W. Dietz in this volume

CONICAL INTERSECTIONS AND ULTRAFAST RADIATIONLESS DECAY

W. Domcke
Institute of Physical and Theoretical Chemistry
Technical University of Munich
D-8046 Garching
West Germany

H. Köppel and L. S. Cederbaum
Theoretical Chemistry Group
Institute of Physical Chemistry
University of Heidelberg
D-6900 Heidelberg
West Germany

ABSTRACT. Theoretical methods to describe the nuclear dynamics on intersecting multidimensional potential energy surfaces are reviewed. It
is pointed out that conical intersections generally cause an ultrafast
electronic population decay on a femtosecond time scale. The spectroscopy and the time-dependent dynamics of the strongly interacting $\tilde{B}^2E_{2g}$
and $\tilde{C}^2A_{2u}$ states of $C_6H_6^+$ are considered to illustrate the general ideas.

1. INTRODUCTION

In polyatomic molecules the radiationless decay of excited electronic
states is an ubiquitous phenomenon. Depending on the size and symmetry
of the molecule and the nature of the electronic states involved in the
process, the transition rates vary over a wide range from about 10^{14} s^{-1}
to about 10^3 s^{-1}. In the present article we are concerned with the very
fastest of these nonradiative decay processes, with lifetimes in the
sub-picosecond range, which may be termed ultrafast relaxation processes.
    When the population of an electronic state decays on a sub-pico-
second timescale, the radiative decay cannot compete, even for strongly
allowed transitions, and the fluorescence quantum yield will be unmea-
surably small. A complete absence of fluorescence (i.e., quantum yield
$\lesssim 10^{-4}$) is indeed observed for many polyatomic radical cations /1/ and
provides at least indirect evidence for ultrafast electronic relaxation
processes in these species. The direct experimental observation of time-
dependent population decay on a timescale of a few femtoseconds is not
possible at present; it should be mentioned, however, that time-resolved
measurements of the population decay of Rydberg states of benzene with
lifetimes as short as 70fs have been reported recently /2/.

R. Lefebvre and S. Mukamel (eds.), Stochasticity and Intramolecular Redistribution of Energy, 217–231.
© 1987 by D. Reidel Publishing Company.

When direct dissociation of the system can be excluded, i.e., when the energy of the initially populated states is below the dissociation threshold of the corresponding electronic potential energy surface, ultrafast decay requires a very strong non-Born-Oppenheimer (BO) coupling to some other electronic state. This in turn implies a near - degeneracy of the electronic state under consideration with another electronic state of the same spin multiplicity in at least some region of vibrational coordinate space. Recent work has shown that conical intersections /3/ of multidimensional electronic potential energy surfaces provide a mechanism for such ultrafast relaxation processes /4-6/. The important vole of conical intersections in the unimolecular decay of a number of radical cations has been pointed out by Lorquet and coworkers /4,7-9/. Köppel and coworkers have treated the quantum dynamics of simple conical-intersection models representing small polyatomic systems and have demonstrated the ultrafast electronic population decay in these systems /5,6/. These are the first attempts to calculate radiationless decay rates in polyatomic molecules on a microscopic basis, i.e., starting from ab initio calculated electronic potential energy surfaces and wave functions. Here we give a brief review of these ideas and report preliminary results of calculations for a conical intersection in a somewhat larger polyatomic system, the benzene cation.

2.   CONICAL INTERSECTIONS

The general concept of conical intersections of adiabatic electronic potential energy surfaces has been introduced by Herzberg and Longuet-Higgins /3/. They have pointed out, in particular, that the occurence of an intersection is not necessarily a consequence of symmetry as in the case of the well-known Jahn-Teller (JT) effect /10/. Conical intersections may exist also in completely nonsymmetric systems such as a triatomic molecule composed of three different atoms /3/.

The simplest way to understand the topology of conical intersections is to consider the following example of a partially symmetry-induced conical intersection. Consider two nondegenerate electronic states $|\Phi_g\rangle$, $|\Phi_u\rangle$ with symmetry species Γ_g, Γ_u in the molecular point group. (We use here the labels "g" and "u" as a mnemotechnical device to indicate the different symmetry of these states; in practice, the irreducible representations of the electronic states may differ by any other symmetry label.) The corresponding potential energy surfaces V_g, V_u in the space of totally symmetric coordinates Q_{gi}, $i=1,2,...,$ may cross along any of the Q_{gi}, since the states are of different symmetry. Consider next the set of vibrational coordinates Q_{uj}, $j=1,2...,$ with symmetry species Γ_Q such that

$$\Gamma_g \times \Gamma_Q \times \Gamma_u \supset \Gamma_A , \qquad (1)$$

where Γ_A denotes the totally symmetric representation. Eq.(1) is the necessary condition that the matrix element

$$\lambda_j = <\phi_g | (\partial H_{el}/\partial Q_{uj}) | \phi_u> \tag{2}$$

is nonzero, H_{el} being the electronic Hamiltonian. This implies that a possible degeneracy of V_g and V_u is lifted in first order in Q_{uj}. Any crossing of V_g and V_u along a totally symmetric mode Q_{gi} is thus converted into a conical intersection in Q_{gi}, Q_{uj} space. We may call the Q_{uj} the coupling modes, since they are responsible for the vibronic coupling according to eq.(2), and the Q_{gi} the tuning modes, since they "tune" the energy separation of $|\phi_g>$ and $|\phi_u>$ and provide for the possibility of exact degeneracies.

A general classification of potential-surface intersections, in particular for triatomic systems, has been given by Carrington /11/ and Davidson /12/. A variety of examples of conical intersections found by ab initio electronic structure calculations is given in ref. /13/.

3. NUCLEAR MOTION ON INTÉRSECTING SURFACES: SPECTROSCOPY AND DYNAMICS

The nuclear dynamics associated with conically intersecting surfaces is genuinely a multi-mode phenomenon, involving at least two, in practice often many more, vibrational degrees of freedom. The treatment of multi-mode molecular dynamics on coupled electronic surfaces clearly presents a challenging computational problem. The problem is strongly nonseparable in the vibrational modes and excludes the application of vibronic perturbation theory, since both small as well as large energy gaps of the interacting states are involved. One way of attack is to replace the problem by a model problem which is sufficiently simplified to allow an efficient numerical solution, but still contains the essential physics of the original problem. Such a model is obtained with the following simplifications, which are fairly standard in the theory of vibronic coupling of polyatomic molecules (see /6/ and references therein): (i) the model hamiltonian is constructed in a diabatic electronic basis, i.e., we consider potential-energy coupling terms, but neglect residual couplings via the nuclear kinetic energy operator T_N; (ii) the harmonic approximation is adopted for the diabatic potentials, and (iii) only linear (in the normal coordinates) interstate coupling terms are retained. A less essential, but convenient, approximation is to assume identical vibrational frequencies in the different unperturbed electronic states.

Assuming that the interacting states $|\phi_g>$, $|\phi_u>$ are excited (or ionized) states of the molecule and employing (dimensionless) normal coordinates of the electronic ground state $|\phi_o>$, the situation described in section 2 is represented by the following model hamiltonian

$$\mathcal{H} = H_o \underline{1} + \begin{pmatrix} E_g + \sum\limits_i \kappa_j^{(g)} Q_{gi} & \sum\limits_j \lambda_j Q_{uj} \\ \\ \sum\limits_j \lambda_j Q_{uj} & E_u + \sum\limits_i \kappa_i^{(u)} Q_{gi} \end{pmatrix} \tag{3}$$

where $\underline{1}$ denotes the 2x2 unit matrix and $H_o = T_N + V_o$ is the multidimensional harmonic oscillator describing the normal vibrations v_{gi}, v_{uj} in the electronic ground state

$$H_o = \sum_i (-\frac{\omega_{gi}}{2} \frac{\partial^2}{\partial Q_{gi}^2} + \frac{\omega_{gi}}{2} Q_{gi}^2) +$$

$$+ \sum_j (-\frac{\omega_{uj}}{2} \frac{\partial^2}{\partial Q_{uj}^2} + \frac{\omega_{uj}}{2} Q_{uj}^2) \tag{4}$$

E_g and E_u are the vertical excitation (or ionization) energies of the electronic states. The parameters $\kappa_i^{(g,u)}$ are given by the gradients of the excitation energies with respect to the totally symmetric modes and may be termed intra-state electron-vibrational coupling constants, while the λ_j are the inter-state vibronic coupling constants /6/.

It is important to realize that the parameters of the model hamiltonian (3), in particular the coupling constants κ_i and λ_j, can be directly obtained from ab initio calculations. Indeed, by diagonalizing (3) in the fixed-nuclei limit, $T_N = 0$, we obtain the excited-state potential-energy surfaces V_1, V_2 of the model, which may be fitted to ab initio calculated surfaces. A simple analysis shows /6/ that the coupling constants are given by

$$\kappa_i^{(k)} = (\partial V_k / \partial Q_{gi})_{Q=0}, \quad k = g, u \tag{5}$$

$$\lambda_j = (\sqrt{\frac{1}{2} \frac{\partial^2}{\partial Q_{uj}^2} (V_1 - V_2)^2})_{Q=0} \tag{6}$$

Note that $V_{g,u} \equiv V_{1,2}$ for $Q_u = 0$.

To determine the vibronic energy levels and eigenstates of (3), we expand the v-th vibronic state $|\Psi_v \rangle\rangle$ in a direct-product basis of electronic ($|\phi_g\rangle, |\phi_u\rangle$) and harmonic oscillator ($|v_{gi}\rangle, |v_{uj}\rangle$) basis states (we use double kets to designate states in the full (electronic plus vibrational) Hilbert space, while single kets designate states in the respective subspaces)

$$|\Psi_v \rangle\rangle = \sum c^v_{g, v_{g1} \ldots v_{u1} \ldots} |\phi_g\rangle |v_{g1}\rangle \ldots |v_{u1}\rangle \ldots$$

$$+ \sum c^v_{u, v_{g1} \ldots v_{u1} \ldots} |\phi_u\rangle |v_{g1}\rangle \ldots |v_{u1}\rangle \ldots \tag{7}$$

The vibronic energy levels and states are then given by the eigenvalues and eigenvectors, respectively, of the secular problem

$$(\mathbf{\mathcal{H}} - E_v \underline{1}) \underline{c}^v = 0 \tag{8}$$

where $\underline{\underline{\chi}}$ is the matrix representation of (3) in the direct-product basis. Since each direct-product basis state in (7) possesses overall g or u symmetry, the secular problem (8) decouples into two sub-problems, corresponding to vibronic states of g and u symmetry, respectively.

Once (8) has been solved for all eigenvalues and eigenvectors, we have complete information about the spectroscopy of the model system. As an example, let us consider the absorption spectrum, i.e., the spectrum corresponding to an optical transition from the unperturbed electronic and vibrational ground state $|\phi_o>|\underline{0}>$ into the manifold of vibronic states. Here $|\underline{0}>$ stands for the product of the ground states of the individual vibrational modes. The line positions in the spectrum are determined by the vibronic energy levels E_ν. The corresponding line intensities are (assuming the Condon approximation in the diabatic representation)

$$I_\nu = |\tau_{og}|^2 \, |<\underline{0}|<\phi_g|\Psi_\nu>>|^2 +$$
$$+ \, |\tau_{ou}|^2 \, |<\underline{0}|<\phi_u|\Psi_\nu>>|^2 \qquad (9)$$

where τ_{og} und τ_{ou} are the electronic transition moments from $|\phi_o>$ to $|\phi_g>$ and $|\phi_u>$, respectively. It is seen from (9) that the <u>first compo-nent</u> of the eigenvectors of (8) determines the line intensity in the absorption spectrum.

The absorption spectrum contains the information on the nonradiative decay of the excited state only indirectly via the "linewidth" of the BO vibrational levels. It is therefore of interest to look directly at the time evolution of the wave function of the system. Such a time-dependent description requires the definition of an initially prepared excited state $|\Psi(0)>>$ which is not an eigenstate of the Hamiltonian. This preparation depends, of course, on the experiment under consideration. Being primarily interested in the dynamics of cations, we consider photoionization as a conceptually simple and experimentally easily realizable process. In photoionization with high-energy photons (excluding thus threshold phenomena and autoionization processes) without energy analysis of the photoelectrons, the prepared state is a coherent superposition of the optically accessible electronic states multiplied by the initial vibrational wave function, in our case

$$|\Psi(0)>> = \tau_{og}|\phi_g>|\underline{0}> + \tau_{ou}|\phi_u>|\underline{0}> \qquad (10)$$

With energy analysis of the photoelectrons and assuming non-overlapping electronic bands in the photoelectron spectrum, we may prepare either of the cationic states, i.e.

$$|\Psi(0)>> = \tau_{og}|\phi_g>|\underline{0}> \text{ or } |\Psi(0)>> = \tau_{ou}|\phi_u>|\underline{0}>. \qquad (11)$$

After preparation, the system propagates on the coupled electronic surfaces of the cation and the time-dependent vibronic wave function is

$$|\Psi(t)>> = e^{-i\underline{\underline{\chi}}t}|\Psi(0)>>. \qquad (12)$$

Other preparation processes defining a different $|\Psi(0)\rangle\rangle$ are of course conceivable, e.g. narrow-band laser excitation from the ground state of the cation into one of the excited states. It should be stressed, however, that the usual concept of preparation of BO vibrational levels of the excited state /14/ is not applicable for conical intersections, since such levels cannot be identified above the minimum energy of the locus of intersection.

An obviously useful measure to test the time dependence of $|\Psi(t)\rangle\rangle$ is the autocorrelation function

$$\chi(t) = \langle\langle\Psi(0)|e^{-i\mathcal{H}t}|\Psi(0)\rangle\rangle, \tag{13}$$

i.e., the time-dependent overlap of the vibronic wave packet with its initial form. Since the Fourier transform of $\chi(t)$ is the absorption spectrum, $\chi(t)$ gives no information which is not already contained in the former. The "experimental" determination of $\chi(t)$ by Fourier transformation of the measured absorption spectrum has been discussed by Lorquet et al. /15/.

In a photoionization experiment with the inital-state preparation (10) or (11) the autocorrelation function does not directly provide information on the radiationless decay, since the inital vibrational state $|0\rangle$ is in general not a stationary state on either excited-state potential energy surface. The corresponding wave packet will thus start to oscillate and $\chi(t)$ contains the information both on this vibrational motion (which occurs also in the absence of vibronic coupling) as well as the population decay of the excited electronic states.

More direct information on the nonradiative decay is obtained from the population probability $P_k(t)$ of the k-th (diabatic) electronic state

$$P_k(t) = \langle\langle\psi(t)|\Phi_k\rangle\langle\Phi_k|\psi(t)\rangle\rangle. \tag{14}$$

If we have prepared the cation in $|\Phi_k\rangle$ at t=0 with unit probability, eq.(14) gives the survival probability of this population at time t. Clearly $P_k(t) \equiv 1$ in the absence of vibronic coupling and radiative decay. Under certain conditions, $P_k(t)$ is a directly measurable quantity. Assume, for example, that the electronic ground state $|\Phi_o\rangle$ of the cation is energetically well separated and not involved in the vibronic coupling. Assume, furthermore, that fluorescence from $|\Phi_k\rangle$ to $|\Phi_o\rangle$ is allowed, the other electronic transitions being dipole-forbidden. Then $P_k(t)$ gives the rate of fluorescence (normalized to unity at t=0) at time t /31/.

The most efficient way to calculate $\chi(t)$ and $P_k(t)$ for short times (i.e., a few vibrational periods) is to solve directly the time-dependent Schrödinger equation. This may be done by representing $\mathcal{H}$ and $|\Psi(t)\rangle\rangle$ in the vibronic direct-product basis introduced above and using a finite-difference method for propagation in time, see, e.g., /16/.

These concepts and numerical methods have been applied in great detail to the X-A conical intersection in the ethylene cation /5,6,16/ which represents a nonseparable three-mode problem (two tuning modes, one coupling mode). It has been shown that this intersection leads to a population decay of the $\tilde{A}$ state on a 10fs time scale. The rather good agreement of the calculated photoelectron spectrum with the experimental spectrum /17,18/ indicates that the resulting picture, in particular the ultrafast nonradiative decay, is basically correct.

A particularly interesting recent development is the application of the Miller-McCurdy-Meyer classical analog model /19/ to the ultrafast nonradiative decay problem by Meyer /20/. It has been shown that this classical model, in which the electronic degrees of freedom are described by classical action-angle variables, reproduces the population probability of the full quantum calculation for $C_2H_4^+$ with excellent accuracy /20/. Since the numerical effort involved in the classical calculation grows much less with the number of vibrational degrees of freedom than the effort of the quantum calculation, Meyer's result indicates the possibility of accurate dynamical calculations for multi-mode conical intersection problems. The classical analog model has recently also been applied to the ExE Jahn-Teller problem by Ezra and coworkers /21/.

4. THE $\tilde{B}$-$\tilde{C}$ CONICAL INTERSECTION IN THE BENZENE CATION

Over the years we have investigated the nuclear dynamics associated with conical intersections in a variety of polyatomic cations ($C_nH_4^+$ with n=2,3,4,5, HCN^+, BF_3^+) and open-shell molecules (NO_2) /6/. These are still rather small systems and it suggests itself to inquire wether the concepts and numerical methods outlined above can also be fruitfully applied to larger polyatomic systems. We have selected the benzene cation for this purpose as an interesting and prototypical system. We report here preliminary results of quantum dynamical calculations for the intersecting $\tilde{B}^2E_{2g}$ and $\tilde{C}^2A_{2u}$ states of $C_6H_6^+$.

The three lowest electronic states of $C_6H_6^+$ are the $\tilde{X}^2E_{1g}$, $\tilde{B}^2E_{2g}$ and $\tilde{C}^2A_{2u}$ states, in conventional notation /22/. It is known from photoelectron spectroscopy that the $\tilde{B}$ and $\tilde{C}$ states are close in energy and that the $\tilde{C}$ band is completely diffuse /23/ which may be taken as an indication of strong vibronic interaction with the $\tilde{B}$ state. Of particular interest is the fact that $C_6H_6^+$ is non-fluorescent (quantum yield $<4\times10^{-5}$ /22/) although the $\tilde{C}$-$\tilde{X}$ transition is strongly allowed (the $\tilde{C}$-$\tilde{B}$ and $\tilde{B}$-$\tilde{X}$ transitions are dipole forbidden) and the lower part of the $\tilde{C}$ band is not predissociated /22/. This implies the existence of a very efficient nonradiative and non-dissociative relaxation mechanism which depopulates the $\tilde{C}$ state. Since sym-$C_6F_3H_3^+$ and $C_6F_6^+$, where the order of the two states is reversed with respect to $C_6H_6^+$, are fluorescent /24/, it is reasonable to assume that the $\tilde{C}$-$\tilde{B}$ interaction plays a decisive role in the radiationless relaxation process in $C_6H_6^+$.

Braitbart et al. /22/ have postulated on the basis of Franck-Condon factors observed in the photoelectron spectrum and bonding properties of the molecular orbitals of benzene that the potential energy curves of the $\tilde{C}$ and $\tilde{B}$ states should intersect along the totally symmetric C-C and C-H stretching coordinates. This conjecture is confirmed by more quantitative ab initio SCF molecular orbital gradient calculations. Figure 1 shows the harmonic potential energy curves of

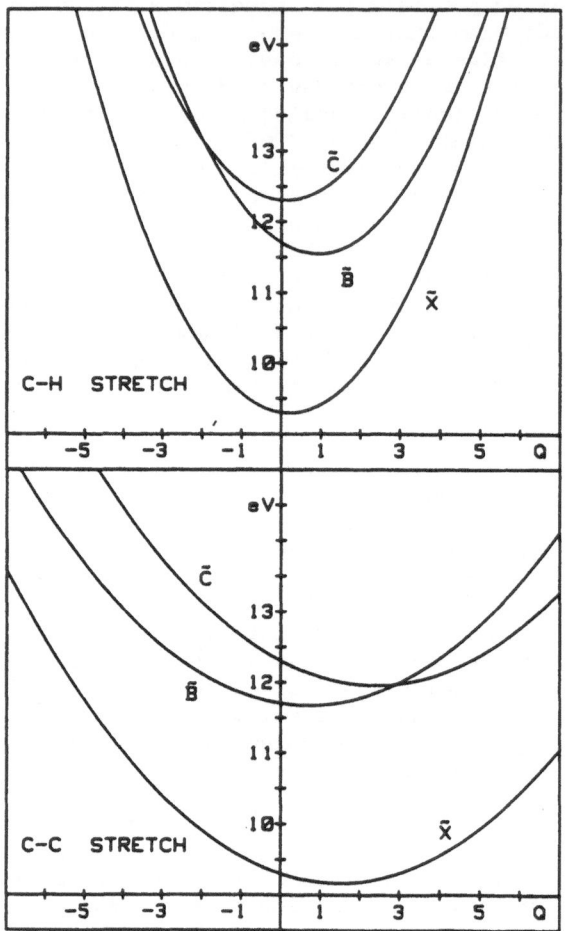

Figure 1.   Potential energy curves of the three lowest electronic
            states of the benzene cation as a function of the C-H
            stretch and C-C stretch normal coordinates.

the $\tilde{X},\tilde{B},\tilde{C}$ states as a function of the C-C and C-H stretch normal coordinates constructed from <u>ab</u> <u>initio</u> SCF electron-vibrational coupling
constants $\kappa_i^{(k)}$ of ref. /25/. These coupling constants have also been
determined by Lipari et al. /26/ using semiempirical (CNDO/S) methods
and by Pulay et al. /27/ using <u>ab</u> <u>initio</u> SCF methods. It is seen that
stretching of the C-C normal coordinate leads to an intersection of
the $\tilde{B}$ and $\tilde{C}$ states very close to the minimum of the latter.
        The vibrational modes which can couple the $\tilde{B}$ and $\tilde{C}$ states accor-
ding to the symmetry selection rule (1) are the degenerate out-of-plane
modes ν_{19},ν_{20} (Herzberg notation) of E_{2u} symmetry. The inter-state

vibronic coupling involving degenerate modes is usually termed pseudo-Jahn-Teller (PJT) effect. Introducing, as usual, polar coordinates r_j, ϕ_j for the degenerate modes, the PJT hamiltonian for the $\tilde{B}-\tilde{C}$ system reads (see, e.g., /6/)

$$\mathcal{H}_{PJT} = H_o \underset{\sim}{1} + \begin{pmatrix} E_g + \sum\limits_i \kappa_i^{(g)} Q_{gi} & \sum\limits_j \lambda_j r_j e^{i\phi j} & 0 \\ \sum\limits_j \lambda_j r_j e^{-i\phi j} & E_u + \sum\limits_i \kappa_i^{(u)} Q_{gi} & \sum\limits_j \lambda_j r_j e^{i\phi j} \\ 0 & \sum\limits_j \lambda_j r_j e^{-i\phi j} & E_g + \sum\limits_i \kappa_i^{(g)} Q_{gi} \end{pmatrix}$$

(15a)

$$H_o = \sum_i \frac{\omega_{gi}}{2}\left(-\frac{\partial^2}{\partial Q_{gi}^2} + Q_{gi}^2\right) + \sum_j \frac{\omega_j}{2}\left(-\frac{1}{r_j}\frac{\partial}{\partial r_j}\frac{1}{r_j}\frac{\partial}{\partial r_j} - \frac{\partial^2}{\partial \phi_j^2} + r_j^2\right)$$

(15b)

where E_g and E_u are the vertical ionization potentials at the $\tilde{B}^2E_{2g}$ and $\tilde{C}^2A_{2u}$ states, respectively, and $\underset{\sim}{1}$ denotes the three-dimensional unit matrix. The i-sum runs over the totally symmetric modes ν_1, ν_2, the j-sum over ν_{19}, ν_{20}. The inclusion of the coupling modes ν_{19}, ν_{20} converts the curve crossings in figure 1 into a multi-dimensional conical intersection of the adiabatic potential energy surfaces.

In addition to the PJT coupling we have to take account of the fact that the $\tilde{B}$ state is degenerate and that the vibrational modes $\nu_{15}-\nu_{18}$ of E_{2g} symmetry are JT active in this state. We have to supplement, therefore, the hamiltonian (15) by the JT hamiltonian

$$\mathcal{H}_{JT} = H_o' \underset{\sim}{1} + \begin{pmatrix} 0 & 0 & \sum\limits_k g_k r_k e^{i\phi_k} \\ 0 & 0 & 0 \\ \sum\limits_k g_k r_k e^{-i\phi_k} & 0 & 0 \end{pmatrix}$$

(16a)

$$H_o' = \sum_k \frac{\omega_k}{2}\left(-\frac{1}{r_k}\frac{\partial}{\partial r_k}\frac{1}{r_k}\frac{\partial}{\partial r_k} - \frac{\partial^2}{\partial \phi_k^2} + r_k^2\right)$$

(16b)

where the k-sum runs over $\nu_{15}, \nu_{16}, \nu_{17}, \nu_{18}$.

The JT effect in the $\tilde{B}$ state and the $\tilde{B}-\tilde{C}$ PJT effect are not independent problems, since the JT splitting of the $\tilde{B}$ state affects the energetic separations of the component states interacting via the PJT effect. This is formally reflected by the non-commutativity of $\mathcal{H}_{JT}$ and $\mathcal{H}_{PJT}$. It should also be noted that the JT interactions via the four E_{2g} modes cannot be treated independently, but have to be considered as a four-mode ExE JT effect (the JT hamiltonians for the individual modes

do not commute with each other). We have therefore, in the harmonic
and linear coupling approximations, altogether 14 nonseparable vibrational degrees of freedom which may contribute to the $\tilde{B}$–$\tilde{C}$ vibronic
interaction.

The overall hamiltonian, $\mathcal{H} = \mathcal{H}_{PJT} + \mathcal{H}_{JT}$, contains 8 coupling constants: two for the totally symmetric modes, two for the PJT coupling
and four for the JT coupling. As mentioned above, values for the $\kappa_i^{(g,u)}$
are available in the literature. Values for the remaining coupling constants can also be found in the literature or determined from published
data. The JT coupling constants g_k have been determined by Lipari et al.
/26/ from CNDO/S molecular orbital calculations. The PJT coupling constants λ_j can be determined from the second derivatives of CNDO/S molecular orbital energies given in /26/ according to eq.(6). The CNDO/S
calculations predict that only two JT couplings (ν_{16} and ν_{18}) are significant in the $\tilde{B}$ state. Dropping the modes ν_{15} and ν_{17}, the problem
reduces to ten vibrational degrees of freedom: $\nu_1,\nu_2(A_{1g})$, $\nu_{19},\nu_{20}(E_{2u})$,
$\nu_{16},\nu_{18}(E_{2g})$. Since the vibrational frequencies of C_6H_6 in its electronic ground state are well known and the vertical ionization energies
E_g, E_u can be estimated from the photoelectron spectrum, all parameters
contained in the model hamiltonian (15,16) are thus determined.

A full quantum mechanical calculation for three (because of the
degeneracy of the $\tilde{B}$ state) electronic states and ten nonseparable vibrational degrees of freedom is a formidable computational problem,
even for a simplified model as considered here. As a preliminary step
towards the solution of the problem we have artificially separated the
JT effect from the PJT effect. The PJT conical intersection problem
(15) has been solved by expansion in a direct-product vibronic basis
with the following number of vibrational basis functions: $N(\nu_1)=6$,
$N(\nu_2)=17$, $N(\nu_{19})=9$, $N(\nu_{20})=4$. This results in a dimension of 24570 of
the two matrices $\mathcal{H}_g, \mathcal{H}_u$ to be diagonalized. The Lanczos method /28/
has been used to exploit the sparsity of the matrices. 7000 Lanczos
iterations /28/ were sufficient to converge the envelope of the calculated spectrum. The two-mode JT problem has been solved in the standard manner /6/ using 12 vibrational basis functions for both ν_{16} and
ν_{18}. This results in a dimension of 819 of the hamiltonian matrix. 500
Lanczos iterations were performed to obtain the spectrum. The spectra
resulting from the JT and PJT calculations for the $\tilde{B}^2E_{2g}$ state were
then convoluted with each other (this amounts to the neglect of the
noncommutativity of (15) and (16)) and the spectrum of the $\tilde{C}^2A_{2u}$ state
added. More details on the numerical calculations are given elsewhere
/29/.

The resulting theoretical photoelectron spectra for the overlap-
ping $\tilde{B}$ and $\tilde{C}$ bands are shown in figure 2 together with the experimental
spectrum of ref. /23/. The calculated stick spectrum has been convo-
luted with a Lorentzian of 40 meV fwhm to account for the limited
experimental resolution. The photoionization cross sections for the
$\tilde{B}$ and the $\tilde{C}$ states have been assumed to be of equal magnitude.

It is seen that the calculation explains the diffuse appearance
of the "$\tilde{C}$ band" between about 12 and 13 eV, in contrast to the earlier
calculation of ref. /25/ where the vibronic coupling with the $\tilde{B}$ state
has been neglected. The present calculation predicts a dense and nearly

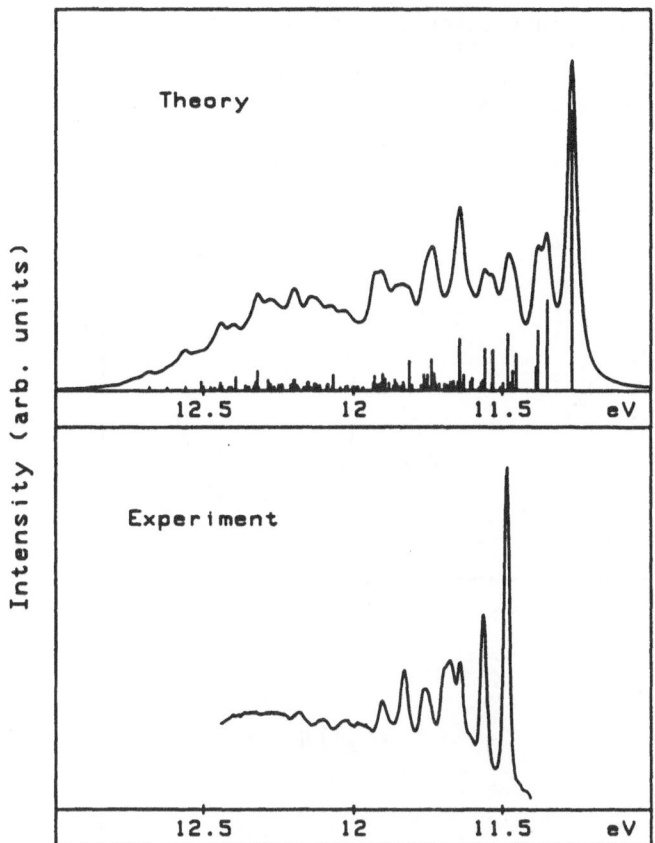

Figure 2.   Calculated (upper frame) and experimental /23/ (lower frame)
            photoelectron spectrum of $C_6H_6^+$ between 11 and 13 eV binding
            energy.

structureless quasi-continuum of vibronic lines in the high-energy part
of the spectrum. The observed density of states is essentially that
of the lower lying $\tilde{B}$ state at this energy. This implies that the
$\tilde{C}^2A_{2u}$ state "dissolves" into the $\tilde{B}^2E_{2g}$ state owing to very strong vibronic coupling caused by the conical intersection of the two states.
     The calculation explains at least to some extent also the ob-
served line structure in the lower part of the $\tilde{B}$-$\tilde{C}$ spectrum. It is
interesting to note that according to the calculation most of the apparently well resolved lines in the low-energy part of the spectrum are
actually composed of numerous individual vibronic lines. A high-reso-
lution spectroscopy of this part of the spectrum of $C_6H_6^+$ would be of
great interest. The remaining discrepancis between the calculated and
experimental spectra in figure 2 are probably due to (i) the use of

vibronic coupling constants taken from CNDO/S calculations which are
only of qualitative (±50%) validity /26/ and (ii) the artificial separation of the PJT and JT vibronic coupling problems. Work an more accurate calculations is in progress /29/.

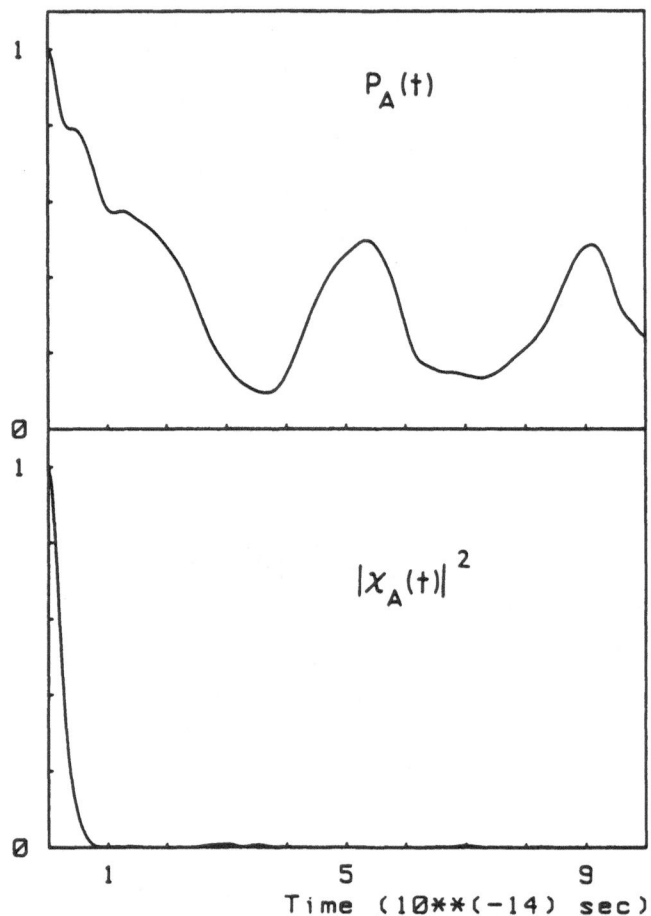

Figure 3.   Time-dependent population probability $P_A(t)$ of the $\tilde{C}^2A_{2u}$
state of $C_6H_6^+$ (upper frame) and autocorrelation func-
tion $\chi_A(t)$ for the same state (lower frame).

Figure 3 shows some first results on the time-dependent description of the dynamics of the system. In the lower part of figure 3 the
absolute square of the autocorrelation function (13) is shown, assuming
initial preparation of the wave packet in the diabatic $\tilde{C}^2A_{2u}$ state. The
autocorrelation function $\chi_A(t)$ exhibits a rapid initial decay within a

few femtoseconds which reflects the dynamics of the C-C stretching mode ν_2. In the absence of vibronic coupling ($\lambda_j=0$), $|\chi_A(t)|^2$ would exhibit periodic recurrences reflecting the regular motion on the harmonic surface of the $\tilde{C}$ state. The complete absence of recurrences on a 100 fs timescale in figure 3 shows that the vibrational motion in the coupled system is highly irregular. The absence of recurrences is not only a consequence of the radiationless decay of the $\tilde{C}$ state, but also reflects the mode-mode coupling in the $\tilde{C}$ state owing to the anharmonity of the adiabatic potential energy surface.

The time-dependent population probability $P_A(t)$ of the $\tilde{C}^2A_{2u}$ state is shown in the upper part of figure 3. It is seen that $P_A(t)$ decays on a timescale of about 20 fs, significantly slower than the decay time of the autocorrelation function, but still extremely fast compared to the radiative decay time of the $\tilde{C}$ state. At larger times $P_A(t)$ exhibits quasiperiodic partial recurrences, i.e., the population oscillates between the $\tilde{C}$ and $\tilde{B}$ electronic states. The period of these quantum beats coincides approximately with the period of the C-C stretch mode, which is the most active tuning mode of the system.

The conclusion to be drawn from these preliminary calculations is that the $\tilde{C}^2A_{2u}$ state of $C_6H_6^+$ dissolves into the vibronic line structure of the $\tilde{B}^2E_{2g}$ state and looses its identity as a separate electronic state in the sense of the BO approximation. The strong mixing of the two states may be traced back to the existence of a conical intersection of the adiabatic potential energy surfaces. In the time-dependent picture, the population of the $\tilde{C}$ electronic state exhibits an ultrafast decay on a timescale of about 20 fs. Since the $\tilde{B}$ state is dark in emission, the fluorescence from the $\tilde{C}$ state is strongly quenched (Douglas effect /30/). To understand the observed very low quantum yield of fluorescence /22/, we probably have to consider also the radiationless decay of the $\tilde{B}$ state into the $\tilde{X}$ state. This decay is expected to occur on a much longer timescale than the $\tilde{C}$-$\tilde{B}$ decay, since the $\tilde{B}$ and $\tilde{X}$ potentials do not intersect in the accessible range of coordinates (see figure 1).

5. CONCLUSIONS

We have reviewed recent quantum mechanical model calculations which show that conical intersections of potential energy surfaces may cause a population decay of electronic states of polyatomic molecules on a femtosecond timescale. Although only a few specific systems, mainly cations, have been investigated so far, there are reasons to believe that conical intersections are a very common phenomenon, at least in radical cations, and that they determine to a large extent the nonradiative relaxation of these systems. The ubiquity of conical intersections may explain the observation that fluorescence is the exception rather than the rule in polyatomic radical cations.

The investigation of ultrafast nonradiative intramolecular decay processes is interesting from a fundamental point of view. Since ultrafast decay requires the existence of strong and specific interactions involving few electronic states and relatively few vibrational modes, a first-principles calculation of the decay rate should be possible and the first steps in this direction have been taken. To test these ideas and the accuracy of the calculations, direct time-resolved measurements of the predicted femtosecond decay times in gas-phase cations would be of great interest.

ACKNOWLEDGEMENTS

The work surveyed here has been supported by the Deutsche Forschungsgemeinschaft and the Fonds der Chemischen Industrie. Computing time has generously been provided by the Universitätsrechenzentrum Heidelberg.

REFERENCES

1. J. P. Maier, in Kinetics of Ion-Molecule Reactions, P. Ausloos, ed., Plenum Press, New York, 1979, p. 437.
2. J. M. Wiesenfeld and B. I. Greene, Phys.Rev.Lett. 51, 1745 (1985).
3. G. Herzberg and H. C. Longuet-Higgins, Disc.Faraday Soc. 35, 77 (1963); H. C. Longuet-Higgins, Proc.Roy.Soc. (London) A 344, 147 (1975).
4. M. Desouter-Lecomte, D. Dehareng, B. Leyh-Nihant, M. T. Praet, A. J. Lorquet and J. C. Lorquet, J.Phys.Chem. 89, 214 (1985).
5. H. Köppel, L. S. Cederbaum, W. Domcke and S. S. Shaik, Angew.Chem. Int.Ed. 22, 210 (1983).
6. H. Köppel, W. Domcke and L. S. Cederbaum, Adv.Chem.Phys. 57, 59 (1984).
7. M. Desouter-Lecomte, C. Galloy and J. C. Lorquet, J.Chem.Phys. 71, 3661 (1979).
8. C. Sannen, G. Raseev, C. Galloy, G. Fauville and J. C. Lorquet, J.Chem.Phys. 74, 2402 (1981).
9. D. Dehareng, X. Chapuisat, J. C. Lorquet, C. Galloy and G. Raseev, J.Chem.Phys. 78, 1246 (1983).
10. E. Teller, J.Phys.Chem. 41, 109 (1937).
11. T. Carrington, Disc.Faraday Soc. 53, 27 (1972); Acc.Chem.Res. 7, 20 (1974).
12. E. R. Davidson, J.Am.Chem.Soc. 99, 397 (1977).
13. E. R. Davidson and W. T. Borden, J.Phys.Chem. 87, 4783 (1983).
14. M. Bixon and J. Jortner, J.Chem.Phys. 50, 3284 (1969).
15. A. J. Lorquet, J. C. Lorquet, J. Delwiche and M. J. Hubin-Franskin, J.Chem.Phys. 76, 4692 (1982); D. Dehareng, B. Leyh, M. Desouter-Lecomte, J. C. Lorquet, J. Delwiche and M. J. Hubin-Franskin, J.Chem.Phys. 79, 3719 (1983).
16. H. Köppel, Chem.Phys. 77, 359 (1983).
17. P. M. Dehmer and J. L. Dehmer, J.Chem.Phys. 70, 4574 (1979).

18. J. E. Pollard, D. J. Trevor, J. E. Reutt, Y. T. Lee and D. A. Shirley, J.Chem.Phys. $\underline{81}$, 5302 (1984); J. E. Reutt, L. S. Wang, J. E. Pollard, D. J. Trevor, Y. T. Lee and D. A. Shirley, J.Chem. Phys. $\underline{84}$, 3022 (1986).
19. W. H. Miller and C. W. McCurdy, J.Chem.Phys. $\underline{69}$, 5163 (1978); H.-D. Meyer and W. H. Miller, J.Chem.Phys. $\underline{70}$, 3214 (1979).
20. H.-D. Meyer, Chem.Phys. $\underline{82}$, 199 (1983).
21. J. W. Zwanziger, R. L. Whetten, G. S. Ezra and E. R. Grant, Chem. Phys.Lett. $\underline{120}$, 106 (1985); R. L. Whetten, G. S. Ezra and E. R. Grant, Ann.Rev.Phys.Chem. $\underline{36}$, 277 (1985).
22. O. Braitbart, E. Castellucci, G. Dujardin and S. Leach, J.Phys. Chem. $\underline{87}$, 4799 (1983).
23. L. Karlsson, L. Mattson, R. Jadrny, T. Bergmark and K. Siegbahn, Physica Scripta $\underline{14}$, 230 (1976).
24. J. P. Maier and F. Thommen, Chem.Phys. $\underline{57}$, 319 (1981).
25. W. von Niessen, L. S., Cederbaum and W. P. Kraemer, J.Chem.Phys. $\underline{65}$, 1378 (1976).
26. N. O. Lipari, C. B. Duke and L. Pietronero, J.Chem.Phys. $\underline{65}$, 1165 (1976).
27. P. Pulay, G. Fogarasi and J. E. Boggs, J.Chem.Phys. $\underline{74}$, 3999 (1981).
28. J. K. Cullum and R. A. Willoughby, Lanczos Algorithms for Large Symmetric Eigenvalue Computations, Vol. I, Birkhäuser, Basel, 1985.
29. H. Köppel and W. Domcke, to be published.
30. A. E. Douglas, J.Chem.Phys. $\underline{45}$, 1007 (1966).
31. H.-D. Meyer and H. Köppel, J.Chem.Phys. $\underline{81}$, 2605 (1984).

THEORETICAL STUDIES OF PHOTODISSOCIATION DYNAMICS IN LARGE CLUSTERS AND
IN SOLIDS

(a)R. Alimi, (b)A. Brokman, and (a)R.B. Gerber
(a)Department of Physical Chemistry
(a)The Fritz Haber Research Center for Molecular Dynamics and
(b)Department of Materials
The Hebrew University of Jerusalem
Jerusalem 91904, Israel

ABSTRACT. The photodissociation of diatomic molecules such as Cl_2 and
HI embedded in large clusters and in solid matrices of Ne, Ar and Xe is
studied by classical trajectories. The clusters studied are of $N \stackrel{\sim}{\sim} 10^2$
to 10^3 atoms, the initial temperature in the calculations was 0°K. The
methods used are: (1) Molecular Dynamics simulations including all particles; (2) A scheme which combines Molecular Dynamics for many particles with a harmonic treatment of atoms far from the reaction site.
This study is the first simulation of reaction dynamics in solids. Specific results include: (i) The $Cl_2[Ne]_N$ clusters lose several Ne atoms
within <1 psec following photodissociation, while the corresponding Ar
clusters have essentially infinite lifetimes. This is due to a shock
wave generated by the photodissociation in the Ne case, while in the Ar
cluster the energy rapidly randomizes. The behavior is dominated by mass
ratios and potential range parameters; (ii) Recombination is incomplete
in $Cl_2[Ne]_N$. The Cl atoms produced are trapped at interstitial sites
within several Å of the photodissociation site; (iii) Comparison of
theory and experiment for $HI[Xe]_N$ suggests that in low-energy photodissociation the H exit from the cage is entirely by tunneling.

1.   INTRODUCTION

There has been intensive activity in recent years directed towards microscopic understanding of molecular reaction dynamics in condensed
matter. The theoretical studies in this field were hitherto confined
to liquid solutions[1]. There are, however, strong reasons for pursuing
molecular reaction dynamics also in the solid state. The much greater
simplicity of crystalline matter in comparison with liquid solutions
suggests that first-principles, microscopic understanding of molecular
reactions should be more easily attained for the former class of systems. The present study is a first attempt towards a systematic understanding at the atomic level of reaction dynamics in solid. The present
subject bears some relation to the topic of vibrational relaxation of
excited impurity molecules in host solid matrices, the theory of which
was pursued by several groups in recent years[2].

R. Lefebvre and S. Mukamel (eds.), Stochasticity and Intramolecular Redistribution of Energy, 233–244.

The processes considered here are photodissociation of diatomic molecules such as Cl_2 and HI embedded in a rare gas matrix. Among the main questions investigated is the physical mechanism whereby the photofragments leave the solvent cage which surrounds them, and also the nature of their migration dynamics following photodissociation, i.e., whether the atoms produced show diffusive motion or kinematic propagation, etc. Another topic central to this study is the question how large must a finite solvent matrix be for the system to exhibit the behavior of an extended solid. Thus we shall explore photodissociation of diatomics in large rare gas clusters (of sizes in the range of $\sim 10^2$ to $\sim 10^3$ atoms) and focus on the question whether the same behavior as in solids is already attained in such systems.

The structure of the article is as follows: In Sec. 2 the theoretical methods used in this study are briefly described. Sec. 3 deals with results on cluster fragmentation, and their relation to the question of finite size effects. A very large and interesting difference is found between $Cl_2[Ar]_N$ and $Cl_2[Ne]_N$, and considerable insight is gained from analysis of these cases. Sec. 4 deals with the dynamics of product exit from the cage and the nature of subsequent migration. The case of $HI[Xe]_N$, in which an important role for tunneling is suggested here, is discussed in Sec. 5.

2. METHODS

2.1. Calculation of structure

The equilibrium structure of the clusters and solids prior to the photodissociation was obtained by straightforward energy minimization calculations. Both for Cl_2 and HI, the impurity molecule occupies a substitutional site in the host rare gas crystal (or cluster). However, the presence of the impurity causes forces that distort the system in comparison with the corresponding pure rare gas crystal or cluster. In the calculations we used pairwise forces between the rare gas atoms, available from gas phase data to great accuracy. We also made the (less satisfactory) assumption that the interaction between the guest molecule AB and the host crystal atoms can be taken approximately as a sum of potentials between A and the surrounding atoms, and between B and the surrounding atoms. Typical shifts in atomic positions in the minimization of the total energy are, for the nearest neighbors the molecule, of the order of 0.1Å (from the position in the corresponding pure rare gas system). Since this aspect does not enter in a qualitatively important way in the discussion of the photodissociation dynamics, we shall not elaborate here further on this point.

2.2. The initial state for photodissociation

In all calculations reported here, the rare gas atoms and the molecular c.m. were taken initially at their classical equilibrium values. This amounts to taking the cluster to be at T = 0K, and, moreover, to neglecting the zero point energies in the vibrations of the atoms and the

molecular c.m. We estimate that the effect of this on the photodissociation dynamics is extremely small, if the rare gas atoms are heavy (Ar,
Xe), and if the cluster temperature is indeed close to 0K. The distribution of initial conditions for the photodissociation is thus, in this
model, due to different possible orientations of the molecule in the
solid. Consider the case of Cl_2 in solid Ar. Using the pairwise potentials discussed above, the calculations show that there are six equivalent energy minima for the Cl_2 corresponding to different orientations
of the molecule in the fcc cage structure of the solid. There is a high
barrier (about 30 meV) which completely hinders rotation at low temperatures. Also, transitions between any two minima by tunneling are extremely slow for this system, so as to be ignored for our purposes here.
A contour map of the potential for Cl_2 rotation in solid Ar is shown in
Fig. 1.

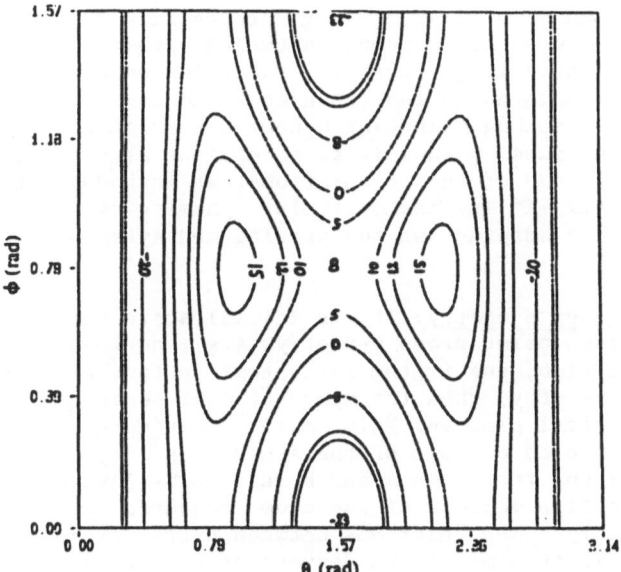

Figure 1: Potential contour plot for Cl_2 rotation in fcc solid Ar. The
energy is in meV, θ is the polar angle, ϕ the azimuthal one.

The potential function in this case is of the typical Devonshire form[3]
for molecular libration in an fcc crystal environment. In each of the
potential wells of Fig. 1 in which the molecule may be trapped, the molecule vibrates (librates) at T = 0K. The initial orientations of the
molecule about each minimum were thus sampled according to $|\psi_o(\theta,\phi)|^2$,
the librational ground state (where θ and ϕ are the orientation angles
of the molecular axis). For Cl_2 in Ar the amplitude of molecular libra-
tions in the ground state is $\sim 7°$ about each minimum. Trajectories of
photodissociation for different initial orientations within the above
mentioned range can differ greatly in behavior, so the orientational

initial-state sampling is extremely important.

The situation for HI in Xe is very different. There is considerable experimental evidence (that is backed by calculations) that hydrides such as HI or CHl are essentially free rotors in a matrix[4]. Therefore in this case we applied uniform sampling for the initial orientations of the molecule in the cluster, or solid.

We modelled the photodissociation event classically, by assuming a vertical transition at the photodissociation frequency from the ground state potential energy curve of the impurity molecule to the pertinent excited-state potential energy curve. Radiative recombination and non-adidatic transitions were not included in the present model.

2.3. Simulation methods for the dynamics

Most of the results reported here were obtained by "numerically exact" solution of the classical equations of motion for the cluster systems involved. Once an observable property (e.g. probability for exit from the cage per absorbed photon) is found converged with respect to cluster size for sufficiently large clusters, the value obtained for that property can be extrapolated to the extended solid case. The calculations reported here are thus Molecular Dynamics simulations for solid slabs[5] of various sizes, with initial conditions as described in the previous Section. In some of the calculations, we introduced the following methods for simplifying the Molecular Dynamics simulations, which are particularly adapted for the problem and systems considered here:

Use of point group symmetry: If the initial state geometry of the cluster has a rigorous point group symmetry, e.g., an inversion center or a plane of reflection, and if the interatomic potentials assumed cannot break that symmetry, then the symmetry must be maintained throughout the dynamical process. This implies in general, linear relations between coordinates of some of the atoms, e.g., $r_A(t) + r_B(t) = 0$ for all times t for any two atoms A and B related by inversion symmetry (r_A, r_B are the position vectors of the atoms measured from the inversion center). Symmetry-based relations between different coordinates can be used to eliminate some of the degrees of freedom in the equations of motion. Such a reduction of the equation by symmetry is important for the problem considered here because we assumed all the center of masses to be initially at their equilibrium values, and because the corresponding structures have high symmetry. Such symmetry-adapted simplification plays a much reduced role in cases where the zero-point vibration of the cluster atoms is considered, or at finite temperature T. In the systems considered here, reductions of the number of dynamical equations by factors 2-4 were typically the case.

Hybrid Molecular Dynamics-Lattice Dynamics method: Consider a photodissociation process of an isolated impurity molecule embedded in a solid or in a very large cluster. One expects that at sufficiently large distances from the reaction site the energy released will disperse to such an extent that each of the atoms that far from the site

will have at most a small amount of excitation energy. It should therefore be justified to treat such sufficiently far atoms in the harmonic
approximation, while "exact" classical dynamics is used for the atoms
in the "near" region. Indeed, at the reaction site and its vicinity
"hard" collisions between atoms occur as consequence of the photodissociation impact, and highly anharmonic interactions are involved. We de-
note by $\underline{r}_a$ the position vector of the atoms in the near region, by $\underline{u}_h$
the displacement vector from equilibrium of the far atoms. The dynamical
equations can then be reduced to the form:

$$\underline{\underline{m}}_a \underline{r}_a = \underline{F}_a(\underline{r}_a) + \underline{\underline{F}}_h(\underline{r}_a)\underline{u}_h \tag{1}$$

$$\underline{\underline{m}}_h \underline{u}_h = \underline{\underline{K}} \ \underline{u}_h + \underline{G}_a(\underline{r}_a) \tag{2}$$

where a matrix notation is used. The force constant matrix $\underline{\underline{K}}$, and the
force terms $\underline{F}_a(\underline{r}_a)$, $\underline{\underline{F}}_h(\underline{r}_a)\underline{u}_h$ etc. are all obtained from the interaction
potential. The linear equations for the $\underline{u}_h$ can be solved in terms of the
$\underline{G}_a(\underline{r}_a)$ by the Lattice Dynamics method. Thus, one is left with equations
of much reduced dimensionality for the $\underline{r}_a$ variables only. Even simpler,
one can solve for the $\underline{u}_h$ by using a "soft" fast integrator for eq. (2),
while Eq. (1) is treated by a more stable but less efficient (e.g.,
Gear), as that equation necessarily required. The possibility of dividing the solid into a limited anharmonic and a much larger harmonic
region makes it feasible to treat an extremely large number of atoms.
However, one must carefully verify that the system is large enough for
the above treatment to be valid. For $N \simeq 10^3$ size clusters this was the
case for most, but not all, the systems studied here.

3.   CLUSTER FRAGMENTATION DYNAMICS

In this section we consider the clusters $Cl_2[Ne]_N$, $Cl_2[Ar]_N$ for $N = 90$
and $N = 1098$. The clusters with $N = 1098$ have a cubic shape, with six
fcc rare gas unit cell along each of the directions x, y, z. The center
of mass Cl_2 is at the center of the cube. The $N = 90$ clusters are
prisms of dimensions $2a:2a:3a$ where a is the unit cell length of the
gas lattice. The c.m. positions of the Cl_2 is placed at the center of
the short (x, y) axes of the prism, its z-position divides that axis of
the cluster by a 1:2 ratio.
      We consider now the question whether photodissociation results in
fragmentation of the cluster, and on which time scale. In all cases the
photodissociation energy suffices for detaching rare gas-atoms from the
cluster, hence ultimately the clusters must undergo partial fragmentation. However, if such a process were to involve substantial energy redistribution and randomization, the predicted lifetimes (by RRKM) should
be extremely long. The classical trajectory simulations show a very interesting difference in the behavior of the Ar and the Ne clusters. In
$Cl_2[Ar]_N$, after the first impact by each Cl atom products on neighboring Ar atoms, the energy rapidly redistributes and no fragmentation
is found throughout the calculated trajectory (until $t \gtrsim 20$ psec). It
appears that a statistical behavior is obtained. The lifetime calculated

by RRKM for $Cl_2[Ar]_{90}$ is $\tau = 1.75$ μsec, that for $Cl_2[Ar]_{1098}$ may be
considered practically infinite. On the other hand, $Cl_2[Ne]_N$, upon photodissociation, loses Ne atoms on the scale of $\tau \approx 0.2$ psec! The fragmentation channels in the two cases are:

$$Cl_2[Ne]_{90} \xrightarrow{h\nu} Cl_2[Ne]_{81} + 9 \ Ne \tag{3}$$

$$Cl_2[Ne]_{1098} \xrightarrow{h\nu} Cl_2[Ne]_{1090} + 8 \ Ne \tag{4}$$

The Ne clusters evidently dissociate by an impulsive mechanism, extremely unlike the RRKM mechanism of the Ar case. The initial structure of
the (relevant part of the) $Cl_2[Ne]_{90}$ cluster, shown in Fig. 2, is helpful in discussing the fragmentation dynamics.

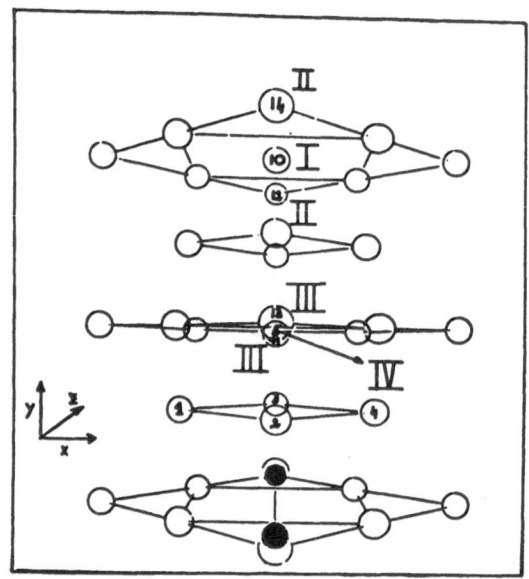

Figure 2. Structure of part of the $Cl_2[Ne]_{90}$ cluster. The Cl atoms are
shown in black. The role of the Ne atoms labelled by numbers is discussed in the text.

Upon photodissociation the "upper" Cl atom hits the window consisting
of atoms (1)-(4), gives some energy to the latter, but due to its high
mass has sufficient inertia to hit also atom 11 on top of the Cl_2 molecule. That atom passes through a second 4 Ne window and hits atom 10,
which leaves with considerable velocity. The first stage thus consisted
of a cascade of coherent, "hard" collisions resulting in fragmentation.
Having lost atom 10, atoms (12) and (14) lose binding energy and leave
the cluster (group II of fragments). Also when atom (11) was struck by
Cl it moved upward to create a temporary vacancy: The atoms (5) and
(13) lost binding forces, and they leave the cluster with low velocity.
The cascade of coherent collisions leading to the fragmentation can be

<u>viewed as a shock wave</u>[6]. The same qualitative behavior is found also
in the $Cl_2[Ne]_{1098}$ cluster, although the details are more complicated.
Fig. 3 shows the Ne location of the Ne clusters that are ejected by the
shock wave in this case, and their directions of flight. It is essential

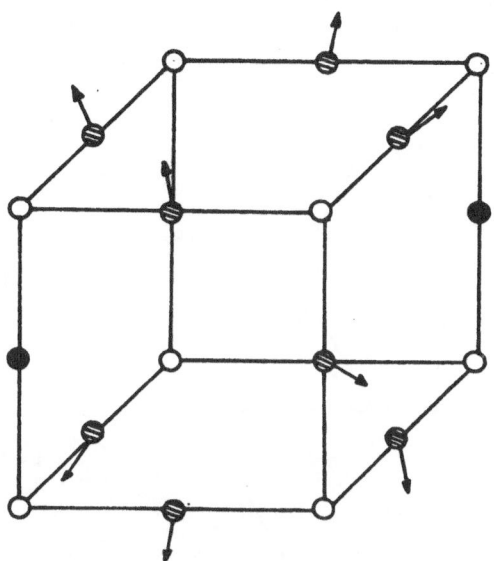

Figure 3. Photodissociation-induced fragmentation of $Cl_2[Ne]_{1098}$. The
Ne atom ejected by the shock waves and their directions of flight are
indicated by arrows.

to stress that the atoms ejected are not necessarily those of lowest
binding energy (the atoms shown in black have almost the same energy),
another aspect of the non-RRKM nature of the mechanism. The propagation
in time of the shock along a certain axis, (001), of $Cl_2[Ne]_{1098}$ is
shown in Fig. 4. Simple considerations of mass ratio and potential determine that the same behavior cannot happen in the Ar-clusters: Going
back to Fig. 2, the almost equal Cl and Ar masses, and the longer-range
repulsive Ar-Cl potential dictate an almost total loss of the Cl energy
to the first "window" of atoms (1)-(4). The shock wave is not formed,
and the energy begins to disperse and "randomize".

4.   SEPARATION OF THE PRODUCT ATOMS, AND RECOMBINATION

We proceed now to discuss the chemical reaction itself, that is the separation of the products and their recombination. Fig. 5 shows Cl-Cl
distances in $Cl_2[Ne]_{90}$ after ∿10 psec, a stage after which any further
separation of the atoms is very slow. As Fig. 5 shows, there is no recombination in this case. Under the impact of the initial impulse, the
Cl atoms seem to reach interstitial near-equilibrium positions, and do

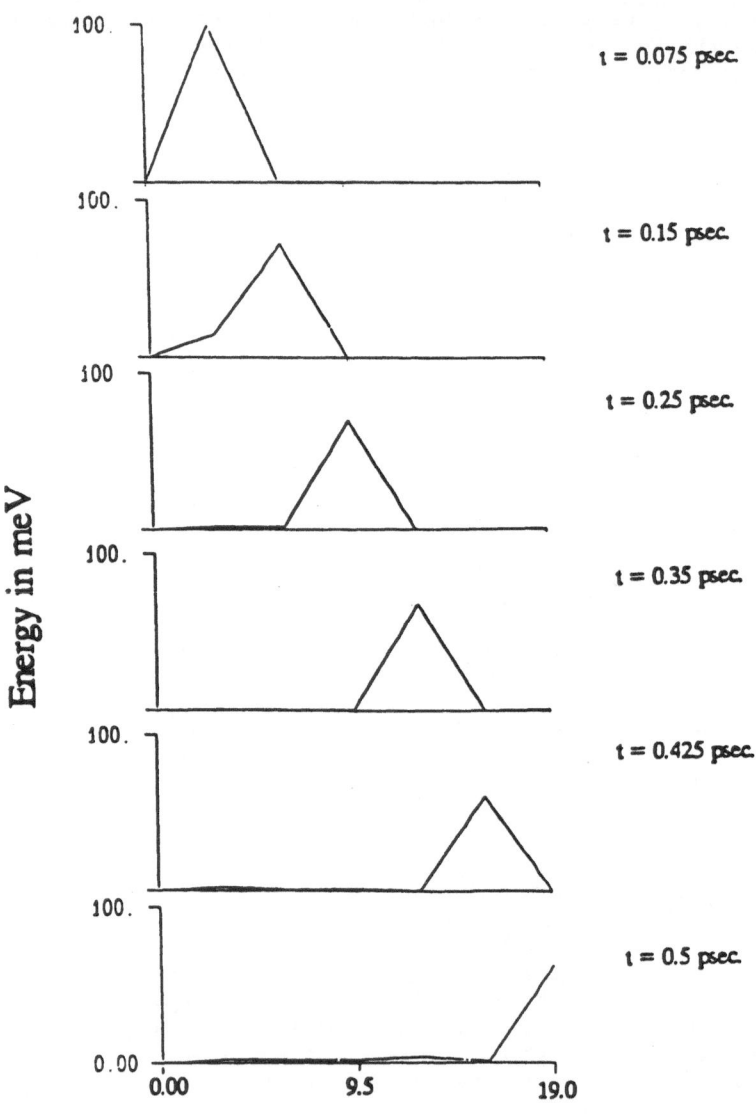

Figure. 4: Energy in the $Cl_2[Ne]_{1098}$ cluster along the (011) axis. The propagation of the pulse shape in time is shown.

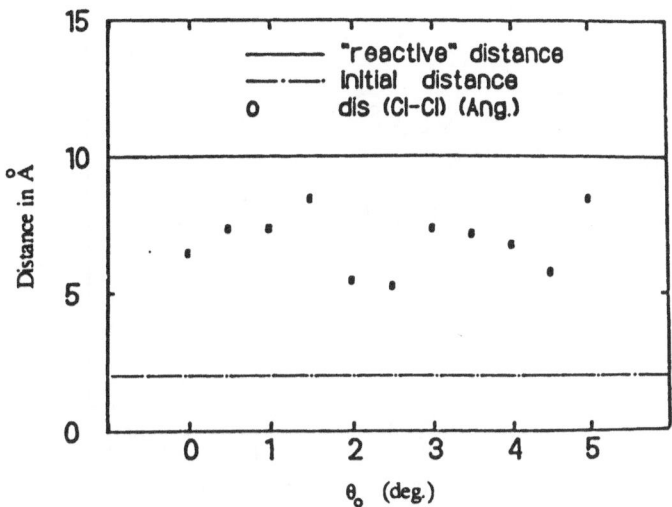

Figure 5: Final (∼10 psec) Cl-Cl distances after photodissociation of Cl$_2$[Ne]. The distances are shown for various initial Cl$_2$ orientation angles in the cluster.

not have sufficient energy to go back and recombine. Clearly somewhat different interstitial sites are reached for different orientations. The Cl-Cl distance vs. time is shown in Fig. 6. It is clear that after the initial "impulse" step in which the hard Cl-Ne collisions occurred the average motion is very slow. There are oscillations, corresponding to Cl vibrations about the interstitial site. Postulating a relation $R^2(t) \approx 4$ Dt, one can extract a "diffusion" constant. Physically, this corresponds not to direct migration of the Cl, but to the fact that the cluster,"softened" by the release of photodissociation energy, undergoes slow structural changes and the interstitial site itself moves in the cluster. This is a new, and potentially very interesting migration mechanism in soft clusters and in matrices[6].

We do not show here results for the Ar clusters, but in this case recombination is complete. As we saw, the Cl/Ar $\approx$ 1 mass ratio leads to extensive energy transfer from the Cl atoms to atoms (1)-(4) in Fig. 1. Therefore the Cl cannot pass through the first window, and the two fragments do not reach the interstitial sites as in the Ne case, in which they can remain un-recombined.

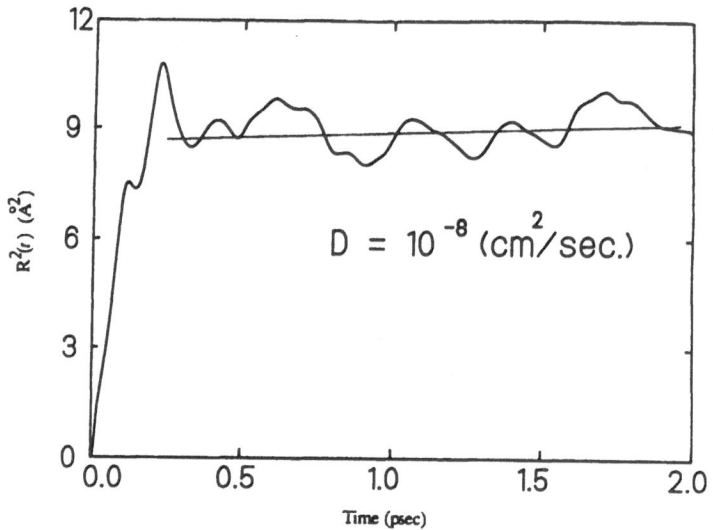

Figure 6: The time-dependence of the square of the distance of the Cl product in $Cl_2[Ne]_{90}$ from its initial position.

5. H ATOM MIGRATION IN CLUSTERS

Photodissociation of a hydride molecule in a rare gas cluster results in a product atom much lighter than the atoms of the cluster in which it may migrate. This large mass scale separation introduces qualitatively different features in the photodissociation behavior of hydrides than those encountered for Cl_2. The system that will briefly be discussed here is $HI[Xe]_{90}$. The position of the HI c.m. in the cluster is similar to that mentioned in a previous Section for $Cl_2[Ar]_{90}$. However, as stressed in Sec. 2, the axis of HI is not locked in a narrow range of orientations as in the Cl_2, but rather this impurity molecule rotates freely in the cluster. Our calculations have shown that depending on the initial orientation of the HI upon photodissociation, the cluster may lose one Xe atom within $\tau \gtrsim 0.1$ psec, lose two Xe atom on such a time scale, or remain intact (the total probability for which is roughly ≈ 0.5) for a very long duration of time. At high photodissociation energies ($E \gtrsim 2.0$ eV) H leaves the cage, just by surmounting the essentially static potential barrier due to the forces exerted by the heavy Xe atoms. However, for photodissociation energy $\lesssim 1.9$ eV we find complete recombination. Experimental evidence, although a very tentative one, points to H atom formation upon photodissociation of HI in Xe at 4 K[7]. The evidence is that at least part of the H atoms are not in a geminate pair situation with an I neighbor, thus exit from the cage must have taken place. Preliminary estimates that we carried out[6], suggest that this difference between the calculations and the experiment cannot be due to the finite, 4 K temperature in the latter. On the other hand, our estimates suggest that H tunneling through the barrier could account for the experimental effect. The present study thus suggests that low energy photodissociation of hydrides in low-temperature matrices, to yield non-geminate H atom products may be a pure tunneling reaction. Experimental study of this conjecture should be of considerable interest.

V. CONCLUSIONS

In this study, several classical simulations of photodissociaition in large crystalline clusters and in matrices were carried out. Several interesting effects were found. One of these is that following photodissociation, large clusters may exhibit rapid distribution of energy and almost infinite lifetimes (e.g., $Cl_2[Ar]_{1098}$), or may undergo very rapid partial decomposition due to a shockwave induced by the primary process (as in $Cl_2[Ne]_{1098}$). Thus, when such a shock wave occurs even clusters with $N > 10^3$ atoms will exhibit strong finite size effects. Another result of interest is that photodissociaition products in a soft cluster (or matrix) may migrate after the initial process by relaxation and shape changes in the cluster induced by the energy released. This is actually a migration of the site in the cluster at which the product is located, rather than a diffusion of an atom in a stiff framework. This mechanism of mass transport may prove important in molecular crystals. Finally, another result suggests that the formation of "free" H atom radicals in a matrix is, at low photodissociation energies and solid temperature, a pure tunneling process. We believe that experimental test of the above findings should be of very considerable interest. The fact that these early simulations predict several pronounced effects for reaction dynamics in solids and in large clusters suggests that this should be a fertile field for experimental explorations and more extensive, further theoretical studies.

Acknowledgement: The Fritz Haber Center is supported by the Minerva Gesellschaft fur die Forschung, Federal Republic of Germany.

REFERENCES

(1) J.T. Hynes in 'Theory of Chemical Reaction Dynamics', edited by M. Baer (CRC Press, Florida, 1985), Vol. 4.

(2) See, for instance, M. Berkowitz and R.B. Gerber, Chem. Phys. 37, 369 (1979).

(3) V.E. Narrayanaramurti and R.O. Phol, Rev. Mod. Phys. 42, 201 (1970).

(4) V.E. Bondybey, J. Chem. Phys. 65, 5138 (1976).

(5)   R.W. Hockney and J.W. Eastwood, 'Computer Simulations Using Particles', (McGraw-Mile, New York, 1981).

(6)   R. Alimi, A. Brokman and R.B. Gerber, to be published.

(7)   K. Kinugawa, T. Miyazaki and H. Hase, J. Phys. Chem. $\underline{82}$, 1697
(1978).

A HYPERSPHERICAL COORDINATE DISSOCIATIVE CORRELATION SCHEME FOR H_3^+

R. Pfeiffer and M.S. Child
Theoretical Chemistry Department,
1 South Parks Road,
Oxford OX1 3TG.

ABSTRACT. A finite difference method is used to determine the eigenvalues for motion in hyperspherical angular variables as a function of
the hyperspherical radius. Degeneracies in the resulting correlation
diagram are taken to infer the nature of the angular motions at different radii. It is concluded that the potential of Burton et al gives
rise to well characterised H^+/H_2 motions at proton-molecule separations
of R>4.00 a.u.

1. INTRODUCTION

The recent H_3^+ vibrational predissociation excitation experiments of
Carrington and Kennedy [1] have stimulated us to develop an interpretation in terms of a high angular momentum H^+/H_2, ion-rotor, complex
[2, 3]. The model employed however begs two important questions.
First it ignores the possibility of exchange between the three possible
protons; and secondly it assumes that the vibrations of the H_2 molecule
can be isolated from the remaining degrees of freedom.
   As a means to investigate the validity of these restrictions we
offer here an adiabatic correlation scheme between the vibrations of
the parent ion, H_3^+, and the motions of the predissociation fragments,
$H^+ + H_2$. The group theoretical aspects of such correlations have of
course been fully explored [4-9], but only a quantitative calculation
will answer the questions central to our investigation. At what point
along the correlation can the possibility of nuclear exchange be
ignored, as evidenced by the accidental degeneracies associated with
the fragment states and similarly at what point can the vibrational
motions of H_2 be recognised?
   The required correlation is carried through in terms of hyperspherical coordinates [10], using the radius R, which is determined by
the moment of inertia of the system, as the correlation coordinate.
The remaining motions are visualised as two dimensional angular motions
on the surface of a sphere and the corresponding quantum-mechanical
eigenvalues are determined by a finite-difference method.
   Details of the coordinate system are given in section 2. The

R. Lefebvre and S. Mukamel (eds.), Stochasticity and Intramolecular Redistribution of Energy, 245–252.
© *1987 by D. Reidel Publishing Company.*

finite difference method is outlined in section 3. Results are given
in section 4 and conclusions collected in section 5.

2.   COORDINATE SYSTEM

The hyperspherical variables are expressed in terms of mass reduced
Jacobi coordinates R_1 and r_1, where for example R_1 is the mass reduced
vector between atom 1 and the centre of mass of 2 and 3 while r_1 is
the mass reduced vector from atom 2 to atom 3. The new variables
(ρ,θ,ϕ), which are similar to those of Johnson [10], are defined in
such a way that

$$
\begin{aligned}
R_1 &= (\rho \cos \theta/2 \sin \phi/2, \; \rho \sin \theta/2 \cos \phi/2, \; 0)\\
r_1 &= (\rho \cos \theta/2 \cos \phi/2, \; \rho \sin \theta/2 \sin \phi/2, \; 0),
\end{aligned}
\tag{1}
$$

with θ and ϕ lying in the ranges $-\pi/2 < \theta < \pi/2$ and $0 < \phi < 2\pi$.
   It then follows that

$$
\begin{aligned}
r_1^2 &= 0.5 \; \rho^2(1 + \cos \theta \cos \phi)\\[4pt]
r_2^2 &= 0.5 \; \rho^2(1 + \cos \theta \cos(\phi-2\pi/3)\\[4pt]
r_3^2 &= 0.5 \; \rho^2(1 + \cos \theta \cos(\phi+2\pi/3).
\end{aligned}
\tag{2}
$$

Consequently equilateral geometries, with $r_1 = r_2 = r_3$, correspond to
$\theta = \pm \pi/2$, which may be visualised as the poles of a sphere. Similarly
there are three points on the equator, $\theta = 0$, with $\phi = 0, \pm 2\pi/3$ at
which two atoms coalesce.
   The molecular motions at small ρ values, applicable to the parent
equilateral ion, therefore favour regions close to $\theta = \pm \pi/2$, whereas
those of the fragments show an energetic preference for $\theta = 0, 2\pi/3$ or
$-2\pi/3$. The evolution of the potential contours of Burton et al [11]
from the equilibrium value $\rho = \rho_e = 1.65$ a.u. to $\rho = 2.4\rho_e$ are shown
in Fig. 1.

3.   ADIABATIC ANGULAR EIGENVALUES

   The angular part of the vibrational hamiltonian may be expressed as

$$
\hat{H}_{ang} = \frac{\hbar^2}{m\rho^2} \left[\frac{1}{\cos\theta} \frac{\partial}{\partial\theta}\left(\cos\theta \frac{\partial}{\partial\theta}\right) + \frac{1}{\cos^2\theta} \right] + V(\rho,\theta,\phi),
\tag{3}
$$

where m is the mean of a hydrogen atom. The eigenfunctions $\psi_\alpha(\rho,\theta,\phi)$,

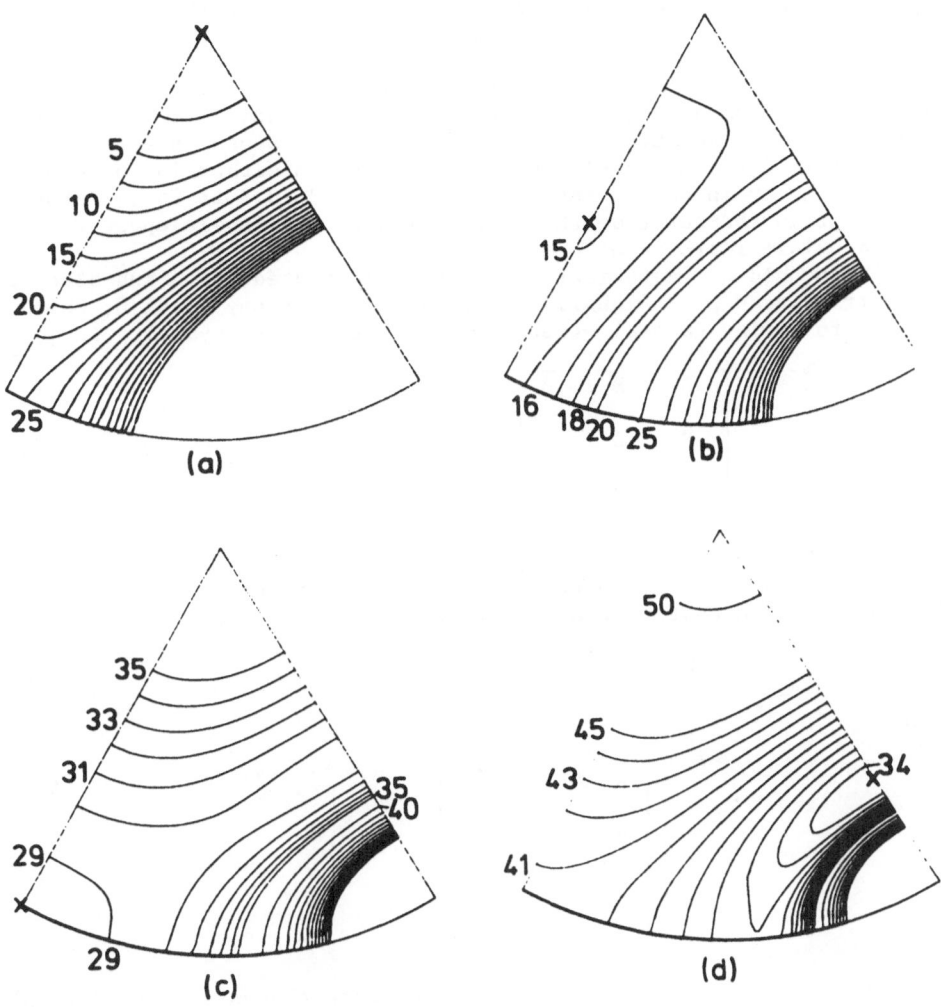

Fig. 1   Contours of the Burton et al [11] potential at (a) $\rho = \rho_e$,
(b) $\rho = 1.5\ \rho_e$,   (c) $\rho = 2.0$ and   (d) $\rho = 2.4\ \rho_e$. The upper vertex in
each diagramm corresponds to the pole of the upper hemisphere and the
left and right hand sides to the lines $\phi = \pi/3$ and $\phi = 2\pi/3$ respectively. The crosses indicate the position of the potential minima.
Contours are shown in multiples of 1000 cm^{-1}.

at given ρ were determined on a (θ,ϕ) grid, using seven point finite difference formulae for the derivatives. This implies in effect a delta-function basis, so that the potential contribution to the hamiltonian matrix is diagonal.

The matrix was factorised before diagonalisation according to the permutational symmetry of the potential $V(\rho,\theta,\phi)$ which is isomorphic with D_{3h}; cyclic permutations of the nuclei correspond to three-fold rotations in ϕ and non-cyclic ones to reflection either in the equational plane or in one of the planes $\phi = 0$, $2\pm2\pi/3$. It is therefore convenient to characterise the eigenstates by the labels appropriate to irreducible representations of D_{3h}.

The method of factorisation may be illustrated for a sub-block of the matrix based on 3n equally spaced ϕ values at any given θ, which has the following general structure by virtue of the symmetry in ϕ;

$$\tilde{h} = \begin{pmatrix} A & B & \overline{B} \\ \overline{B} & A & B \\ B & \overline{B} & A \end{pmatrix} \tag{4}$$

here A, B and $\overline{B}$ are nxn matrices and $\overline{B}$ is the transpose of B. Now consider the unitary transformation

$$h = U^{+} h U , \tag{5}$$

where

$$U = 3^{-\frac{1}{2}} \begin{pmatrix} I & I & I \\ I & w & w^{*} \\ I & w^{*} & w \end{pmatrix}, \tag{6}$$

and I, w and w^{*} are diagonal nxn matrices with elements 1, $\exp(2\pi i/3)$ and $\exp(-2\pi i/3)$ respectively. It is readily verified that $\tilde{h}$ reduces to the block diagonal form

$$\tilde{h} = \begin{pmatrix} A + B + \overline{B}, & 0 , & 0 \\ 0 , & A + wB + w^{*}B, & \\ 0 , & 0 , & A + w^{+}B + w\overline{B} \end{pmatrix}. \tag{7}$$

The eigenstates of the first block have symmetry A and those of the second and third are mutually degenerate in pairs (symmetry E) because

the two matrices are complex conjugates. A further factorisation of
the A block may be achieved by choosing an even number, $n = 2m$ of grid
points between 0 and $2\pi/3$ and ensuring that the plane $\phi = \pi/3$ bisects
the mth and $(m + 1)$th points. Symmetry or antisymmetry about $\phi = \pi/3$
then gives rise to the A_1 or A_2 representations, but particular care is
required to ensure that the correct behaviour is preserved in crossing
the poles. Finally a further factorisation may be made according to
the symmetry or antisymmetry with respect to reflection in the equational plane, the resulting representations being labelled Γ' or Γ''
respectively.

4.   RESULTS

The correlation diagram obtained by this scheme is shown in Figs.
2 and 3. Note that only the curves belonging to the Γ' representations
have been plotted in Fig. 2. These are degenerate with the corresponding Γ'' curves for $\rho < 1.4\rho_e$ at which point, as seen from Fig. 1, the
equilibrium geometry starts to move away from the pole towards the
equator and penetration across the equatorial plane becomes energetically allowed. Hence the Γ', Γ'' degeneracies are removed. At the same
time however the opportunities for exchange between the three protons,
as represented by relatively free ϕ motion in Fig. 1(a), become progressively hindered leading to the development of three fold degeneracies between the lower A_1', E' or A_2', E' pairs of curves in Fig. 2.
Similar three fold degeneracies also develop between the higher curves
and becomes essentially complete for $\rho > 2.4$.

Fig. 3 illustrates the development of another type of degeneracy
associated with motion around the roughly circular valley in Fig. 1(d).
At low energies the degeneracies occur between A_1' and A'' states or A_2'
and A'' which may be verified to imply vibrational motion along the
valley, around the equilibrium point. These degeneracies are however
lifted at higher energies and new A_1', A_2'' or A_2, A_1'' degeneracies occur,
which are characteristic of free rotation around the valley. These
patterns indicate a restricted rotation of the H_2 molecule with respect
to the departing proton.

Motion perpendicular to the valley in Fig. 1(d) corresponds on the
other hand to vibrational motion of the H_2 molecule, and Fig. 3 shows
one high energy A_1', A_1' pair for $\rho > 2.4\rho_e$ which may be assigned to the
first excited vibrational state of H_2.

5.   CONCLUSIONS

A practical scheme to determine the correlation between 'molecular'
and 'fragment' motions of H_3^+ has been developed. The correlation is
treated as adiabatic in the hyperspherical radius ρ, and the formulation adopted is well suited to future studies of the strength of non-
adiabatic oupling.

When implemented in terms of the potential of Burton et al [11],
the scheme indicates well developed fragment states for $\rho > 2.4\rho_e$

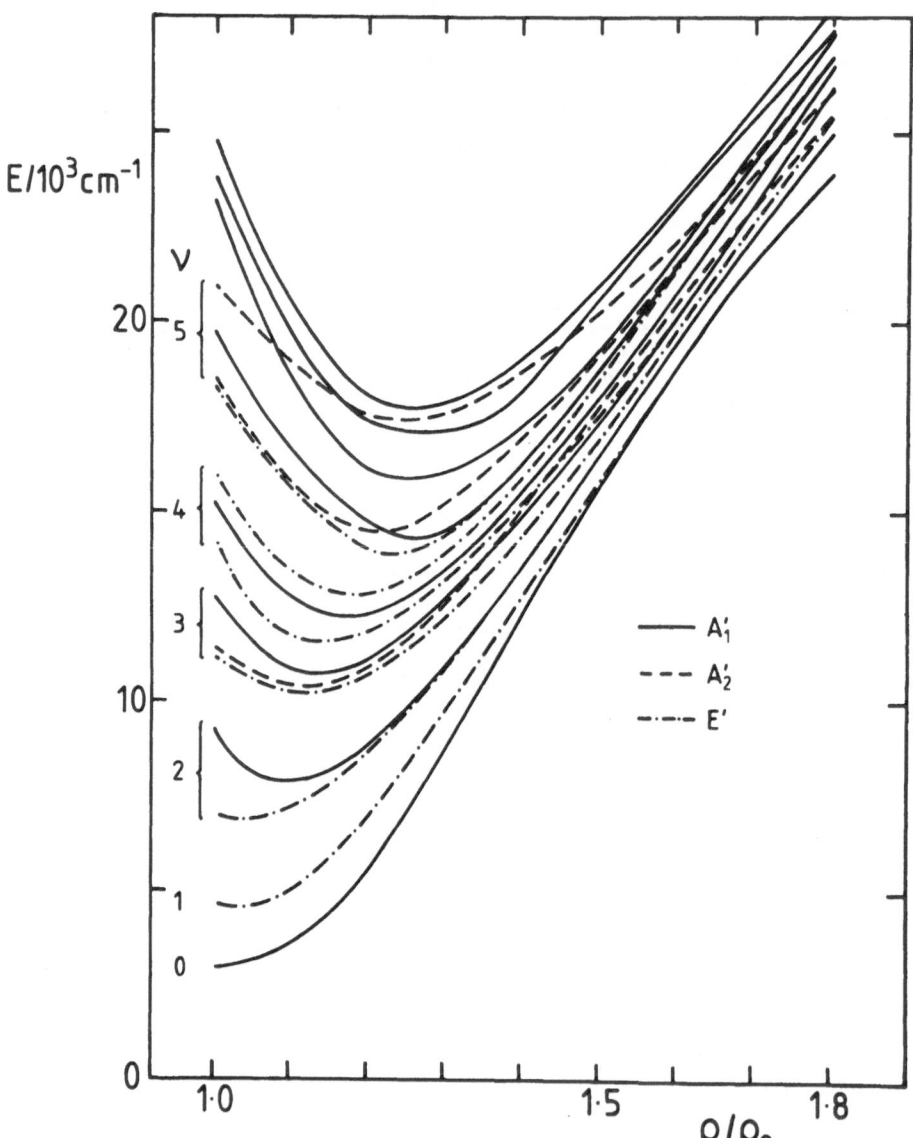

Fig. 2. The correlation of A'_1 and A'_2 and E' angular states for $\rho_e < \rho < 1.8\,\rho_e$.

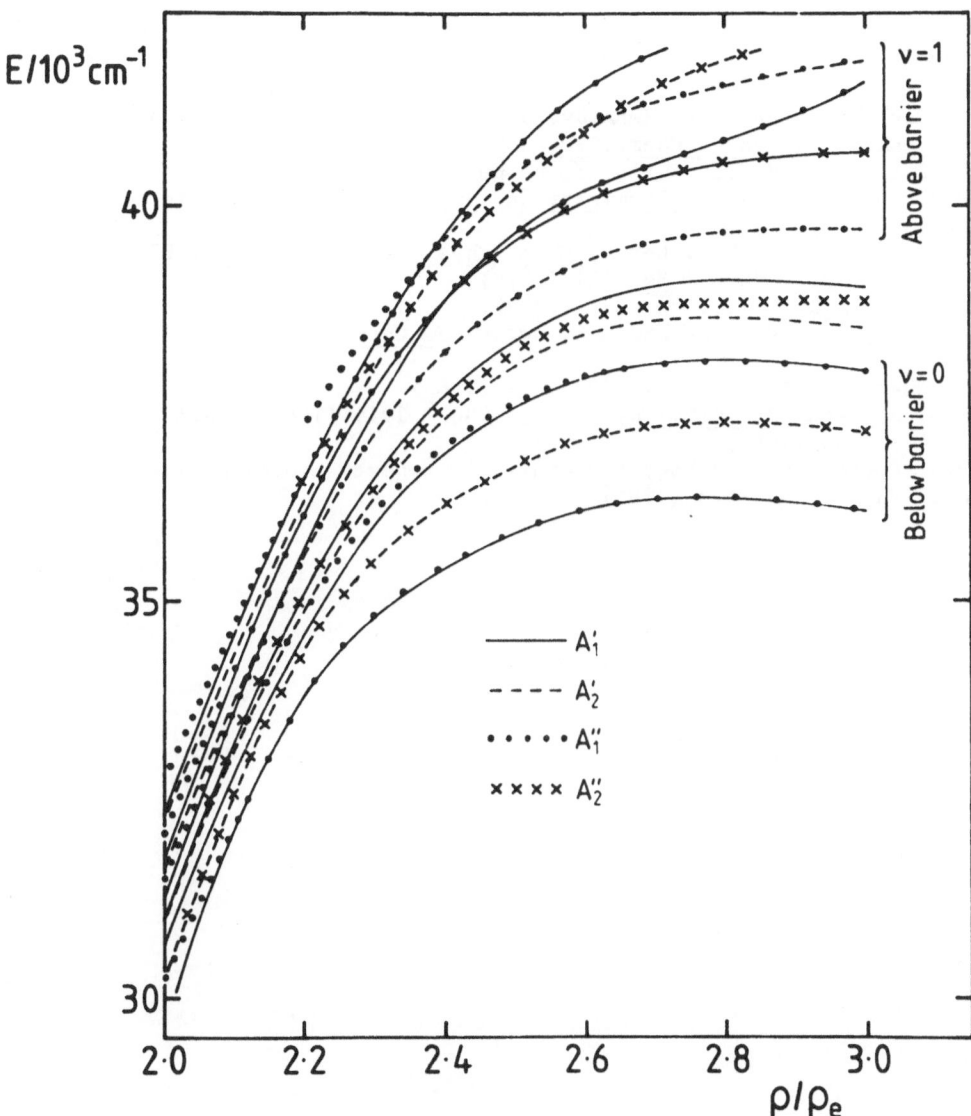

Fig. 3.   The correlation of A'₁, A'₂, A"₂ states for $2\rho_e < \rho < 5\rho_e$.

which translates in terms of real variables to a proton/molecule sep-
atation R > 4.00 a.u. At this distance proton exchange can be ignored
and the molecule undergoes well defined vibrational and restricted
rotational motions.

REFERENCES

1. A. Carrington and R.A. Kennedy, J. Chem. Phys., 81, 91 (1984).
2. M.S. Child, J. Phys. Chem., 90, 3595 (1986).
3. R. Pfeiffer and M.S. Child (to be published).
4. P.R. Bunker, Molecular Symmetry and Spectroscopy, (Academic
    Press, 1979).
5. M.E. Kellman and R.S. Berry, Chem. Phys. Lett., 42, 327 (1976).
6. M. Quack, Mol. Phys., 34, 477 (1977).
7. G. Ezra, Mol. Phys., 38, 863 (1979).
8. F. Amar, M.E. Kellman and R.S. Berry, J. Chem. Phys., 73, 2387
    (1980).
9. M. Quack, J. Chem. Phys., 82, 3277 (1985).
10. R.E. Johnson, J. Chem. Phys., 79, 1906 - 1915 (1983); ibid 79,
    1916 - 1935 (1983).
11. P.G. Burton, E. von Nagy-felsobuki, G/ Doherty and M. Hamilton,
    Mol. Phys., 55, 527 (1985).

SPECTROSCOPY OF PREDISSOCIATING MOLECULES

Veronica Vaida
Department of Chemistry and Biochemistry
University of Colorado
Boulder, CO 80309 USA

ABSTRACT. The excited state structure and dynamics of predissociative electronic states of model three and four atom molecules, namely CS_2, OCS and NH_3, is discussed. Experimental investigations rely on jet cooled absorption spectroscopy. The data is used in conjunction with results of theoretical calculations to model the excited electronic surface and open the way for exploration of reaction dynamics.

I. INTRODUCTION

The complete understanding of polyatomic molecular potential energy surfaces and their chemistry is becoming a realistic goal. Particularly suited for such a study are molecular systems where the characteristic reaction is a light induced unimolecular process. Exploration of the photodissociation process requires a detailed knowledge of the potential energy surfaces that control the reaction; such information can, in principle, be obtained from high resolution optical spectra [1] and calculated from ab-initio theory. In practice, efficient chemical channels shorten the electronic state lifetime and lead to diffuse spectra with little or no rotational structure. The short excited state lifetime generally precludes emission and ionization rendering these sensitive techniques useless. As a consequence, spectra of dissociative molecules contain little of the type of information obtainable from bound-bound transitions. The theoretical treatment of photodissociative molecules is complicated by these problems and can no longer be reduced to a simple rotor-vibrator model hamiltonian. Understanding of these complex reactive molecules must come from a close combination of experimental results and theoretical modeling.

Excitation of the molecular models discussed here (CS_2, OCS and NH_3), induces a large change in geometry (either linear to bent geometry as in CS_2 and OCS or pyramidal to planar as in NH_3). Efficient photodissociation is known to occur on the excited surface of CS_2, OCS and NH_3, to CS + S, CS + O and NH_2 + H, respectively. Effects of vibrational level coupling on the energy flow and on their

R. Lefebvre and S. Mukamel (eds.), Stochasticity and Intramolecular Redistribution of Energy, 253–261.

photochemistry are manifested as spectroscopic perturbations obtainable
from the spectra.

Presently, the spectroscopy of CS_2, OCS and NH_3 will be discussed
in light of the information available for their excited potential energy
surface and the chemical dynamics on this excited surface.

II. METHODS

The excited state structure and dynamics of reactive polyatomic systems
is probed in our laboratory by direct absorption of jet-cooled samples
[2]. This technique and its relationship to other modern experimental
methods available to study dissociation is discussed elsewhere [2,3].
In this work, characterization of the upper state relies on describing
the excited electronic state by comparison with the ground state of the
molecule via a Franck-Condon analysis of observed intensity profiles of
spectroscopic features. This can be done in certain cases if accurate
experimental itensities are available. The direct absorption experiment
we use relies on low intensity light sources and yields corrected
relative vibronic band intensities. At the same time, the samples are
cooled so that inhomogeneous contributions are eliminated and
homogeneous linewidths are obtainable from the spectra. For the
dissociative molecules studied, the homogeneous vibronic linewidth
increases with vibronic level. Evaluation of theoretical methods by
comparison of predicted and experimental spectra is best performed when
an optimum choice of lineshapes is employed in the synthesis of model
spectra.

Recently, experimental information about reactive excited states of
polyatomic molecules has been obtained using resonance fluorescence
(resonance Raman) techniques [4]. The combination of resonance
fluorescence, absorption and theoretical modeling was employed [4] to
study the excited states of CH_3I and O_3. Resonance Raman data is
available [5,6] for NH_3 and CS_2 and is used as a complement to the
absorption data and theoretical results describing the excited
electronic states of these molecules.

III. RESULTS

The examples chosen for discussion involve small molecules with
efficient dissociation occurring on an excited electronic state. From
the spectra, we attempt to extract information about the upper state
structure and the dissociation dynamics on this excited surface.

CS_2

The $^1\Sigma g^+ \rightarrow {}^1B_2(^1\Sigma u^+)$ transition of CS_2 occurring at ~ 50000 cm^{-1}
provides an interesting model where electronic excitation induces a
large change in geometry and leads to efficient dissociation. The
chemistry of this excited electronic state has been studied extensively
[7], yet a dynamical picture is yet to be proposed. A prerequisite to a

quantitative understanding of the reaction dynamics is the proper
description of the excited state structure. The structural information
obtained for the $^1B_2$ state of CS_2 comes mainly from high resolution work
[8-13] in the hot band region of the spectrum. This work led to a
geometry for the low vibrational levels of the upper $^1B_2$ state with the
C-S bond lengthening from the ground state value of 1.55 to 1.66 Å in
the excited state and the S-C-S angle decreasing from 180° to 153°.
Both progressions in the symmetric stretch ν_1 and bend ν_2 are therefore
expected in the spectrum. Detailed analysis of most of the spectrum was
precluded by large homogeneous spectral linewidth and congestion caused
by similar frequencies for ν_1', ν_2' and hot bands in ν_2''. Much of this
congestion could be eliminated in the jet cooled absorption spectrum
studied in our laboratory [14]. The spectrum observed in the jet
(Figure 1) is characterized by clusters of bands separated by 400-450

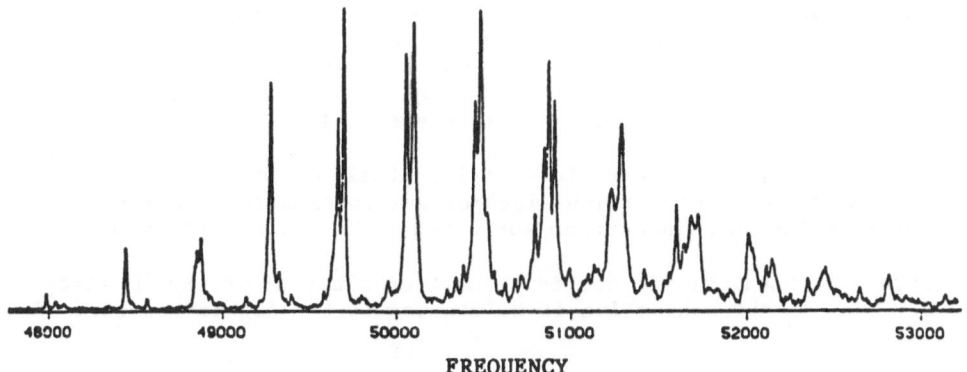

FREQUENCY

Figure 1. Absorption spectrum of the $^1\Sigma u^+ \leftarrow {}^1\Sigma g^+$ transition of CS_2
cooled in a supersonic expansion with Ar.

cm^{-1}. In the low energy region of this spectrum the vibronic bands
observed display irregular spacings and appear to belong to levels below
the barrier to linearity. In contrast, at the Franck-Condon maximum, a
simple interval pattern is observed. At these energies the K-level
structure converges toward that expected for the quasilinear molecule.
At higher energies, enhancement of the weaker bands within each cluster
suggests significant resonance effects between multiply excited ν_1' and
ν_2'.

The structure of the $^1B_2$ state was further probed by jet cooled
spectra [15] of $^{12}C^{34}S_2$ and $^{13}C^{32}S_2$. The observed isotopic shifts of
vibronic bands as a function of energy suggest that the active mode seen
in the spectrum is the symmetric stretch ν_1 perturbed by interaction
with the bending mode ν_2.

In spite of the extensive spectroscopic data now available for this
excited state of CS_2, the vibrational structure remains unsolved. This
is particularly noteworthy in the simple case of a triatomic with only
three normal modes present. That an uncoupled mode (ν_1 and ν_2) picture
does not explain the spectrum is obvious from the information at hand.

The proper way to look at this spectrum, whether making use of normal modes or using a completely different approach, is yet to be established.

OCS

The ground $^1\Sigma^+$ electronic state of OCS is known to be linear [1] with three normal modes of vibration, the "C-O" stretch at 2062 cm^{-1}, the "C-S" stretch at 859 cm^{-1} and the bending mode at 527 cm^{-1}. Excitation at ~58000 cm^{-1} leads to a large change in geometry to a bent $^1\Pi$ excited state [1,16]. On photodissociation, OCS gives rise to CO + S. The internal energy distribution of the photoproducts has been extensively studied following excitation to the $^1\Sigma^+$ and $^1\Delta$ excited electronic state [17]. The room temperature vapor spectrum [16] consists of a continuum superimposed on which are two progressions with an average spacing of 513 cm^{-1}, separated from each other by 270 cm^{-1}; both have a Franck-Condon maximum approximately 13 quanta away from the origin. The two observed progressions have been assigned to the two component electronic transitions which arise from Renner-Teller splitting of the bent $^1\Pi$ electronic state.

The incomplete structural information available for the excited electronic surface as well as the lack of any information about the reaction dynamics on this surface motivated our spectroscopic study [18].

Figure 2 shows the jet-cooled absorption spectrum of the $^1\Pi$ state of OCS. Much of the congestion is eliminated at the low temperature of

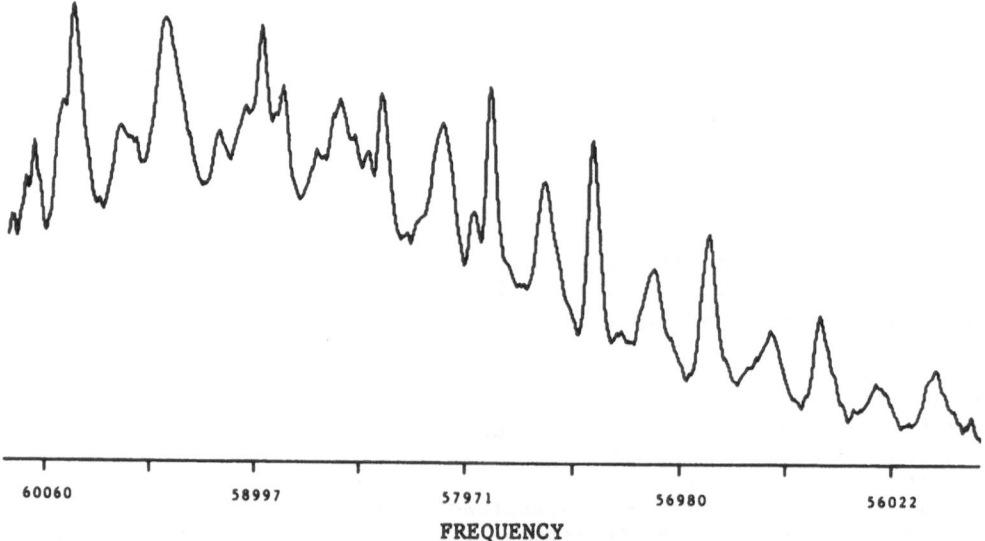

| 60060 | 58997 | 57971 | 56980 | 56022 |

FREQUENCY

Figure 2.   Absorption spectrum of the $^1\Pi \leftarrow {}^1\Sigma^+$ transition of OCS cooled in a supersonic expansion with Ar.

the jet revealing an observed spectrum with a much more complex pattern
than the one seen at room temperature. Every quantum of vibration is
allowed in this transition, unlike the case of CS_2 discussed above,
because in OCS of $C_{\infty v}$ point group, the parity selection rule is
relaxed. At low energy, the observed pattern consist of two
progressions which each split into two at ~58500 cm^{-1}. This pattern
is indicative of the molecule becoming quasilinear above this energy.
Transitions below 58500 cm^{-1} correspond to two progressions, with an
average spacing of 520 cm^{-1} and 504 cm^{-1}, close to the value of the
bending mode in the ground state. In the cooled spectrum, the width of
vibronic features corresponding to these two progressions is different
by about a factor of three. As the rotational profiles in this molecule
at the temperature of the jet cannot account for this discrepancy, we
attribute the observed bandwidth to the dissociation dynamics. The data
then suggests that the progression with large fwhm is predominantly the
dissociative stretch seen in combination with quanta of the bending
mode. The second progression arises presumably from a progression in
the pure bend, coupled less efficiently to the dissociative coordinate.

   The structural details of this electronic state and the exact
coupling schemes responsible for the dissociation are yet to be
quantitatively determined, yet much of the information needed can be
read directly from the absorption spectrum of the cold, isolated
molecule.

NH_3

Ultraviolet radiation of approximately 47000 cm^{-1} excites the pyramidal
ground state of ammonia to an electronic state of planar geometry [19].
NH_3 in this A^1A_2'' state dissociates to NH_2 and H with a near unity
quantum yield [20]. Extensive experimental studies [21] were performed
on the spectroscopy and the photochemistry of the A state of this
molecule. Problems concerning both the structure and dynamics of this
excited state were the starting point for our investigation [21-24].
The challenge from an experimental point of view in answering these
questions came from the fact that the efficient dissociation shortens
the excited state lifetime giving rise to diffuse spectra with no
rotational structure. As chemistry competes efficiently with emission,
no fluorescence was observed on excitation of the A state of NH_3,
although fluorescense was seen following excitation of ND_3 [25]. For
the same reasons, multiphoton ionization spectra of the A state [26-28]
are weak and complicated by predissociation which occurs faster than the
up-pumping needed for MPI.

   We investigated the A^1A_2'' state of ammonia by a combination of
spectroscopic [21,22] and theoretical techniques [23,24].
Experimentally we relied on jet-cooled absorption spectra of NH_3 and ND_3
which yielded accurate band intensities and homogeneous band width [21,
22]. The relative oscillator strength could then be used to compare
with results of an MCSCF calculation of the A^1A_2'' + X transition [23].
The potential energy surface relevant to the NH_3 + NH_2 + H dissociation

was calculated and this model was used to interpret the dynamical
effects observed in the spectrum [24].

Figure 3 shows the ultraviolet absorption spectrum of NH_3 obtained
in a supersonic jet. A long progression spaced by ~ 900 cm^{-1} is

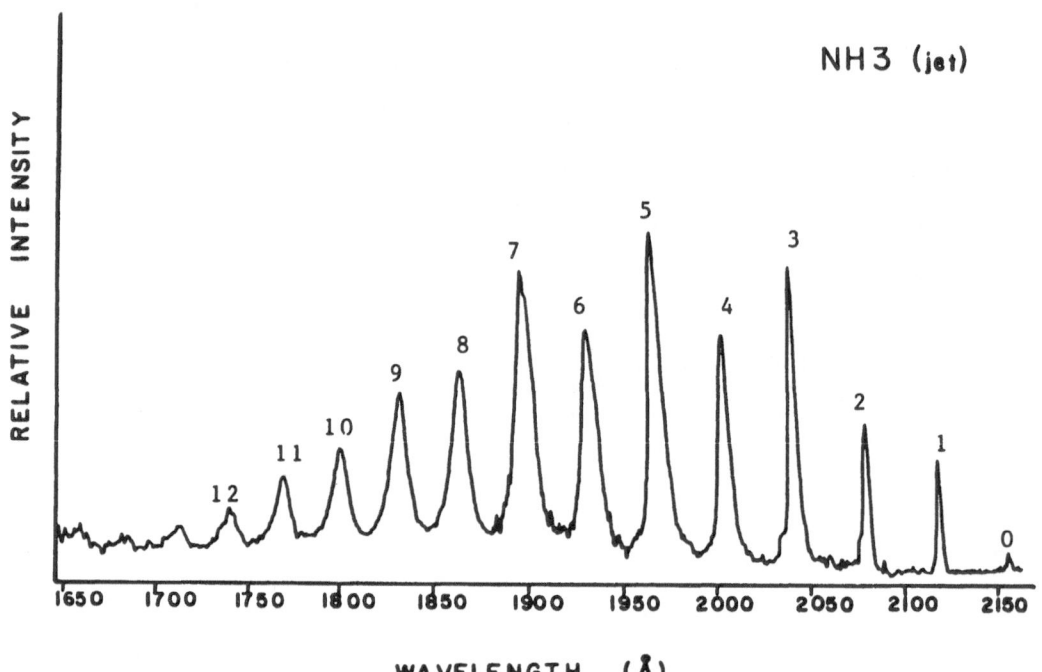

Figure 3.   Absorption spectrum of the A ← X transition of NH_3
            cooled in a supersonic expansion with Ar. The ν_2'
            quantum number is shown above each vibronic band.

observed and assigned to the umbrella bending mode ν_2. As seen from the
spectrum of the cold molecule, band intensities for odd quanta of the
inversion mode are enhanced at the expense of the intensities of bands
of even quanta. This temperature dependent alternation of band
intensities is a result of selection rules on nuclear permutation
symmetry for protons in K = 0 levels. A quantitative analysis gave a
temperature of 15 K for NH_3 cooled in the supersonic expansion [22,23].

The main problem in extracting the structure of the A state of
ammonia from the spectrum has been the lack of observable intensity in
the ν_1 (symmetric stretch) progression. The ground state of ND_3 was
determined [19] to be pyramidal with a N–D bond length of 1.012 Å. The
rotational structure of the ν_2' = 0,1 bands in the A state indicated
[19] a planar electronic state with a N–D bond length of 1.08 Å. In
this transition, both the D–N–D angle and the N–D bond length change
considerably. The Franck–Condon analysis using two uncoupled harmonic

modes predicts intense progressions in both modes. Several ideas were
presented [29] to explain the lack of observable ν_1 progression in the
room temperature vapor spectrum of NH_3, but none could explain the
jet-cooled spectra of NH_3 and ND_3. This discrepancy between predictions
of the separable mode harmonic Franck–Condon model and the experimental
absorption spectra motivated an ab-initio calculation of the ammonia X
and A surfaces [23]. CASSCF and CEPA electronic wavefunctions were used
to obtain potential energy, electric dipole moment and electronic
transition moment surfaces for the A and X states. Franck–Condon
factors and A–X radiative transition probabilities for the symmetrical
stretching and bending modes were evaluated. The symmetric stretching
mode in the A state is calculated to have only small intensity in the
A–X absorption spectrum due to anharmonic coupling of ν_1 and ν_2. The
relative intensities calculated for the absorption spectrum at room and
low temperatures now agree well with the experimental data.

   At the low rotational temperatures of ammonia in the jet, the
observed spectral bandwidth, deconvoluted from rotational and instrument
factors, reflect accurate homogeneous width [21]. The fwhm increase
with vibrational quantum number from 30 to 300 cm^{-1}, corresponding to
lifetimes of 0.15 to 0.02 psec.

   To model the photodissociation of ammonia, potential energy
surfaces for the lowest dissociative pathways of the singlet X and A
states yielding $NH_2(X^2B_1, A^2A_1) + H(^2S)$ products have been calculated
using ab-initio wavefunctions [24]. The results indicate that the A
state dissociation proceeds via a planar barrier, calculated to be 3226
cm^{-1} high. This barrier increases with increased out of plane angle in
ammonia. The minimal energy path through the X–A conical intersection
is seen to follow planar geometries.

   With help from this model surface, it should be possible to
understand the level coupling involved in funneling energy into the
dissociative mode and to calculate the dissociation dynamics consistent
with experimental data for the A state of ammonia.

IV. DISCUSSION

Experimental studies of reactive surfaces of polyatomic molecules can be
performed using absorption spectroscopy of samples free of inhomogeneous
effects. This technique, along with recently developed resonance Raman,
photofragment spectroscopy and theoretical calculations can provide the
necessary information to develop a comprehensive picture of reactive
excited electronic surfaces. The A state of ammonia illustrates that
ab-initio theoretical methods can successfully be used to describe
excited state structure and predict spectra even in cases where
efficient mode coupling occurs and simple models fail.

   From a potential surface obtained by ab-initio or semiempirical
electronic structure theory, which properly account for the spectra,
dynamical calculations can be used to predict many properties of
reactive molecules. The aim of this work is to develop, from experiment
working in concert with theory, a complete excited state picture for

dissociative molecules such as the three and four atom model systems discussed here.

ACKNOWLEDGEMENT

The studies described here were performed in collaboration with Dr. D. G. Leopold, Dr. R. J. Hemley, Dr. J. L. Roebber, M. I. McCarthy, Dr. P. Rosmus, Dr. H. J. Werner, Dr. P. Botschwina and Dr. P. C. Engelking whose contributions I gratefully acknowledge.
    Financial support for this work was provided by the National Science Foundation.

REFERENCES

1. Herzberg, G., Electronic Spectroscopy of Polyatomic Molecules (Van Nostrand, New York 1966).
2. Vaida V. Acc. Chem. Res. 19, 114 (1986) and references therein.
3. Recently, level and time-resolved experiments have been developed and used to study, in microscopic detail, the elementary photodissociation process. Several reviews have discussed these methods: Leone, S. R., Adv. Chem. Phys. 1982, 50, 255. Simons, J. P. J. Phys. Chem. 1984, 88, 1287. Bersohn, R. J. Phys. Chem. 1984, 8, 5145.
4. Imre D.; Kinsey, J.; Field, R.; Katayama, D. J. Phys. Chem. 1982, 86, 2564.
5. Desiderio, R. A.; Gerrity, D. P.; Hudson, B. S. Chem. Phys. Lett. 1985, 115, 29.
6. Ziegler, L. D.; Hudson, B. J. Phys. Chem. 1984, 88, 1110.
7. Gallear, A. B. Proc. R. Soc. London A 1963, 276, 401; Yang, S. C., Freedman, A., Kawasaki, M.; Bersohn, R. J. Chem. Phys. 1980, 72, 4058; De Sorgo, M., Yarwood, A. J., Strauz, O. P., Gunning, H. E. Can. J. Chem. 1965, 43, 1886; Butler, J. E., Drozdoski, W. S., McDonald, J. R. Chem. Phys. 1980, 50, 413; Okabe, H. J. Chem. Phys. 1972, 56, 4381.
8. Price, W. C.; Simpson, D. M. Proc. R. Soc. London A. 1938, 165, 272
9. Ramasastry, C.; Rao, K. R. Indian J. Phys. 1947, 21, 313.
10. Walsh, A. D. J. Chem. Soc. 1953, 2266.
11. Douglas, A. F.; Zanon, I. Can. J. Phys. 1964, 42, 627.
12. Rabelais, J. W.; McDonald, J. M.; Scherr, V.; McGlynn, S. P. Chem. Rev. 1971, 71, 73.
13. Greening, F. R.; King, G. W. J. Mol. Spectrom. 1976, 59, 312/
14. Hemley, R. J.; Leopold, D. G.; Roebber, J. L.; Vaida, V. J. Chem. Phys. 1983, 79, 5219.
15. Roebber, J. L.; Vaida, V. J. Chem. Phys. 1985, 83, 2748.
16. Rabalais, J. W.; McDonald, J. M.; V. Scherr, V. and McGlynn, S. P. Chem. Rev. 1971, 71, 73.
17. Sivakumar, N., Burak, I. Cheung, W.-Y., Houston, P. L. J. Phys. Chem. 1985, 89, 3609.
18. McCarthy, M. and Vaida, V. to be published.

19. Douglas, A. F. Discuss. Faraday Soc. 1963, 35, 158.
20. Groth, W. E., Schurath, U., Schindler R. N. J. Phys. Chem. 1968,
    72, 3914; Hertzberg, G., Longuet-Higgins, H. C. Discuss. Faraday
    Soc. 1963, 35, 77.
21. For a review of experimental work on ammonia see Vaida, V.;
    McCarthy, M.; Engelking, P. C.; Rosmus, P.; Werner, H. J. and
    Botschwina, J. Chem. Phys. submitted for publication.
22. Vaida, V.; Hess, W.; Roebber, J. L. J. Phys. Chem. 1984, 88, 3397.
23. Rosmus, P.; Botschwina, P.; Werner, H. J.; Engelking, P. C.;
    McCarthy, M. and Vaida, V. J. Chem. Phys. submitted for
    publication.
24. McCarthy, M.; Rosmus, P.; Botschwina, P.; Werner, H. J. and Vaida,
    V. J. Chem. Phys. submitted for publication.
25. Koda, S.; Hackett, P. A. and Back, R. A. Chem. Phys. Lett. 1974,
    28, 532.
26. Glownia, J. H.; Riley, S. J. and Colson, S. D. J. Chem. Phys. 1980,
    73, 4296.
27. Ashfold, M. N., Bennett, C. L., Dixon, R. N. Chem. Phys. 1985, 93,
    293; Ashfold, M. N., Dixon, R. N., Rosser, K. N., Stickland, R. J.
    and Western, C. M. Chem. Phys. 1986, 101, 467.
28. Kay, B. D.; Grimley, A. J.; Raymond, T. D. SPIE vol. 540 Southwest
    Conference on Optics 330 (1985).
29. Harshbarger, W. R. J. Chem. Phys. 1970, 53, 903; Avouris, P.,
    Rossi, A. R., Albrecht, A. C. J. Chem. Phys. 1981, 74, 5516.

A SCHRÖDINGER EQUATION ANALOG TO THE GENERALIZED LANGEVIN EQUATION OF CLASSICAL MECHANICS, WITH APPLICATION TO REACTIVE FLUX CORRELATION FUNCTIONS

William H. Miller
Department of Chemistry, University of California
and Materials and Molecular Research Division
Lawrence Berkeley Laboratory
Berkeley, California 94720 U.S.A.

ABSTRACT. Within the time-dependent self consistent field (TDSCF) approximation it is shown that a Schrödinger equation can be derived which is analogous to the generalized Langevin equation (GLE) of classical mechanics. Application of this to evaluate reactive flux autocorrelation functions is discussed.

1. INTRODUCTION

The model of a "system", consisting of a few interesting degrees of freedom, coupled to a "bath" of many (less interesting) degrees of freedom is ubiquitous in chemistry and physics. It is obviously a common situation in statistical mechanics, where the bath may consist of 10^{23} degrees of freedom, but it is also a very relevant and useful point of view in the field of reaction dynamics where the bath may consist of relatively few (e.g., 2-10) degrees of freedom, though still too many to be able to treat the complete system plus bath without approximation.

The specific system-bath model that has concerned my group in recent years arises from the reaction path (or reaction surface) Hamiltonian description of a reactive process in a polyatomic molecular system.[1] The reaction coordinate (or coordinates) constitute the "system", and the remaining degrees of freedom, which are locally harmonic vibrations perpendicular to the reaction path (or reaction surface), are the "bath". In this paper, though, I consider the simpler generic system-bath model, which is characterized by the Hamiltonian

$$H(p_s, s, \underline{P}, \underline{Q}) = \frac{p_s^2}{2m} + V_0(s) + \sum_k \frac{P_k^2}{2m} + \frac{1}{2} m\omega_k^2 Q_k^2 - \sum_k Q_k f_k(s) \quad . \qquad (1)$$

The "system" here is the reaction coordinate s. The potential $V_0(s)$ has the topology of the chemical process being described; an intramolecular H-atom transfer process, for example, would be

263

R. Lefebvre and S. Mukamel (eds.), Stochasticity and Intramolecular Redistribution of Energy, 263–271.
© *1987 by D. Reidel Publishing Company.*

characterized by a double-well potential function. The bath consist
of harmonic oscillators, which are linearly coupled to the system.
    The universal strategy for dealing with a system-bath situation
is to find some way, exact or approximate, to eliminate the bath and
then to deal accurately with the system. Within the framework of
<u>classical mechanics</u> there is a very simple and elegant way to
accomplish this: writing out the Newtonian form of the classical
equations of motion,

$$m\ddot{s}(t) = -\frac{\partial V}{\partial s} = -V_0'(s) + \sum_k f_k'(s)Q_k \tag{2a}$$

$$m\ddot{Q}_k(t) = -\frac{\partial V}{\partial Q_k} = -m\omega_k^2 Q_k + f_k(s) , \tag{2b}$$

one is able to solve Eq. (2b) exactly for the oscillator trajectory
(in terms of the "system" trajectory s(t))

$$Q_k(t) = Q_k^{(0)}(t) + \int_0^t dt' \, \sin[\omega_k(t-t')]f_k(s(t'))/m\omega_k \tag{3a}$$

where $Q_k^{(0)}(t)$ is the trajectory for the unperturbed oscillator,

$$Q_k^{(0)}(t) = Q_k^{(0)}\cos(\omega_k t) + (P_k^{(0)}/m\omega_k)\sin(\omega_k t) , \tag{3b}$$

$(Q_k^{(0)}, P_k^{(0)})$ being the initial (t=0) coordinates and momenta of the
"bath". Substituting $Q_k(t)$ from Eq. (3a) into Eq. (2a) then gives the
following closed equation – the generalized Langevin equation (GLE) –
for the trajectory of the "system",

$$m\ddot{s}(t) = -V_0'(s) + \sum_k f_k'(s)Q_k^{(0)}(t)$$

$$+ \int_0^t dt' \, \sum_k f_k'(s(t))f_k(s(t'))\sin[\omega_k(t-t')]/m\omega_k . \tag{4}$$

The three terms of the right hand side of Eq. (4) are referred to as
the local force, the random force (because the initial conditions
$(Q_k^{(0)}, P_k^{(0)})$ are usually averaged over), and the non-local friction,
respectively. The identification with friction is more obvious if one
integrates the last term in Eq. (4) by parts, whereby the right hand
side of Eq. (4) becomes

$$-V'_{eff}(s) + \sum_k f'_k(s)\tilde{Q}_k^{(0)}(t)$$

$$- \int_0^t dt'\, \dot{s}(t') \sum_k \frac{f'_k(s(t))f'_k(s(t'))}{m\omega_k^2} \cos[\omega_k(t-t')]; \qquad (5a)$$

$\tilde{Q}_k^{(0)}(t)$ differs from $Q_k^{(0)}(t)$ in that $Q_k^{(0)}$ is replaced by $Q_k^{(0)}$ $-f_k(s(0))/m\omega_k^2$, and

$$V_{eff}(s) = V_0(s) - \sum_k \frac{f_k(s)^2}{2m\omega_k^2} . \qquad (5b)$$

The last term in Eq. (5a) now clearly has the form of a non-local friction.

One would like very much to have a quantum mechanical analog to the generalized Langevin equation i.e., a closed quantum mechanical equation for only the "system" degrees of freedom which correctly incorporates the effect of the "bath" of it. Feynman path integral methodology[3] is one way in which this can actually be accomplished: the path integral over the "bath" degrees of freedom can be evaluated analytically, so that one is left with a path integral (i.e., a quantum calculation) for only the "system". This path integral, however, must in general be evaluated numerically, and unfortunately numerical (i.e., Monte Carlo) path integration is poorly behaved for real time dynamics, i.e., the propagator $e^{-iHt/\hbar}$. We,[4] and others,[5-7] have made some progress in this regard, but the prognosis is not very encouraging. (For evaluating the Boltzmann operator $e^{-\beta H}$, on the other hand, Monte Carlo path integration is very attractive.)

Much easier than doing quantum mechanics by evaluating a path integral would be to solve a Schrödinger equation. Unfortunately, though, there seems to be no Schrödinger-like equation for the "system" which incorporates the effect of the "bath" exactly. The purpose of this paper, however, is to show that within the time-dependent self consistent field (TDSCF) approximation one actually can obtain a Schrödinger equation analog of the generalized Langevin equation.

Section 2 first summarizes the TDSCF approximation, which is then applied to the system-bath Hamiltonian, Eq. (1), in Section 3. Section 4 concludes with a brief discussion of how this overall scheme applies to the evaluation of reactive flux correlation functions.

2. THE TIME-DEPENDENT SELF CONSISTENT FIELD APPROXIMATION

The TDSCF approximation has been well-described a number of times.[8] It is obtained from the Dirac[9] time-dependent variational principle,

$$\delta S = 0 \, , \tag{6a}$$

where S is the functional

$$S[\psi] = \int dt \int dq \, \psi(q,t)^* (i\hbar\frac{\partial}{\partial t} - H)\psi(q,t) \, , \tag{6b}$$

(q denotes the coordinates for all the degrees of freedom) when the SCF _ansatz_ is chosen for the trial wavefunction:

$$\psi(q,t) = e^{i\alpha(t)} \prod_{i=1}^{N} \phi_i(q_i,t) \, . \tag{6c}$$

If the Hamiltonian has the form

$$H = \sum_{i=1}^{N} h_i + V(q) \, , \tag{7}$$

then the TDSCF procedure leads to the following coupled one-dimensional Schrödinger equations for the one particle wave functions (the "orbitals") ϕ_i

$$i\hbar\frac{\partial}{\partial t}\phi_i(q_i,t) = [h_i + V_i^{eff}(q_i,t)]\phi_i(q_i,t) \tag{8}$$

where the time-dependent effective potential for the ith degree of freedom is given by

$$V_i^{eff}(q_i,t) = \int dq_i' \prod_{\substack{j=1 \\ j \neq i}}^{N} |\phi_j(q_j,t)|^2 V(q) \, ; \tag{9}$$

$\int dq_i'$ denotes an integral over all the coordinates except q_i. The overall phase $\alpha(t)$ is given by

$$\alpha(t) = (N-1)\hbar^{-1} \int_0^t dt' \, \bar{V}(t') \tag{10a}$$

where

$$\bar{V}(t) = \int dq \prod_{i=1}^{N} |\phi_i(q_i,t)|^2 V(q) \, . \tag{10b}$$

An important feature of the time-dependent SCF equations (8)-(10) is that once the initial conditions are specified - i.e., the initial "orbitals" $\phi_i(q_i,0)$ - the equations may be integrated forward in time non-iteratively. Also important, of course, is that these are each <u>one-dimensional</u> Schrödinger equations.

3.   TDSCF FOR THE SYSTEM-BATH HAMILTONIAN

For the system-bath Hamiltonian of Eq. (1) the equations of Section 2 simplify. The time-dependent wavefunction is of the SCF form

$$\psi(s,\underline{Q},t) = e^{i\alpha(t)}\chi(s,t) \prod_k \phi_k(Q_k,t) , \qquad (11)$$

and the coupled one-dimensional Schrödinger equations for the orbitals are

$$i\hbar \frac{\partial}{\partial t} \chi(s,t) = [h_s - \sum_k Q_k(t)f_k(s)]\chi(s,t) \qquad (12a)$$

$$i\hbar \frac{\partial}{\partial t} \phi_k(Q_k,t) = [h_k - f_k(t)Q_k]\phi_k(Q_k,t) , \qquad (12b)$$

where

$$h_s = \frac{p_s^2}{2m} + V_0(s) \qquad (13a)$$

$$h_k = \frac{p_k^2}{2m} + \tfrac{1}{2}m\omega_k^2 Q_k^2 \qquad (13b)$$

and

$$Q_k(t) = \langle\phi_k(t)|Q_k|\phi_k(t)\rangle \qquad (14a)$$

$$f_k(t) = \langle\chi(t)|f_k|\chi(t)\rangle . \qquad (14b)$$

The phase $\alpha(t)$ is given by

$$\alpha(t) = -\frac{1}{\hbar}\int_0^t dt' \sum_k f_k(t')Q_k(t') . \qquad (15)$$

Progress can be made because Eq. (12b) is recognized to be the

Schrödinger equation for a linearly forced time dependent harmonic
oscillator, which can be solved exactly:

$$\phi_k(Q_k,t) = \int_{-\infty}^{\infty} dQ_k' \, K(Q_k,Q_k';t) \, \phi_k(Q_k',0) \; , \tag{16a}$$

where K is the Feynman kernel;[10]

$$K(Q_k,Q_k';t) = \left(\frac{m\omega_k}{2\pi i \hbar \sin(\omega_k t)}\right)^{1/2} \exp\left\{\frac{im\omega_k}{2\hbar\sin(\omega_k t)} \left[\cos(\omega_k t)(Q_k^2 + Q_k'^2)\right.\right.$$

$$- 2Q_k Q_k' + \frac{2Q_k}{m\omega_k} \int_0^t dt' \, f_k(t')\sin(\omega_k t')$$

$$+ \frac{2Q_k'}{m\omega_k} \int_0^t dt' \, f_k(t')\sin[\omega_k(t-t')]$$

$$\left.\left.- \frac{2}{m^2\omega_k^2} \int_0^t dt' \int_0^{t'} dt'' f_k(t') f_k(t'')\sin(\omega_k t'')\sin[\omega_k(t-t')]\right\} \; . \tag{16b}$$

Using Eq. (16) one can evaluate Eq. (14a), giving

$$Q_k(t) = Q_k^{(0)}\cos(\omega_k t) + (P_k^{(0)}/m\omega_k)\sin(\omega_k t)$$

$$+ \frac{1}{m\omega_k} \int_0^t dt' \, f_k(t')\sin[\omega_k(t-t')] \; , \tag{17a}$$

where

$$Q_k^{(0)} = \langle\phi_k(0)|Q_k|\phi_k(0)\rangle \tag{17b}$$

$$P_k^{(0)} = \langle\phi_k(0)|P_k|\phi_k(0)\rangle \; . \tag{17c}$$

Note that the similarity of Eq. (17) to the classical Eq. (3).

Substituting Eq. (17a) into Eq. (12a) gives the desired closed
Schrödinger equation for the "system" wavefunction $\chi(s,t)$, which is
summarized here:

$$i\hbar\frac{\partial}{\partial t}\chi(s,t) = \left\{h_s - \sum_k f_k(s)\left(Q_k^{(0)}\cos(\omega_k t) + \frac{P_k^{(0)}}{m\omega_k}\sin(\omega_k t)\right)\right.$$

$$- \sum_k \frac{f_k(s)}{m\omega_k} \int_0^t dt' <\chi(t')|f_k|\chi(t')> \sin[\omega_k(t-t')]\}\chi(s,t) \ . \quad (18)$$

Once Eq. (18) is solved for $\chi(s,t)$, the phase $\alpha(t)$ is evaluated via Eq. (15).

The Schrödinger Equation, Eq. (18), for the "system" wavefunction $\chi(s,t)$ is clearly seen to be directly analogous to the classical generalized Langevin equation, Eq. (4). Even though this Schrödinger equation for χ is non-local in time - because of the memory term, the last term on the right hand side of Eq. (18) - it can be integrated non-iteratively because the memory term involves only past and not future times.

4.   CONCLUDING REMARKS

With the TDSCF approximation, therefore, one obtains a Schrödinger equation that is analogous to the generalized Langevin equation of classical mechanics. It is a much simpler calculation to integrate this one-dimensional Schrödinger equation, even with the non-local frictional effects due to coupling to the bath, than it is to evaluate the one-dimensional path integral than results in the path integral approach.

One of the important usages envisioned for this TDSCF system-bath model is to evaluate reactive flux autocorrelation functions.[4] If the "dividing surface" between reactants and products is at s=0, then the thermally averaged flux-flux autocorrelation function is

$$C_f(t) = tr[F \ e^{iHt_c*/\hbar} \ F \ e^{-iHt_c/\hbar}] \ , \quad (19)$$

where $t_c = t - i\hbar\beta/2$, $\beta = (kT)^{-1}$, and F is the flux operator

$$F = \frac{1}{2} [\frac{p_s}{m} \delta(s) + \delta(s) \frac{p_s}{m}] \ . \quad (20)$$

The time integral of $C_f(t)$ from 0 to ∞ is the thermal rate constant for the reaction.

Introducing the basis $\chi_i(s) \prod \phi_{\xi_k}(Q_k)$ to evaluate the trace, and also using the TDSCF approximation for the time-evolution operators, gives

$$C_f(t) = \sum_\xi \sum_{i,i'} F_{i,i'}(0) F_{i',i}(t_c) \rho_\xi \ , \quad (21)$$

where

$$F_{i',i}(t_c) = \frac{\hbar}{2im} \left(\chi_{i'}(s,t_c)^* \frac{\partial \chi_i(s,t_c)}{\partial s} \right.$$

$$\left. - \frac{\partial \chi_{i'}(s,t_c)^*}{\partial s} \chi_i(s,t_c) \right) \Big|_{s=0} , \tag{22}$$

$$\rho_\xi = \prod_k \int dQ_k \; |\phi_{\xi_k}(Q_k,t_c)|^2 . \tag{23}$$

The novel feature here is that one must integrate Eq. (18) for the system wavefunction $\chi_i(s,t_c)$ from t=0 to the complex time $t_c = -i\frac{\hbar}{2} \beta + t$, but this presents no essential difficulties: it is simplest to integrate Eq. (18) first from t=0 to t = $-i\frac{\hbar}{2}\beta$ - for which the equation is completely real - and then to use the function $\chi(s,-i\frac{\hbar}{2}\beta)$ as the initial condition for integrating in the real time direction.

We envision, therefore, that this TDSCF system-bath model will be extremely valuable for computing these reactive flux correlation functions. All one-dimensional quantum effects (e.g., barrier tunneling) are obviously incorporated exactly within this approach, the main feature to be tested being how well the TDSCF approximation describes coupling between the degrees of freedom. The important practical feature of the approach is that one must solve numerically a one-dimensional time-dependent Schrödinger equation, Eq. (18), no matter how large the molecular system.

ACKNOWLEDGMENTS

This work has been supported by the National Science Foundation Grant CHE84-16345.

REFERENCES

1. For reviews, see (a) W. H. Miller, J. Phys. Chem. 87, 3811 (1983); (b) W. H. Miller, in The Theory of Chemical Reaction Dynamics, ed. D. C. Clary, Reidel, Boston, 1986, pp. 27-45.
2. For example, S. A. Adelman and J. D. Doll, J. Chem. Phys. 61, 4242 (1974); 64, 2375 (1976).
3. R. P. Feynman and A. R. Hibbs, Quantum Mechanics and Path Integral, McGraw-Hill, N.Y., 1965.
4. (a) W. H. Miller, S. D. Schwartz, and J. W. Tromp, J. Chem. Phys. 79, 4889 (1983); (b) R. Jaquet and W. H. Miller, J. Phys. Chem. 89, 2139 (1985); (c) K. Yamashita and W. H. Miller, J. Chem. Phys. 82, 5475 (1985).

5. (a) D. Thirumalai and B. J. Berne, J. Chem. Phys. 79, 5029
    (1983); (b) D. Thirumalai, E. J. Bruskin, and B. J. Berne, ibid.
    79, 5063 (1983); (c) D. Thirumalai and B. J. Berne, ibid. 81,
    2512 (1984).
6. J. D. Doll, J. Chem. Phys. 81, 3536 (1984).
7. E. C. Behrmann, G. A. Jongeward, and P. G. Wolynes, J. Chem.
    Phys. 79, 6277 (1983).
8. See, for example, (a) A. K. Kerman and S. E. Koonin, Ann. Phys.
    (N.Y.) 100, 332 (1976); (b) J. W. Negele, Rev. Mod. Phys. 54, 913
    (1982).
9. P. A. M. Dirac, Proc. Camb. Phil. Soc. 26, 376 (1930).
10. Ref. 3, p. 64.

THEORETICAL STUDIES OF OVERTONE-INDUCED CHEMICAL REACTIONS

T. Uzer, School of Physics,
Georgia Institute of Technology, Atlanta, GA 30332 USA

James T. Hynes, Department of Chemistry and Biochemistry
University of Colorado, Boulder, CO 80309 USA

ABSTRACT. Overtone-induced chemical reactions are studied theoretically via a combination of classical trajectories and nonlinear resonance analysis. The examples discussed are 1) a model unimolecular isomerization (akin to the HCN → HNC reaction) subsequent to a CH bond excitation, 2) the dissociation of hydrogen peroxide to hydroxyl radicals after an OH bond excitation and 3) the model dissociation of a bond remote from an initial excitation site, as influenced by an intervening metal atom "blocker". In all these examples, emphasis is given to the mechanism, time scales and possible "nonstatistical" character of the energy flow to the reaction site.

I. INTRODUCTION

The course and rate of the intramolecular flow of energy leading to reaction in a vibrationally excited molecule are clearly fundamental – but largely poorly understood – essentials in the molecular level mechanism of a unimolecular (and bimolecular) chemical reaction [1]. Such overtone excitation, which is in some sense selective, can help reveal how the total internal energy, the excitation state, and the various internal molecular intermode couplings effect and affect the unimolecular reaction induced by the excitation. In consequence the number of experimental overtone induced reaction studies is waxing at a rapid pace [2].

Recent theoretical research has revealed the essential role played by nonlinear resonances in effecting intramolecular energy transfer in certain nonreactive systems [3-6]. Here the critical first step of the overall energy flow from e.g. an initially excited CH bond is often energy transfer from the excited stretching mode to the adjacent bending mode. This transfer is especially effective when there is a Fermi resonance -- a 2:1 frequency ratio between the stretch and bend motions. Most notably, the intramolecular relaxation of excited CH stretches in benzene [4] and in other hydrocarbons [5] has been successfully described using this resonance picture. In other cases

R. Lefebvre and S. Mukamel (eds.), Stochasticity and Intramolecular Redistribution of Energy, 273–283.

[6], it is instead 1:1 resonant frequency ratios for CH stretches and other molecular modes that are crucial for the flow.

Here we will focus on theoretical studies of chemical reactions induced by overtone excitation. The central issues will revolve around how and how fast the energy leaves an initially excited CH or OH stretch, how and how fast it flows into the reaction coordinate, and how direct and nonstatistical the process is. The unimolecular reactions studied range from simple models to attempted realistic treatments. They include overtone-induced isomerization [7], dissociation of HO$_2$H [8], and dissociation of a remote bond whose separation from the excitation site includes a heavy metal atom as a potential energy flow blocker [9]. The theoretical techniques employed are classical trajectories and nonlinear resonance analysis. A central theme of our studies is that while the energy flow out of an excited overtone may be rapid — so that energy localization in the overtone is difficult -- the path out of the excited bond is highly specific and thus potentially useful in directing the course of a reactive event.

II. OVERTONE-INDUCED ISOMERIZATION MODEL

Some real isomerization reactions proceed by, or at least heavily involve, large amplitude internal bending motions. The best known example is the unimolecular conversion of HCN to HNC, which is observed to occur in hot HCN [10]; another example is the methyl cyanide isomerization [11].

Perhaps the simplest possible model system (Fig. 1) for the investigation of how nonlinear stretch-bend Fermi resonance effects intramolecular energy transfer from a nonreactive vibrational mode into a reactive mode was constructed and analyzed by Uzer and Hynes [7].

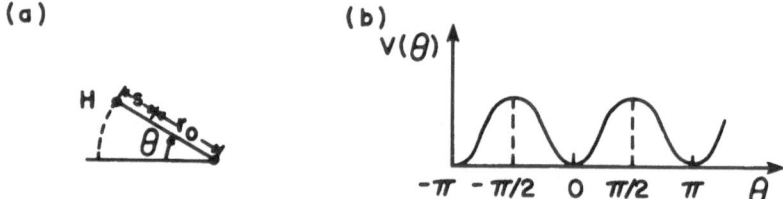

Figure 1. a) The model isomerization model and b) the isomerization potential in the bending reaction coordinate.

The model is a classical two dimensional system (very roughly patterned after the HCN → HNC process, but not intended as a faithful description). The stretching coordinate is described as a Morse oscillator

$$V(s) = D\left(1 - e^{-\alpha s}\right)^2 ;$$

$$E_v(cm^{-1}) = 3157\left(v + \frac{1}{2}\right) - 114\left(v + \frac{1}{2}\right)^2 , \qquad (1)$$

with the same unperturbed frequency as, but twice the anharmonicity of, the local CH stretch in benzene [4]. The bending motion is anchored at an infinitely heavy atom and the moving particle at the bond terminus is assigned the hydrogen atom mass m_H. The isomerization potential energy $V(\theta) = V_0(1-\cos 2\theta)/2$ in the deviation θ from the reactant configuration equilibrium angle is assigned a barrier height $V_0 = 2$ ev (not too dissimilar to the HCN barrier). The bending motion Hamiltonian

$$H_b = g_s(\theta)p_\theta^2/2 + V(\theta) \; ; \qquad g_s(\theta) = [m_H(r_0 + s)^2]^{-1} \qquad (2)$$

couples the stretch and bend through the variation of the g-matrix element with the bond extension s about the equilibrium separation r_0. This coupling arises from the dependence of the bend moment of inertia on the instantaneous bond extension s, and is most effective – and the energy transfer between the modes is the most rapid – when there is a 2:1 Fermi resonance condition: $\omega_s = 2\omega_b$. Indeed the leading order coupling arising from eq. (2) is $\propto sp_\theta^2$, explicitly exhibiting the 2:1 characteristics.

Since the stretching frequency ω_s declines from its harmonic value with increasing energy due to the anharmonicity (as does ω_b), the Fermi resonance condition can be "tuned into" by variation of the initial stretch overtone energy. We thus anticipate nonstatistical behavior in the reaction arising from this resonance specificity of the flow from the excited stretch into the reaction coordinate θ.

The results (Fig. 2) of the classical trajectory calculations [7]

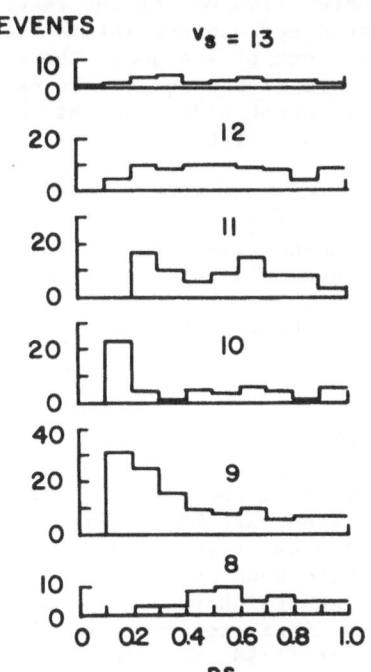

EVENTS

$v_s = 13$

Figure 2. The distribution of reaction times for 200 trajectories in 1 ps for various excitations v_s of the stretching motion.

bear out our expectation. This figure shows the reaction patterns for
various initial stretch quanta and with the Transition State Theory
assumption that passage over the barrier is sufficient for reaction (see
below). The stretch-bend Fermi resonance occurs at approximately v_s =
9, and the results show that reaction is most likely at this resonance
with many fast reactions occurring with reaction times less than 0.2
ps. When the resonance is detuned either by a decrease or increase in
v_s, fast reactions become progressively less likely. This
nonmonotonic pattern, which is easily understood via our resonance
picture, is in complete contrast to the monotonic increase in rate (and
the shape of individual v_s reactivity patterns) predicted by the
statistical RRKM theory.

Similar nonstatistical features were found [7] for other aspects.
For example, reaction following the direct and very high excitation of
the bend ($v_b \approx 20$) is direct and more rapid than for the equal energy
case of excitation in the stretch. Since here the bend is far off
resonance with the stretch, there can be no loss of energy to the
nonreactive stretch mode to compromise the direct character of the
reaction.

It is important to note that the Transition State Theory definition
of the rate is not in fact the true rate; it ignores recrossings of the
barrier [12]. For example, at v_s = 9, about 50% of the 60% of the
trajectories which crossed the barrier towards products in 1 ps were
stabilized in the product well on this time scale. The remainder either
returned to reactants or kept "winding" around the anchor atom. This
feature highlights the important general question of the stabilization
of the products by appropriately times energy flow out of the reaction
coordinate. This should be greatly assisted by resonant interactions of
the bend in the product well with other degrees of freedom. The catch
here is that such resonant interactions would also likely drain energy
from the bend reaction coordinate on the reactant side, and this will
diminish the pronounced mode specific and nonstatistical behavior we
have observed.

In this way, one can begin to imagine that the RRKM regime might be
more or less rapidly approached with more internal degrees of freedom.
Clearly a critical remaining task is to examine this question for more
internal degrees of freedom and assorted resonance possibilities,
perhaps involving molecular rotation.

(For interesting subsequent studies of the actual HCN
isomerization, see Ref. [13]).

III. PHOTODISSOCIATION OF HYDROGEN PEROXIDE

In a gas phase room temperature experiment, Crim and coworkers [14] have
used v_{OH} = 6 overtone excitation to photodissociate HO_2H in its ground
electronic state to OH radicals. The mechanism, time scale and possible
nonstatistical character of this process have been examined in a
classical trajectory study by Uzer, Hynes and Reinhardt [8] whose
central results are now briefly described. (This reference should be
consulted for details of the molecular force field employed.)

The first question to be asked is: how does the energy find its way from the anharmonic OH stretch to the dissociative OO bond? Since the OH frequency ω_{OH} = 2615 cm^{-1} at v_{OH} = 6 shows no simple integer relation to the OO frequency ω_{OO} = 986 cm^{-1} at v_{OO} = 0 and there is no strong direct coupling of the OH and OO bonds, a rapid <u>direct</u> transfer is precluded. Instead an <u>indirect</u> path is a possibility: an initial 2:1 Fermi resonance of the $\overline{v_{OH} = 6}$ OH stretch and the HOO bend of harmonic frequency 1374 cm^{-1}, followed by an approximate 1:1 resonance between the bend and the fairly strongly coupled OO stretch. Our trajectory studies reveal in fact that aspects of this last scenario operate, but that the full mechanism of the OO bond rupture subsequent to OH overtone excitation is rather more complex. Separation of the discussion into portions dealing with the short time ($\lesssim$0.5 ps) and longer time ($\gtrsim$0.5 ps) dynamics provides the clearest picture.

On short time scales, energy flow from the OH stretch to the HOO bend is indeed observed. But it is not as strong as would be anticipated from a simple estimate of the HOO bend frequency ω_{HOO}, which locates the 2:1 stretch-bend Fermi resonance at $v_{OH} \approx 5$. This feature has its origin in the dynamic lowering of ω_{HOO} by the increased mass ("moment of inertia") for the bend ensuing from the increased excited OH bond length. The actual stretch-bend resonance center is then located at $v_{OH} \approx 8$ rather than ≈ 5. Nonetheless enough energy transfers to the vicinal HOO bend to in principle subsequently excite the OO bond. But not much of this in fact occurs (Figure 3). Rather,

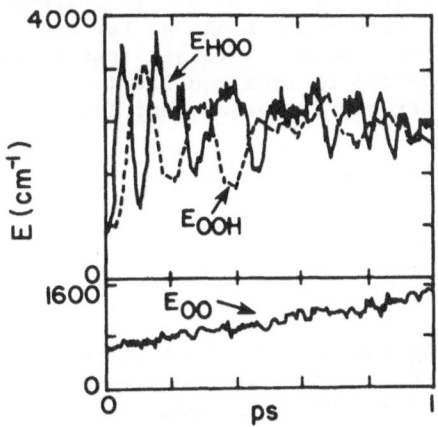

Figure 3. Short time energy flow in the central portion of HO$_2$H, including the vicinal HOO bend, the far OOH bend and the OO stretch. Average is over 50 trajectories.

energy transfers to the opposite OOH bend, with subsequent beats between the two bends, but with only a surprisingly mild accumulation of energy in the central OO bond.

This striking "bond-skipping" effect can be understood [8] in terms of the behavior of three coupled bond oscillators with the central oscillator disparate in frequency from the flanking bonds. Excitation of the left bond corresponds to exciting a linear combination of antisymmetric and symmetric modes composed largely of motions of the terminal oscillators. In HO_2H, this corresponds to excitation of the antisymmetric and symmetric bends, which subsequently beat with little excitation of the central OO bond. Evidently the HOO bend and OO stretch are too far off 1:1 resonance for rapid transfer to the latter leading to dissociation.

The critical feature necessary to the understanding of the longer time scale events resulting in dissociation is the following. As the OO bond lengthens, the HOO bend frequency ω_{HOO} must decline; indeed, in the limit, the bends become free rotations and $\omega_{HOO} \to 0$. Therefore when some energy accrues in the OO bond, it is easier to satisfy a 1:1 resonance condition with the HOO bend, to pick up more energy from that motion, and thus to dissociate.

Figure 4 shows a fairly rapidly dissociating trajectory. The two bends perform very irregular but bounded oscillations; in contrast, the

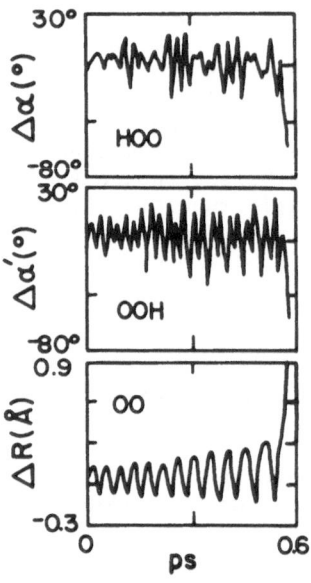

Figure 4. A representative dissociative trajectory in HO_2H. Panels from top to bottom display respectively the angular deviations from equilibrium $\Delta\alpha$ and $\Delta\alpha'$ in degrees for the vicinal HOO bend and far OOH bend as they suddenly rotate inward, as the deviation ΔR from equilibrium of the OO stretch increases rapidly.

OO stretch coordinate shows remarkably regular periodic oscillations with slowly and steadily increasing amplitude. The picture remains the same until about 0.05 ps before dissociation. Both the bends suddenly rotate inwards and the OO bond is extended to dissociation. This is a typical pattern. Energy flows between the initially excited OH stretch, the two bends and the OO bond. After a gradual accumulation of energy in the OO bond, evidently favorable configurations and phasing are attained at a certain moment. Then the bonds simultaneously rotate inward and the OO bond breaks in a nearly impulsive fashion.

A reaction lifetime histogram for 200 trajectories for $v_{OH} = 6$ dissociating HO_2H yields a 1/e time of about 6 ps. This is comparable to the low end of the 5-50 ps range for estimated statistical RRKM lifetimes [14]. But we observe that the initially excited OH stretch

never becomes excited during the HO$_2$H lifetime: its harmonic frequency
($\approx$3800 cm^{-1}) is simply too far off resonance with any physically
proximate motion in the molecule to pick up any energy. (Its
contribution to an RRKM rate would also be small due to the high
frequency). In this sense, the reaction is nonstatistical.

The width of $\approx$1 cm^{-1} associated with our predicted 6 ps lifetime is
small compared to the experimental v_{OH} = 6 width of 86 cm^{-1}, so that
there is negligible lifetime broadening. Nor is the width predicted to
arise in any significant portion from the internal decay of the v_{OH} = 6
initial state; this decay time is calculated to be about 1.5 ps,
corresponding to a width of $\approx$3 cm^{-1}. Rather we attribute the width to
inhomogeneous rotational broadening, for which our estimates give [8]
$\approx$85 cm^{-1} at 300 K. Our predictions are consistent with the subsequent
very recent jet experimental results on v_{OH} = 6 HO$_2$H of Crim and
coworkers [15]: the spectral width collapses extensively compared to the
room temperature results and the "homogeneous" width corresponds to $\approx$3.5
ps. This success gives us a certain amount of confidence in our
picture; nonetheless, calculations with more sophisticated HO$_2$H force
fields specifically constructed for large amplitude motions would
clearly be of interest.

IV. REMOTE BOND DISSOCIATION

Much attention in recent years has been devoted to the question of the
efficacy of "blockers" to intramolecular energy flow. Experimentally
the possibility seems to exist [16] that energy might be localized in
one reactive portion of an energized molecule by a heavy metal atom;
RRKM behavior involving rapid redistribution and reaction remote from
the bond excitation site is thwarted. (Statistical behavior is,
however, found in some molecules with heavy metal "blockers" [17].
While there are a number of theoretical studies of potential blocking
agents to energy flow for nonreactive model systems [18], there is only
one study of a reaction system (subject to initial chemical excitation)
[19].

Here we present preliminary results of a study by us [9] of remote
weak bond CX dissociation in linear HCMCX model compounds induced by
overtone excitation of a HC bond separated from the weak CX bond by a
potential blocking heavy metal atom M. The model is sketched in Fig. 5,

	H —— C —— M —— C —— X				
ω_e	3100	1821		1635	ω_e
ω_{xe}	57	29		95	ω_{xe}
D_e	43700	29500		7900	D_e

Figure 5. Definition of the collinear remote bond
dissociation model with energy parameters in
E_v(cm^{-1}) $= \omega_e\left(v+\frac{1}{2}\right) - \omega_e x_e\left(v+\frac{1}{2}\right)^2$ for the Morse potential
bonds and the bond dissociation energies. All numbers
in cm^{-1}.

which indicates the parameters of the various Morse oscillator bonds in
the molecule. M is given the mass of C or Hg in the calculations and
the CM frequency is artificially taken to be independent of M. We
initially deposit energy into the HC bond corresponding to a certain
v_{HC} overtone and then examine the resulting energy flow and reactions
via classical trajectories.

The mechanism by which M might act as a blocker relies on two
critical features [3,19]: (a) the kinetic coupling [3,19] between the CM
and MC bonds will tend to be weak for a large mass of M and (b) there
must be sufficient detuning of the 1:1 resonance $\omega_{CM} = \omega_{MC}$ to suppress
the flow. This requires a mismatch in the CM and MC bond energies so
that anharmonicity can lower one frequency compared to the other.

The threshold for dissociation of the weak CX bond is $v_{HC} = 3$.
Table 1 shows a remarkable set of results for HC overtone excitations at

Table 1.   Dissociation patterns for HCMCX after HC
overtone excitation in 100 trajectories.

v_{HC}	13	14	15	16
CM ruptures	39	64	67	69
CX ruptures	61	36	33	31

and above the threshold $v_{HC} = 13$ for dissociating the strong 29500 cm^{-1}
CM bond adjacent to the initially excited HC. Here M has the mass of
Hg. There is a marked tendency for the nonstatistical rupture of the
strong CM bond as opposed to the heavily statistically favored rupture
of the remote weak 7900 cm^{-1} CX bond. No MC bond ruptures are
observed. All of this can be understood as follows. The center of the
1:1 resonance between the HC bond and the adjacent CM bond is at
approximately $v_{HC} = 11$ and considerable energy can flow into CM. But
this influx of energy drastically lowers the CM frequency, detuning it
from that of the adjacent unexcited MC bond. Together with the weak
CM-MC kinetic coupling, this evidently isolates the left hand side of
the molecule sufficiently to cause the striking nonstatistical bond
rupture pattern displayed.

We next focus on the overtone $v_{HC} = 11$. This is at the center of
the HC-CM 1:1 resonance but without of course enough energy to break a
CM bond. Figure 6 displays the reactivity patterns for HCMCX with M

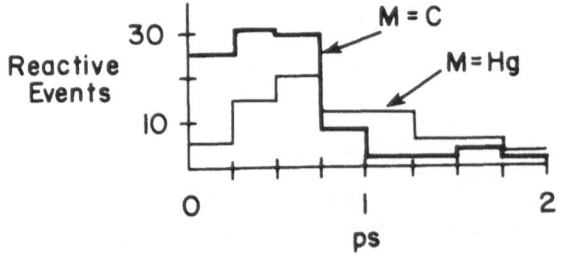

Figure 6.   Reaction lifetime
pattern for 100 trajectories
with initial HC overtone
excitation $v_{HC} = 11$ when M
in HCMCX is C or Hg.

having the mass of Hg, and for HCCCX. The HC–CM(C) 1:1 resonance leads
to an initial rapid flow of energy to the CM(C) bond in the
trajectories. In the HCCCX species, energy continues to flow rapidly
over the entire molecule with negligible apparent blocking ability of
the "light" central C atom. The reactivity histogram (Fig. 6) has an
exponential appearance. But this is not at all the case for the heavy
metal atom species HCMCX with M having the mass of Hg. The reaction
time scale is noticeably longer (Fig. 6), and inspection of the
trajectories indicates that this feature arises in considerable part
from the detuning of the CM and MC bonds which are only weakly
kinetically coupled. The heavy metal atom is acting as a strong
retarding agent of the flow.

These preliminary results indicate strongly that the heavy metal
atom in the linear model compound serves to moderate the rate of energy
transmission through the molecule and to lead to pronounced
nonstatistical behavior. More detailed studies and analysis are in
progress.

V.   CONCLUDING REMARKS

The studies briefly herein described of overtone-induced reactions
indicate, in our view, a number of lessons. First, the initial flow
route of energy out of an excited bond is specific, and that specificity
has important consequences for the reaction. Second, the flow rates and
mechanisms can be comprehended in terms of nonlinear resonance ideas.
Third and perhaps most sobering, we find that with increasing realism
and number of coupled internal degrees of freedom (as in HO_2H), the flow
routes become increasingly complex. Here one begins to see how RRKM
behavior can eventually result. But even in that limit, intramolecular
flow rates are finite and energy transfer paths are not random; their
elucidation in chemically realistic models still presents an open
challenge.

ACKNOWLEDGMENTS

Acknowledgment is made to the Donors of the Petroleum Research Fund, as
administered by the American Chemical Society, for partial support of
this work. This work was also supported by NSF grants CHE 81-13240 and
84-19830 to JTH. We also acknowledge W. P. Reinhardt for his
collaboration on the HO_2H work, which was also supported by NSF grants
CHE-80-14428, PHY 82-00805, CHE 83-10122 and CHE 84-16459 to WPR.

REFERENCES

1.   See e.g. D. W. Noid, M. L. Koszykowski and R. A. Marcus, <u>Annu.
      Rev. Phys. Chem.</u> **32**, 267 (1981); W. L. Hase, in <u>Dynamics of
      Molecular Collisions, Part B</u>, W. H. Miller, ed. (Plenum, New York,
      1986).

2.   See, e.g., F. F. Crim, Annu. Rev. Phys. Chem. 35, 657 (1984); J.
     W. Wong and C. B. Moore, J. Chem. Phys. 77, 603 (1982); J. W.
     Perry and A. H. Zewail, J. Phys. Chem. 86, 5197 (1982); D. W.
     Chandler, W. E. Farneth and R. N. Zare, J. Chem. Phys. 77, 4447
     (1982); K. V. Reddy, D. F. Heller and M. J. Berry, ibid. 78, 2817
     (1983); C. I. Manzanares, N. L. S. Yamasaki, E. Weitz and T.
     Knudtson, Chem. Phys. Lett. 117, 477 (1985); J. E. Baggott, M.-C.
     Chang, R. N. Zare, H. R. Dubal and M. Quack, J. Chem. Phys. 82,
     1186 (1985).
3.   E. L. Sibert, W. P. Reinhardt and J. T. Hynes, J. Chem. Phys. 77.
     3583 (1982); C. Jaffé and P. Brumer, ibid. 73, 5646 (1980); M. S.
     Child and L. Halonen, Adv. Chem. Phys. 57, 1 (1984).   See also
     comments by R. A. Marcus and M. S. Child in Faraday Discuss. Chem.
     Soc. 75, 157 (1983).
4.   E. L. Sibert, W. P. Reinhardt and J. T. Hynes, Chem. Phys. Lett.
     92, 455 (1982); J. Chem. Phys. 81, 1115,1135 (1984).
5.   J. S. Hutchinson, J. T. Hynes and W. P. Reinhardt, J. Phys. Chem.
     90, 3528 (1986); E. L. Sibert, J. S. Hutchinson, J. T. Hynes and
     W. P. Reinhardt, in Ultrafast Phenomena IV, D. H. Auston and K. B.
     Eisenthal, eds. (Springer-Verlag, New York, 1984).   p336.
6.   J. S. Hutchinson, W. P. Reinhardt and J. T. Hynes, J. Chem. Phys.
     79, 4247 (1983); J. S. Hutchinson, J. T. Hynes and W. P. Reinhardt,
     Chem. Phys. Lett. 108, 253 (1984); J. S. Hutchinson, J. Chem. Phys.
     82, 22 (1985); T. A. Holme and J. S. Hutchinson, ibid. 93, 419
     (1985).
7.   T. Uzer and J. T. Hynes, Chem. Phys. Lett. 113, 483 (1985).
8.   T. Uzer, J. T. Hynes and W. P. Reinhardt, J. Chem. Phys. 1986, in
     press.
9.   T. Uzer and J. T. Hynes, unpublished.
10. A. G. Maki and R. L. Sams, J. Chem. Phys. 75, 4178 (1981).
11. K. V. Reddy and M. J. Berry, Chem. Phys. Lett. 52, 111 (1977).
12. D. G. Truhlar, W. L. Hase and J. T. Hynes, J. Phys. Chem. 87, 2664
     (1983).
13. T. A Holme and J. S. Hutchinson, J. Chem. Phys. 83, 2860 (1985);
     Z. Bacic, R. B. Gerber and M. A. Ratner, J. Phys. Chem. 90, 3606
     (1986).
14. T. R. Rizzo, C. C. Hayden and F. F. Crim, Faraday Discuss. Chem.
     Soc. 75, 223 (1983) (see also the discussion remarks by H. R. Dubal
     and M. Quack); J. Chem. Phys. 81, 4501 (1984).
15. L. V. Butler, R. M. Ticich, M. D. Likar and F. F. Crim, J. Chem.
     Phys. 85, 2331 (1986).   The v_{OH}=5 HO_2H photodissociation results
     of N. F. Scherer, F. E. Doary, A. H. Zewail and J. W. Perry,
     ibid. 84, 1932 (1986) are also of interest in this connection.
16. P. J. Rogers, J. I. Selco and F. S. Rowland, Chem. Phys. Lett.
     97, 313 (1983); P. J. Rogers, D. C. Montague, J. P. Frank, S. C.
     Tyler and R. S. Rowland, ibid. 89, 9 (1982).
17. S. P. Wrigley and B. S. Rabinovitch, Chem. Phys. Lett. 98, 363
     (1983); S. P. Wrigley, D. A. Oswald and B. S. Rabinovitch,
     ibid. 104, 52 (1984).
18. K. N. Swamy and W. L. Hase, J. Chem. Phys. 82, 123 (1985).

19. V. Lopez and R. A. Marcus, Chem. Phys. Lett. **93**, 2132 (1982); S. M. Lederman, V. Lopez, G. A. Voth and R. A. Marcus, ibid. **124**, 93 (1986); T. Uzer and J. T. Hynes, J. Phys. Chem. **90**, 3524 (1986); C. Sloane and W. L. Hase, J. Chem. Phys. **66**, 1523 (1977).

LIST OF PARTICIPANTS

Dr. A. Amirav
School of Chemistry
Tel-Aviv University
Tel-Aviv 69978
ISRAEL

Dr. O. Bohigas
Institut de Physique Nucléaire
Université Paris-Sud
91405 Orsay
France

Prof. Mark Child
Theoretical Chemistry Department
1 South Parks Road
Oxford OX1 3TG
England

Dr. Nguyen Dang
Faculté des Sciences
Université de Sherbrooke
Sherbrooke, Québec J1K 2R1
Canada

Dr. W. Dietz
Technische Universität München
Institut für Theoretische Physik T38
D-8046 Garching,
West Germany

Prof. W. Domcke
Institute of Physical and Theoretical
Chemistry
Technical University of Munich
D-8046 Garching
West Germany

Dr. M. Dupré
Service National des Champs Intenses
du CNRS
BP 166X
38042 Grenoble Cedex
France

Dr. Stavros Farantos
Department of Chemistry
University of Crete
          and
Institute of Electronic Structure and Laser
Research Center of Crete
711 10 Iraklion, Crete
Greece

Prof. S. Fischer
Technische Universität München
Institut für Theoretische Physik T38
D-8046 Garching
West Germany

Prof. R.B. Gerber
Fritz Haber Center for Molecular Dynamics
                 and
Department of Physical Chemistry
Hebrew University
Jerusalem
Israel

Prof. Maximo Garcia-Sucre
Instituto Venezolano de Investigaciones
Cientificas
                 and
Universidad Central de Venezuela
Caracas
Venezuela

Prof. James T. Hynes
Department of Chemistry and Biochemistry
University of Colorado
Boulder, Col. 80309
U.S.A.

Dr. R. Jost
Service National des Champs Intenses
du CNRS
BP 166 X
38042 Grenoble Cedex
France

Prof. Jan Kommandeur
Laboratory for Physical Chemistry
The University of Groningen
Nijenborgh 16
9747 AG GRONINGEN
The Netherlands

Dr. Sydney Leach
Laboratoire de Photophysique Moléculaire
du CNRS
Université Paris-Sud
91405 Orsay
France

Prof. Roland Lefebvre
Laboratoire de Photophysique Moléculaire
du CNRS
Université Paris-Sud
91405 Orsay
France

Prof. Donald H. Levy
James Franck Institute
and Department of Chemistry
University of Chicago
Chicago, Ill. 60637
U.S.A.

Dr. M. Lombardi
Service National des Champs Intenses
du CNRS
BP 166 X
38042 Grenoble Cedex
France

Dr. Giorgio Mantica
Dipartimento di Fisica
Universita di Milano
Milano
Italy

Prof. William Miller
Department of Chemistry
University of California
           and
Materials and Molecular Research Division
Lawrence Berkeley Laboratory
Berkeley, Cal. 94720
U. S. A.

Prof. C. B. Moore
Department of Chemistry
University of Berkely
Berkeley, Cal. 94720
U. S. A.

Prof. Shaul Mukamel
Department of Chemistry
University of Rochester
Rochester, N. . Y. 14627
U. S. A.

Dr. P. Papagiannakopoulos
Department of Chemistry
University of Crete
           and
Institute of Electronic Structure and Laser
Research Center of Crete
711 10 Iraklion, Crete
Greece

Prof. Charles S. Parmenter
Department of Chemistry
Indiana University
Bloomington, Ind. 47405
U. S. A.

Prof. Mark A. Ratner
Department of Chemistry
Northwestern University
Evanston, Ill. 60201
U. S. A.

Prof. E. Riedle
Institut für Physikalische und
Theoretische Chemie
Technische Universität München
Lichtenbergstr. 4
D-8046 Garching
West Germany

Prof. Israel Schek
Department of Chemistry
and Institute for Theoretical Chemistry
University of Texas
Austin, Texas 78712
U.S.A.

Dr. Benoit Soep
Laboratoire de Photophysique Moléculaire
du CNRS
Université Paris-Sud
91405 Orsay
France

Prof. H.S. Taylor
Department of Chemistry
University of Southern California
Los Angeles, Cal. 90089
U.S.A.

Dr. André Tramer
Laboratoire de Photophysique Moléculaire
du CNRS
Université Paris-Sud
91405 Orsay
France

Prof. Veronica Vaida
Department of Chemistry and Biochemistry
University of Colorado
Boulder, Col. 80309
U.S.A.

SUBJECT INDEX

A

Absorption spectrum, see CS_2, OCS, NH_3
Action spectrum, 158–159
Adiabatic switching method, 82
Adiabatic angular eigenvalues, 246
Alkyl benzene, 163
Anderson localization, 6
Anharmonic interactions, 193
Anthracene, 172
    absolute fluorescence quantum yield, 173–174
    absorption spectrum of, 173
    fluorescence excitation spectrum of, 173–174
Atom migration, 242
Attractor, 17
Autocorrelation function, 222, 228

B

Bath, 263
Benzene, 171, 193, 203, 273
    cation, 223, 227, 228
Billiards, 2
Born Oppenheimer separation, 23
Bound–free emission, 155, 156
Brody distribution, 33, 66, 68

C

Channel three, 203
Chaos, 1, 15–29
    classical, 1, 16
    quantum, 1, 16, 31, 48
    vibrational, 24
Chemical timing, 164
Chirikov resonance, 4
CH_3I, 254
Classical equations of motion, 236
Clusters, 233–243
    fragmentation dynamics, 237–239
CO_2, 76
Conformer, 141
Conical intersection, 217, 218

W

Z